T0202061

The Metaphysics of Laws of Nature

The Metaphysics of Laws of Nature

The Rules of the Game

WALTER OTT

OXFORD
UNIVERSITY PRESS

OXFORD
UNIVERSITY PRESS

Great Clarendon Street, Oxford, OX2 6DP,
United Kingdom

Oxford University Press is a department of the University of Oxford.
It furthers the University's objective of excellence in research, scholarship,
and education by publishing worldwide. Oxford is a registered trade mark of
Oxford University Press in the UK and in certain other countries

First Edition published in 2022

Published in the United States of America by Oxford University Press
198 Madison Avenue, New York, NY 10016, United States of America

British Library Cataloguing in Publication Data
Data available

Library of Congress Control Number: 2022930784

ISBN 978–0–19–285923–5

DOI: 10.1093/oso/9780192859235.001.0001

Printed and bound in the UK by
Clays Ltd, Elcograf S.p.A.

Contents

Preface   ix
Acknowledgments   xi
Chronology 1600–1900   xiii

1. The Game   1
   1. Starting out   1
   2. Methods   1
   3. Three Axes   3
   4. Structure   5

PART I. OPENINGS

2. The Early Days   11
   1. The Debate   11
   2. The Wrong Criterion   12
   3. The Thin Concept   14
   4. Starting over   21
   5. Descartes   26
      5.1 Laws and Ontology   27
      5.2 The Problem of *Ceteris Paribus* Clauses   31
      5.3 Intentions, not Regularities   33
      5.4 Epistemic Problems   37
   6. Virtues and Vices   38

3. Descartes's Legacy   42
   1. Malebranche: Intervention or Autonomy?   42
      1.1 Efficacious Laws   43
      1.2 Leibniz's Critique   45
   2. Newton   47
      2.1 Methods and Varieties of Explanation   48
      2.2 Ontology   52
   3. Berkeley   56
      3.1 Early Berkeley   56
      3.2 The Web of Laws in *De Motu*   59
   4. Summing up: Top-down Laws in the Modern Period   61

PART II. CONTEMPORARY TOP-DOWN VIEWS

4. Primitivism — 67
   1. Governing without God — 67
   2. Motivations — 68
   3. Mirrors and Spin — 71
   4. The Underdetermination Argument — 76
   5. What is Governing? — 78
      5.1 The Governing Dilemma — 79
   6. The Virtues of Primitivism — 82

5. Universals (I) — 85
   1. Motivations — 85
   2. Armstrong's View — 85
   3. Quiddities: Epistemic Worries — 90
   4. Quiddities: Metaphysical Worries — 92
   5. Epiphenomenalism? — 95
   6. Laws and Explanation — 96
   7. Intra-world Variance — 99

6. Universals (II) — 102
   1. The Inference Problem — 102
      1.1 Legitimacy — 104
      1.2 Mechanism — 107
   2. A Revision: N as an Internal Relation — 109
   3. Applications of the Revised View — 110
   4. Modal Inversion — 112
   5. Conservation Laws — 114
   6. The Ontology of Relations — 115
   7. A Second Revision: Many-to-One — 116
   8. The Price of Revision — 118

PART III. POWERS

7. Origins of the Powers View — 123
   1. The Moderns — 123
   2. Bacon — 124
   3. Spinoza — 128
   4. Euler and Shepherd — 132
   5. Helmholtz — 135
   6. A Contemporary View — 136

8. The Powers Ontology — 141
   1. Whose Powers? — 141
   2. Essential and Invariant — 142
   3. Independent — 145
   4. Intrinsic — 146

  5. Reciprocal   148
  6. Irreducible   149
  7. Intentional   151
  8. Functions   155
  9. Relational Views   157
 10. The Inevitability of Physical Intentionality   159

9. The Arguments for Powers   161
  1. The Conceptual Argument   162
  2. Arguments from Science   164
    2.1 The Missing Categorical Properties   164
    2.2 The Ungrounded Argument   165
    2.3 The Vices of Humility   167
  3. The Metaphysical and Epistemic Arguments   168
    3.1 A Regress?   168
    3.2 Ignorance   170
  4. Varieties of Dispositionalism   171
    4.1 Pan-dispositionalism   172
    4.2 Dual-sided and Neutral Monist Views   176
    4.3 Dualism   177
  5. The Best Powers View   178

10. Facing up to the Moderns   180
  1. Doubts   180
  2. Little Souls   180
    2.1 Holism   183
    2.2 Monism   186
    2.3 The Blower   187
  3. The Moderns' Way out   189
    3.1 Locks and Keys   189
    3.2 Extrinsic and Reducible Powers   192
    3.3 Problems   194

PART IV. THE REGULARITY THEORY AND BEYOND

11. Origins: Hume and Mill   201
  1. A New Family   201
  2. Force and Gravity   203
  3. The Tension in Hume   209
  4. Mill's Web   213

12. Contemporary Best Systems Analyses (I)   218
  1. Two Masters   218
  2. Anthropomorphism and Better Systems for Us   219
  3. Natural Properties   222
  4. Humean Supervenience   223

     5. The Underdetermination Argument Redux   227
     6. Counterfactuals   230

13. Contemporary Best Systems Analyses (II)   235
     1. Mirrors and Nomic Stability   235
     2. Explanation—Case   238
     3. Explanation—Whole   242
     4. What is a Regularity?   244
     5. *Ceteris Paribus* Clauses   247
     6. The Central Tension   251
     7. *Ceteris Paribus* Redux   255

14. The Alternatives   257
     1. The State of Play   257
     2. Conservative Projectivism   257
     3. Contemporary Projectivism   260
     4. Laws as Rules   262
        4.1 The Mismatch Problem   263
        4.2 The Problem of *Ceteris Paribus* Clauses   264
        4.3 Truth and Supervenience   266
        4.4 Counterfactuals and Explanation   269
        4.5 Nomic Stability   271
     5. The Path of Anti-realism   272

PART V. THE ENDGAME

15. Settling up   275
     1. The Rules   275
     2. Three Axes   278
     3. Law Statements to Truthmakers   279
     4. Top-down/Bottom-up   281
     5. Realism/Anti-realism   285
     6. The Hybrids   287
     7. Decisions   289

References   293
Index   307

Preface

In some ways, this book is the natural sequel to *Causation and Laws of Nature in Early Modern Philosophy*, which was published more than a decade ago. That book ended in 1748 with Hume's *Enquiry*, albeit with occasional forays into later figures. I try not to rely on results I argued for there; at the same time, I'm loath to chew my cabbage twice. I don't presuppose any knowledge of the earlier book on the part of the reader. But here and there, I do indicate where fuller arguments for a given point may be found.

In other ways, it's a departure: this is not primarily a work in the history of modern philosophy. So I don't propose to give a systematic treatment of laws of nature from 1748 to the present moment. This book is organized thematically, and although I believe I cover most of the important developments in the years since Hume's *Enquiry*, giving a complete account of them is not my goal.[1]

---

[1] A notable exception is Kant. Earlier drafts included a long section on Kant, which threatened to become a book of its own. Given my purposes here, fighting through the details of his view, and defending my reading against competitors, didn't in the end seem worth the ink.

Acknowledgments

This book would have been much worse without the help of fellow philosophers. I've benefited from exchanges with Tyler Hildebrand, Ben Jantzen, Barry Loewer, Antonia LoLordo, Lydia Patton, and Stathis Psillos. I don't mean to imply that these philosophers would endorse all (or indeed any) of what follows.

I had looked forward to attending Barry's workshop on laws of nature in April of 2020. That date alone will explain why it didn't happen. Barry generously commented anyway on my paper and subsequently on the entire book, for which I'm most grateful. I would also like to thank James Reed, Zachary Veroneau, and Evan Welchance for comments and discussion. Travis Tanner gave the manuscript a final, careful read, which led to many improvements.

I'm indebted to two readers for Oxford University Press for their diligent and perceptive criticisms.

I'm grateful to Doug Juers, Professor of Physics at Whitman College, for helping me with the physics in general and explaining Noether's theorem in particular. Thanks to Michael A. Ciccone for vetting the final version of my physics material. Any mistakes that remain are my own.

In 2017, I presented a paper on the problem of *ceteris paribus* clauses among the moderns at Stefano Di Bella's and Tzuchien Tho's *Causa Sive Ratio* conference in Milan, Italy. For their insightful criticisms, I'm very grateful to Stefano and Tzuchien, as well as the other attendees, especially Katherine Brading, Manuel Fasko, Hylarie Kochiras, Ansgar Lyssy, Jeffrey McDonough, and Andrew Platt.

I'm grateful to Oxford University Press for permission to re-print snippets of '*Leges Sive Natura*: Bacon, Spinoza, and a Forgotten Concept of Law,' from *Laws of Nature* (Oxford, 2018), as well as 'The Case Against Powers,' from Stathis Psillos, Benjamin Hill, and Henrik Lagerlund (eds), *Causal Powers in Science: Blending Historical and Conceptual Perspectives* (Oxford, 2021). My thanks to the editors and publishers of the *Journal of Modern Philosophy* for permission to use part of my 2019 article, 'Berkeley's Best System: An Alternative Approach to Laws of Nature.'

I'm indebted to Peter Momtchiloff, Philosophy Editor at Oxford University Press UK, for his guidance and encouragement. Thanks to Kalpana Sagayanathan for overseeing the final phases of production. I'm very grateful to Jackie Pritchard and Edwin Pritchard for their expert copy-editing. My thanks to Leslie Oakey for the cover art and to the design team at Oxford.

I would like to thank the University of Virginia and my department chair, Antonia LoLordo, for research leave in the fall of 2020.

Finally, I owe a much older debt. In the summer of 2004, I was lucky enough to attend an NEH Institute expertly directed by Steven Nadler and Donald Rutherford. That experience put me on the scent of causal powers and laws of nature, and I've been happily galloping down the path, in fits and starts, ever since.

Charlottesville, Virginia, 1 April 2022

Chronology 1600–1900

1603	Dupleix	*La Physique, ou science des choses naturelles*
1620	Bacon	*Novum Organum*
1632	Galileo	*Dialogo sopra i Due Massimi Sistemi del Mondo*
1638	Galileo	*Discorsi e dimostrazioni matematiche intorno a due nuove scienze*
1644	Descartes	*Principia Philosophiae*
1654	Charleton	*Physiologia Epicuro-Gassendo-Charletoniana*
1655	Gassendi	*Syntagma Philosophicum*
1655	Hobbes	*De Corpore*
1666	Boyle	*The Origin of Forms and Qualities according to the Corpuscular Philosophy*
1669	Huygens	Letter to the *Journal des Sçavans*
1671	Rouhault	*Traité de Physique*
1674–5	Malebranche	*De la recherche de la vérité*
1677	Spinoza	*Ethica*
1678	Cudworth	*The True Intellectual System of the Universe*
1686	Boyle	'A Free Inquiry into the Vulgarly Received Notion of Nature'
1687	Newton	*Philosophiae Naturalis Principia Mathematica*
1688	Malebranche	*Entretiens sur la métaphysique et sur la religion*
1689	Locke	*An Essay concerning Human Understanding*
1696	Sergeant	*The Method to Science*
1704	Régis	*L'Usage de la raison et de la foi*
1710	Berkeley	*A Treatise concerning the Principles of Human Knowledge*
1715	Newton	'An Account of the Book Entitled *Commercium Epistolicum*'
1717	Leibniz and Clarke	Correspondence
1721	Berkeley	*De Motu*
1732	Berkeley	*Alciphron*
1739–40	Hume	*A Treatise of Human Nature*
1740	Du Châtelet	*Institutions de Physique*
1744	Berkeley	*Siris*
1747	s'Gravesande	*Mathematical Elements of Natural Philosophy*
1748	Hume	*An Enquiry concerning Human Understanding*
1750	Euler	*Reflexions sur l'espace et sur les tems*
1768	Euler	*Letters to a German Princess*

1788	Reid	*Essays on the Active Powers of Man*
1818	Brown	*Inquiry into the Relation of Cause and Effect*
1819	Anon. (Shepherd)	*Enquiry Respecting the Relation of Cause and Effect*
1827	Shepherd	*Essays*
1843	Mill	*A System of Logic*
1860	Whewell	*On the Philosophy of Discovery*
1869	Helmholtz	'On the Aim and Progress of Physical Science'
1883	Mach	*Die Mechanik in ihrer Entwickelung*
1889	Venn	*The Principles of Empirical or Inductive Logic*

1

The Game

1. Starting out

I was no good at science in high school. In addition to being a teenaged boy—a condition that comes with its own cognitive handicaps—I resented being told some equation or other was a 'law.' I couldn't work out what that was supposed to mean. It couldn't be a law in the political sense; that requires sentient beings who obey it (or not). So it must be some kind of metaphor. My rare and stumbling efforts to get the occasional adult to tell me what it was supposed to be a metaphor *for* hardly endeared me to them. Where were these laws, and how could they do anything? Laws were presented as if they were cosmic firemen, going around and adjusting, say, the temperature, volume, and pressure of gases, in short, 'governing' events—another political metaphor. Thanks a lot.

What I didn't realize was that my questions, however ill posed and dimly understood by me at the time, were not really scientific questions at all; they were philosophical ones. The teachers had no answers for me, simply because that wasn't the kind of problem they engaged with. To the degree they understood what I was asking for at all, they seemed to think I was confused, slow-witted, or both. Look, if you follow Newton's laws, you can build bridges that don't fall down. What more do you want?[1] Shut up and calculate. Not being inclined to do either, I moved on to other things.

The questions I'm interested in cannot themselves be settled by science. They are at once conceptual and metaphysical questions on which science is, by its nature, silent. How, then, are we to approach them?

2. Methods

A quick word about my subtitle. At one level, it's just another way of referring to the laws of nature: in most contexts, laws function as rules for making

---

[1] This kind of experience is hardly unique to me; cf. Marc Lange's (2002) preface, as well as Mumford (2004, 7).

The Metaphysics of Laws of Nature: The Rules of the Game. Walter Ott, Oxford University Press. © Walter Ott 2022.
DOI: 10.1093/oso/9780192859235.003.0001

calculations and predictions. I'm interested in working out what kinds of things the rules might be, and what the universe would have to be like if it's going to play by them. But at the metaphilosophical level, we can also ask: what are the rules of *this* game, the debate over the laws of nature? What counts as scoring a point? As winning?

Here's one way to do it, familiar from other parts of philosophy: aim for reflective equilibrium. One begins with a set of desiderata or *endoxa*, and then tries to find the view that preserves as many of them as possible, in as coherent a fashion as possible. These desiderata carry different weights: preserving moral responsibility is a reasonable expectation of an account of free will; turning out to be consistent with this or that theory of causation might be an advantage but a less important one.

The dangers of this method are obvious: we might end up with a garbage-in/garbage-out procedure. One's conclusions will only be as good as one's starting points and the various weights one assigns them. In some cases, the danger might not be very great: given some reasonable meta-ethical assumptions, for example, the project of reconciling and preserving intuitions about moral responsibility is likely to yield fruit. In the present case, however, the danger is very real. The concept of a 'law of nature' is not the common property of all agents, derived from sound practical experience. It's a highly artificial notion, introduced by natural philosophers to play a very specific role.

That, at least, is the conclusion of the first part of this book. If I'm right, we need to do some historical work to unearth the original notion of a law of nature. I argue that the right definition is a functional one. What matters is not the name but the role a given proposition plays relative to a particular scientific discipline. Alongside the historical work, of course, we have to do some evaluating: which features of the seventeenth-century concept are essential, and which are mere reflections of the era in which the concept was developed? Armed with such a genealogy, we'll be better able to formulate requirements that are rooted in the job description for laws, and not the results of historical accident. As we move through the historical figures, we'll develop a suite of desiderata we will want to hold later views accountable to.

Another reason to take the long view of the debate over laws is less controversial: we want to avoid re-inventing the wheel or, worse, re-inventing a wheel that didn't work in the first place. Some of these debates have happened before, and we ignore their earlier rounds at our peril. To give just one example: the view that bodies are endowed with powers that allow them to do what they do is experiencing a resurgence. But of course the seventeenth century was a watershed for natural philosophy partly because some of its leading figures—especially Descartes and Newton—ejected powers from their ontology. So before going 'back to Aristotle,' we should be sure we can answer the arguments that turned philosophers away from him, hundreds of years ago.

So I think it would be a mistake to pursue reflective equilibrium from the start. We need to back up a step and work out which features of the original concept are necessary for laws to do the work we expect of them, and which features need to be added. We can then go about awarding points for preserving these necessary features, and more generally weighing virtues and vices.

I've suggested that scientific results themselves, of whatever era, cannot by themselves settle the metaphysical and conceptual questions. But obviously they must be incorporated into the rules of the game. This happens in two ways. First, the scientific results form some of the data the conceptual work must analyze and be faithful to. If we arrive at a functional definition of laws, which takes lawhood to be a matter of playing a special kind of role in a given discipline, we will already be incorporating scientific practice in our theorizing. We'll continue to return to individual laws from different disciplines as a way of testing the philosophical views. Just as important, the deliverances of science are an important constraint on how the game is to be played. To give just one example: if the best going version of physics requires that some quantity—whether energy, momentum, or, at the quantum level, spin—is conserved, we had better not come up with a metaphysics that makes such conservation unintelligible.

3. Three Axes

We now have some sense how the game is to be played. Who are the players? We can sort the present competitors along three axes: top-down/bottom-up; realist/anti-realist; and one-to-one/one-to-many.

The early modern period introduced the notion of a law of nature by adopting what I call the 'top-down' position. On this sort of view, the laws of nature genuinely govern: they help to determine the course events take. You might know all the facts about the properties and positions of all objects in a given world and yet be at a loss to predict what will happen next. For what happens in that world is fixed by the laws of nature that obtain in it. In the modern period, these laws are divine volitions; in the contemporary debate, they are necessitation relations among universals or irreducible features of the universe. What all these views have in common is the idea that laws swing free of the things they govern. In contemporary terms, these views deny that laws supervene on what they govern. You can 'wiggle' the laws without wiggling any of the intrinsic properties of the objects whose behavior they dictate. There are lots of complications here, but the general idea should be familiar enough.[2]

Other views on this axis are 'bottom-up': one way or another, they make the laws supervene on the non-nomic facts. Consider the powers ontology, which the top-down view was designed to supplant. For Aristotle and his fellow-travelers, what counts is the arrangement of properties, at least some of which are causal powers. If there are any laws at all, they merely describe the powers that are responsible for the events we see around us. Such a view is bottom-up in the sense that what determines the course of nature is Nature itself. The powers ontology, declared quite dead by many of the moderns, has found itself back in favor after all these centuries. Whether it has been truly resurrected or merely disinterred is something we shall have to decide.[3]

Both the top-down and bottom-up views mentioned so far are 'realist' in the following, purely stipulative sense: they think something is responsible for making events follow the course they do. Not all accounts of laws accept that. On a broadly Humean position, the fundamental facts are neither powers nor laws, but the mosaic of property instantiations across time and space. To call something a 'law' is just to say that it states a regularity that plays an important role in scientific theorizing. The laws are given by the best system of axioms, where 'best' is defined by a balance between simplicity and strength. Such a view is anti-realist, in my technical sense of 'realism': no powers, necessary connections, or governance of any kind need apply. It is of course not anti-realist

---

[2] One such complication is worth mentioning: someone might argue that bodies inherit dispositions from the laws of nature, and claim that those dispositions should count as intrinsic properties of bodies. If so, then even the top-down view could agree that worlds that are indistinguishable in terms of bodies, their intrinsic properties, and their locations, will have indistinguishable histories and futures (assuming determinism). For my part, I can't see why a disposition foisted on an object by virtue of the laws alone should count as an intrinsic property.

[3] Another position in logical space is a powers view that is neither top-down nor bottom-up, simply because it denies that there are laws at all; see, e.g., Ronald Giere (1999). It's not uncommon to see proponents of the powers view reject laws of nature full stop. In her earlier work, Nancy Cartwright suggests there might be still be room for laws in her system, although they do not govern or necessitate anything; instead, laws 'describe what causes are capable of doing' (1993, 429). But in more recent work, Cartwright seems to have shifted to a more radical position: 'We do not predict and manage the world by tracing out the consequences of scientific laws, but by situation-specific models; and there are no fixed rules for how to build these models nor for how to evaluate their reliability' (2019, 5). There's also an independent movement toward the 'no-laws' view, one that grows out of the philosophy of biology. According to the 'new mechanism,' as it has come to be called, the principle of explanation ought to be 'entities (or parts) whose activities and interactions are organized so as to be responsible for the phenomenon' in question; see Stuart Glennan (2017, 17). See also Glennan (2017, 3–8), Bechtel and Richardson (1993), and Machamer, Darden, and Craver (2000).

in any other sense: which regularities are really out there in the world is a perfectly objective matter.

The top-down/bottom-up and realist/anti-realist axes cut across each other. The anti-realist Humean position is a bottom-up story: the arrangement of properties across space-time fixes the laws, and not the other way around. But it could hardly be more different from its bottom-up companion, the powers theory. While both reject the top-down governance of laws, the Humean also rejects any 'metaphysical glue' that might bind property instantiations. Only one of the four possible combinations is incoherent: no view counts as both top-down and anti-realist.

There is yet a third axis on which accounts of laws can vary: the relationship between law statements and their truthmakers. In some cases, it might seem as if we have a one-to-one law-to-truthmaker relationship. '$F=ma$' might be true in virtue of the fact that $F=ma$. But that is far from obvious. Lots of true propositions, perhaps most of them, cannot be paired with a single state of affairs that is its truthmaker (not, anyway, without a lot mereological gerry-mandering). Consider a proposition such as: 'the average U.S. taxpayer has 2.5 children.' There is no single, fundamental fact, *the average taxpayer has 2.5 children*, that makes this proposition true, for there is no single entity, *the average taxpayer*, to serve as an element in such a fact, nor should there be any such entity as *2.5 children* (absent a Solomonic decree). Instead, the truth-maker involves a hugely complex array of more circumscribed facts: facts about individual humans, their political and economic status (as taxpayers), and their activities. Philosophers from George Berkeley to Gilbert Ryle have rightly warned of assuming that any truth of the form 'x is F' requires x and F to exist as mind-independent individuals. Although some powers theories, for instance, take the one-to-one route and invoke a single power as the truthmaker for each law, more commonly they say only that the powers are responsible for the laws, leaving open the possibility of a one-to-many rela-tionship. So we cannot assume that locating the position of a view on the other two axes fixes its position on this one.

4. Structure

With these axes, we can chop up the territory among three families of views, and proceed one by one. First we have the top-down family; these are by their nature realist positions. Our next realist position is the powers view, which is realist and bottom-up. Finally, we have the anti-realist version of the bottom-up picture: Humeanism, or the 'Best System Analysis' ('BSA'). Within some families, we'll examine siblings that vary along the third axis, one-to-one and one-to-many.

The distinctive—though hardly the sole—early modern conception is a top-down picture, so we begin there (Part I). There is no doubt that the top-down view allows Descartes and then Newton to formulate laws or axioms that are broader in scope and different in kind from the vast majority of earlier so-called 'laws.' Whether *only* the top-down view permits the formulation of such laws is, of course, up for debate. I'll suggest that, although the top-down picture is probably a historical phase through which thinking about laws had to pass, it's not the only viable position.

The particular form the top-down view takes in the early modern period is explicitly theological. Indeed, I'll argue that it depends on the claim that God is the only true cause, apart perhaps from finite minds. Although the theological entanglements of the early modern view make it hard (for me, at least) to take seriously, it has several virtues worth preserving if we can. In particular, although the problem of *ceteris paribus* clauses is never, as far as I can tell, explicitly addressed by the Cartesians, I argue that their view prevents it from arising in the first place.

Next, we turn to the contemporary top-down descendants of that Cartesian picture (Part II). The first takes laws to be primitive, unanalyzable features of the worlds they govern. Although it preserves some of the Cartesian virtues, I argue that primitivism has too little to tell us, not just about what laws are, but what is more important, about what they do: how, exactly, is it that the inverse square law 'governs' the motion of objects? Further chapters investigate the possibility of treating laws as statements of necessitation relations among universals. By making revisions in response to objections, I try to distill the best version of a top-down view. Such a view will need to make the necessitation relations hold in every possible world; they should turn out to be 'internal' relations in the sense that they obtain solely in virtue of the natures of the universals they join. Making the necessitation relation internal is indispensable in replying to the most important objections.

Then we survey the bottom-up views, beginning with powers (Part III). On such a view, events take the course they do because of the nature and arrangement of powers. Whatever laws there are either supervene on, or are identified with, these powers. After assembling what I take to be the most defensible version of the powers theory, I turn back to the moderns. I reconstruct both their most powerful argument against powers as well as the alternative they try to put in its place.

Finally, I turn to the anti-realist positions (Part IV). After defending the Best System Analysis from a number of prominent objections, I pose one of my own, which I think gets at the heart of the problem with such views. I argue that the Best System Analysis goes wrong at the start by insisting that only statements of regularities are candidates for lawhood. Once free of the regularity requirement, we can see our way clear to a more defensible view that, like its predecessor, has no need of metaphysical glue.

I wanted to write a book that would reproduce the back-and-forth of figuring things out, rather than presenting the ossified results and then defending myself against all comers. At one time or another, I've found every view I discuss here plausible, with the exception of theism. So I try to show why the views are attractive. Then I use objections, not to shoot them down, but to refine them. Objections can be used as a cudgel or, as I (mostly) do here, as a sieve: to strain out the weakest versions of a position on the way to finding the strongest. Only at the end do I take sides. That final chapter was written after all the others were in draft, and had been allowed to sit for a considerable time. I felt, but resisted, the impulse to go back and kick over the traces.

Some readers will find this method irritating. I sympathize. On the bright side, such readers—if they are still with me at this point—can skip to the final chapter to find out where I end up standing. I think that would be a shame. Coming as it does at the end, the chapter can seem as if it's meant to be the apex of the whole project. Just the opposite is the case. That I end up endorsing this or that position is at most of biographical interest, something that is, even on my end, non-existent.

What I really care about are the arguments. Some are my own; some ring the changes on others'; some are culled from the literature. In response to those arguments, I've suggested changes to the views they are designed to undermine or support. As we go, I try to mold the best representative of each family before trying to choose among them.

PART I

OPENINGS

2

The Early Days

1. The Debate

Where and when, exactly, did the concept of a law of nature originate?[1] Sifting
through the literature on the subject, one finds a consensus emerging: something happened in the seventeenth century. And that's about it. How important, and how widespread, the change is are matters of dispute. Nor is there
agreement on whether what happened was unprecedented: on some views, the
moderns appropriate a concept that had been around much earlier, maybe as
far back as Ptolemy.[2]

I think two things are responsible for this disagreement. First, and most
obviously: the historians haven't been able to agree on the criteria for the
'modern' concept of law. But I shall also argue that, despite the seemingly
chaotic hunt, most of the pursuers in fact assume that laws must state
regularities, that is, true propositions that report on patterns of property
instantiations. If rigorously applied, that assumption would rule out the
paradigmatic conceptions of the nomic in the early modern period.

---

[1] The anglophone reader searching for seventeenth-century uses of 'laws of nature' must guard
against the inventions of translators. To give just one example: Stillman Drake renders the start of
Galileo's *Dialogue concerning the Two Chief World Systems* (1632) thus: 'Yesterday we resolved to meet
today and discuss as clearly and in as much detail as possible the character and the efficacy of those laws
of nature which up to the present have been put forth by the partisans of the Aristotelian and Ptolemaic
position on the one hand, and by the followers of the Copernican system on the other' (Drake in Galileo
2001, 9). But nothing in the Italian answers to 'laws': the word is plain old *ragioni*. Thomas Salusbury's
translation (1641, 1: 1) renders the phrase as 'natural reasons.' Elsewhere in the *Dialogue*, Galileo does
speak of '*leggi matematiche*,' but not '*leggi della natura*' or anything similar. As David Wootton notes,
'Galileo makes only three references to the laws of nature, on each occasion when he is arguing against
the theological objections to Copernicanism; there are no laws of nature in his more properly scientific
works' (2015, 371).
[2] John Henry (2004) provides a helpful taxonomy of accounts of the development of laws in the
seventeenth century. They fall into three main camps: (1) the view that laws first appear in the work of
Descartes, driven by the development of absolute monarchy (Zilsel (1942)); (2) the view that early
modern laws have their sources in thirteenth-century theology, specifically divine voluntarism (Oakley
(1961), Funkenstein (1986), Crombie (1994)); (3) the view that laws really develop in the medieval
period with mathematical sciences and then come to dominate thinking about physics in the seventeenth century (Ruby (1986)). Very useful overviews include Steinle (2002) and chapter 9 of Wootton
(2015). An excellent recent treatment of the development of the concept of laws is given by Stathis
Psillos (2018).

The Metaphysics of Laws of Nature: The Rules of the Game. Walter Ott, Oxford University Press. © Walter Ott 2022.
DOI: 10.1093/oso/9780192859235.003.0002

2. The Wrong Criterion

When historians search for the origin of our concept of a law of nature, what exactly are they looking for?[3] One feature is uncontroversial: specificity.[4] A law of nature should have some precise content and not simply be a way of lauding the general orderliness of nature, as Cicero and Aquinas both do.[5] Within the context of medieval theism in particular, it is hardly more than a pious platitude to say that the world is governed by God's laws, or that witchcraft, for example, violates (or attempts to violate) the laws of nature.

On the surface, it seems no other criteria have gone unchallenged. For Edgar Zilsel, writing in 1942, Descartes creates the 'modern' concept of law by fusing 'the basic idea of physical regularities and quantitative rules of operation' with the Biblical idea of God legislating for nature.[6] Jane Ruby disagrees: on her view, the theological element is insignificant. Indeed, she takes the distinctive features of laws to be their descriptivity and freedom from divine legislation, and finds the notion in Roger Bacon's optical works in the thirteenth century.[7] More recently, J. R. Milton's work suggests others: in addition to the specificity all parties agree to, he seems to require that laws explicitly play an explanatory role and perhaps be the object of consensus.[8] But there seem to be plenty of pre-modern candidates, including the epicyclic astronomy of Ptolemy, that might meet those criteria.[9]

For all this, most parties to the debate share a remarkable stretch of common ground. For both Zilsel and Ruby, as well as more recent commentators, 'our' concept of law—the one we should be looking for—is the concept of a regularity.[10] This is perhaps not surprising, given the depth and reach of the regularity theory in the twentieth century: one even finds more reflective

---

[3] One might worry that, given the diversity of positions we'll examine in this book, there is no single concept of a law of nature at all; the search for its antecedents is thus pointless. Others entertain this possibility; see Roux (2001, 570). I think the thin concept of laws that I develop below is thin enough to ameliorate this worry.

[4] Daryn Lehoux (2006, 528) makes this point well.

[5] Aquinas, *Summa Theologicae* Part I q. 103, art. 1 (1997, 2: 951) cites both Cicero and Aristotle as agreeing with the claim that the world is governed, and adds elsewhere that this 'government of things in God, the ruler of the universe, has the character of law' (*Summa* Part II, q. 90, art. 1 (1997, 2: 748)). A useful review of occurrences of this 'loose' sense of laws of nature is to be found in Henry (2004).

[6] Zilsel (1942, 265).

[7] For a very helpful discussion of the criteria Zilsel, Ruby, and J. R. Milton (discussed below) apply for the presence of the concept of law, see Lehoux (2006).

[8] See Milton (1998).

[9] See Lehoux (2006). There is more to Milton's account, though, and we'll return to it below.

[10] Consider how Zilsel introduces his inquiry, in one of the first serious efforts at the history of the concept: 'The naturalist observes recurrent associations of certain events or qualities. He is convinced that these regularities, observed in the past, will hold in the future as well, and he calls them "laws of nature," especially if he has succeeded in expressing them by mathematical formulas' (1942, 245). Jane

scientists, such as the 1942 Nobel laureate in physics, Eugene Wigner, noting almost in passing that laws are nothing but 'the regularities of the events.'[11]

The problem is that if we build in the regularity criterion, we at once rule out Descartes, Newton, and the rest of the seventeenth century (or so I'll argue). One would have to come forward at least as far as David Hume to find such an identification. If laws have to be regularities, their inventor in the modern period comes pretty late in the game.

What of the requirement that laws be descriptive, not prescriptive? At first it's a little hard to see what the difference is. The *content* of a law is always descriptive: if God decides it shall be a law *that bodies move in a straight line unless something interferes*, the italicized bit has to be descriptive. The prescriptivity comes in only when we talk about what we might call the force of the law. If one hunts for laws free of this theistic baggage in the paradigmatic early moderns—Descartes, Malebranche, Berkeley, Newton—one is bound to be disappointed. If anything, the notion that laws have prescriptive force—in that they are rules God follows in his own actions—gains ground during the seventeenth and eighteenth centuries. Here are some representative examples:

- When, in 1690, the term (in its non-moral sense) first appears in the *Dictionnaire* of Antoine Furetière, it receives this definition:

LAW, also called rules, and maxims of science, and even of games. The *laws* of optics mean that the angles of incidence are equal to those of reflection. . . . The *law* of the game is that whoever quits the game loses.

In the 1727 edition, as Sophie Roux documents, the entry has changed; rather than purging the definition of its normative cast, the entry exaggerates it:

LAW, also called general rules of nature. . . . God acts always in a manner that is simple, uniform, and according to the general *laws* he has established.

Ruby opens her inquiry by describing the 'modern' concept of law thus: 'the intelligible, measurable, predictable regularities' found in nature (1986, 341). The tradition continues in Sophie Roux's impressive work (2001); she isolates the classical sense of 'law' by pointing to two ingredients: a law 'reveals a natural regularity' (the 'physico-mathematical' sense) and '[t]he laws of nature, which apply without exception to all natural bodies, constitute and determine their entire behavior; laws are literally the laws that a divine legislator has imposed on nature' (the 'metaphysical' sense) (2001, 542, translation mine). John Henry's otherwise excellent paper on the origins of laws treats them as 'regularities in nature which are assumed to reflect an underlying causal necessity' (2004, 74). Daryn Lehoux takes laws to be 'generalizations about natural regularities' (2006, 546).

[11] Wigner (1967, 42).

BAY[LE] God does not break the *laws* of nature except in order to bring about miracles. MAL[EBRANCHE].[12]

- Pierre-Sylvain Régis's *L'Usage de la raison et de la foi* (1704) includes a dictionary that distinguishes between natural laws in the moral and non-moral sense; the latter are 'those [laws] according to which God acts in a manner we can understand.'[13] The laws are laws *for* God's actions; they are laws he himself follows.
- In 1747, Willem s'Gravesande prefaces his popular *vade mecum* of Newton's *Principia* with the lines,

> *What Laws JEHOVAH to himself prescrib'd,*
> *And of his Work the firm Foundation made,*
> *When he of Things the first Design survey'd.*[14]

If we take the purging of divine legislation to be constitutive of 'our' or 'the' concept of laws, the modern period is a low water mark. Compared to prior centuries, the seventeenth shows natural philosophers surprisingly willing to bring in considerations about God; the scholastic period, by contrast, is marked by a sharp and often rigidly policed border between natural philosophy and theology.[15]

In fact, I'll argue that the concept of law in Descartes and Newton crucially relies on divine legislation and therefore on ubiquitous divine action. It is precisely this theological background that untethers the laws from the objects whose behavior they mandate and allows for the thorough-going mathematization of nature. But before we get there, we need a neutral concept of laws that will let us bring into relief all the variations among the substantive positions.

3. The Thin Concept

I propose we wipe the slate nearly clean, retaining only the specificity criterion. Let's try a different strategy: looking at some paradigmatic cases and trying to tease out the minimal features that unite them. We should bracket any areas of controversy and look for a 'thin' concept of laws that will at least allow us to

---

[12] Roux (2001, 565–6, my trans.).   [13] Régis (1996, 959, my trans.).
[14] s'Gravesande (1747, 1).   [15] See Shank (2019).

zero in on our subject matter. Later on, we'll use it to work out what, if anything, is new in the modern's 'thick' concept of laws.

To begin with, then: what are some of the things that *have* in fact been accounted 'laws of nature'? Even this purely descriptive question is ambiguous: are we asking about propositions or statements, on one hand, or states of affairs on the other? The phrase 'law of nature' does duty for both.[16] Let's begin by focusing on law statements, and worry about the ontology later. From this vantage point, the laws of a given science are sentences or propositions that are accorded a certain status and play a certain role in that science. Let's start with an uncontroversial example:

$$F = G\frac{m_1 m_2}{r^2}$$

Any two masses, m_1 and m_2, exert an attractive force F on each other that varies with the distance between them, r. We can calculate the value of this force by multiplying the gravitational constant, G, by the product of the masses and then dividing our result by the square of the distance between them. Gravitational force, then, is inversely proportional to the square of the distance.

Taking Newton's inverse square law as our paradigm, we can say something about what statements of laws typically look like. First, they tend to be general: they do not refer to particular points in space and time. Another way to put this is to say that they range over *all* such points indifferently. Nothing about the inverse square law limits it to any region of space; surely part of Newton's advance over medieval physics lies in his unification of celestial and terrestrial mechanics. A boulder rolling down a mountain and a planet circumnavigating the sun are superficially quite different but ultimately explicable (at least in part) by the operation of the same force whose value is given by the inverse square law.[17]

Second, law statements play what I'll call the 'axiom role' in the science in question.[18] Sometimes this is captured by saying that laws have to be 'fundamental' to the science; they typically shouldn't appear as downstream corollaries

[16] James Woodward (2018, 167) offers a nice discussion of the 'dual or Janus-faced character' of laws.

[17] Note that my thin concept is meant to include, but not necessarily be limited to, what Tim Maudlin calls 'fundamental laws of temporal evolution,' or 'FLOTES': laws that describe 'the evolution of physical states through time' (2007, 12). I certainly agree with Maudlin that many paradigmatic laws are FLOTES, or at least LOTES. But since FLOTES presuppose a direction of time, that category might not include all the laws we would want. For criticism of Maudlin on the direction of time, see esp. Eddy Keming Chen and Sheldon Goldstein (forthcoming, 17).

[18] I realize that the term 'axiom,' in the context of the debate over laws of nature, is closely associated with the Best System Analysis of David Lewis and others. I intend the term here instead in its historically dominant sense.

to other propositions.[19] What is much more important, playing the axiom role means being suited to the derivation of a wide range of states of affairs, symmetrically in time and space.[20] (It does *not* mean 'being called a law' by any particular theory or science.[21]) The basic idea is that laws are defined by what they do: they have to let us make predictions and retrodictions. When the laws have mathematical form, they let us calculate values for the variables they range over. In the early modern period, when the mathematization of nature is well under way, it's quite common for natural philosophers to speak indifferently of 'rules' and 'laws.' Huygens, for example, seems to use the terms interchangeably, when he writes in 1669 of 'the general rule for determining the movement acquired by hard bodies by their direct collision.'[22]

What actually plays the axiom role in a given science is of course relative to the history and state of that science. Some proposition or equation P might count as a law relative to one science at a given time but get displaced by a proposition that allows the derivation of P. From a bird's-eye view, this is what happened to Kepler's laws when Newton came along.

Playing the axiom role enables laws to function as explanations. The philosophy of explanation is of course a subject in its own right.[23] I don't hope to give a reductive or even complete account of explanation. Plenty of explanations don't involve laws at all: to explain how to get to Sesame Street, or the rules of Hanoverian succession, one needn't feel compelled to adduce a law

---

[19] I am grateful to an anonymous referee, who pointed out that sometimes a consequence of a law is itself counted as a law, as when the claim that charged particles repel each other is treated as a law, even though it's a consequence of Coulomb's law. And Gauss's law and Coulomb's law are standardly taken to be derivable from one another. What is important for present purposes, however, is that laws be suitable to play the axiom role; talk of laws not being 'downstream' is just an intuitive way of making this point. I am happy to treat the axiom role as relative to the proposition's role in relation other propositions, such that in some contexts, the same proposition counts as a law and in others not. Others may wish to be less ecumenical, and insist on fundamentality relative to the science as a whole.

[20] John Roberts provides two criteria we might consider for the own 'axiom role.' As he puts it, the 'special role' played by laws has two broad features: (i) 'a typical application of Newton's theory [for example] involves setting up and solving a set of differential equations for a given set of initial conditions; the laws of that theory are the sources from which the differential equations are derived'; (ii) 'the laws of the theory often play a crucial role in establishing the reliability of a method of measuring something; for example, Newton uses the law of universal gravitation to justify his method of estimating the mass ratios of the bodies in the solar system' (2008, 84). Although I agree with Roberts, I wouldn't want to build these into the axiom role, given that the axiom role is meant to define the thinnest possible concept of laws.

[21] Roberts points out that such a functional definition of a law can explain why it is that 'those who write about laws of nature seem to have no trouble identifying what the laws of various theories are,' even when none of them 'has a standard name with the word "law" in it' (2008, 84). So when skeptics about laws point out that you can read vast swaths of scientific literature without encountering the word 'law,' they do nothing to establish that the scientific inquiry in question is operating without laws.

[22] '[L]a règle generale pour determiner le mouvement qu'acquierent les corps durs par leur rencontre directe . . . ' (1669, 29).

[23] A fine recent treatment is Skow (2016); see also Salmon (1984) and (1989).

THE THIN CONCEPT 17

of nature.[24] Where laws *do* play a role in explanation, we can distinguish two very different questions:

(Case)   What is it for a law of nature to explain its instances?[25]

(Whole)   Under what conditions does a theory of laws promise to explain the whole mosaic?

Whether any account of laws ought to provide a whole explanation in the first place is a matter of debate; it's especially controversial in the case of the Best System Analysis. For now, I want to focus on case explanation. I think it's common ground that *if* there are laws at all, they should help explain their instances.[26]

My claim is a conditional one: *if* a law of nature is involved in an explanation, it should at least unify cases, and show how the present case is similar to others. Armed with our laws of free fall, our explanation of the motion of a dropped bowling ball should allow us to see what the behavior of the ball has in common with the motion of other unsupported objects near the Earth's surface.[27] Again, my claim is not that this kind of assimilation is sufficient for explanation generally, or even for explanation involving laws. I do think it's a minimal, necessary condition for laws to function as *explanans*. What kind of law of free fall would we have, if it couldn't show us what dropping a bowling ball on your foot has in common with parachuting?

This broad-stroke notion of assimilation is the only kind of case explanation I want to build into the thin concept. It's part of the axiom role laws have been expected to play since the beginning. Let me just flag a second possible element of case explanation. It seems plausible that a nomic explanation should point to relations of counterfactual dependence. If we cite laws to explain the expansion of a metal object in a room with increasing ambient temperature, we're at least committed to the claim that, had that temperature been different, the metal would have behaved differently. The basic idea is that laws state relationships among variables, and allow us to see how results depend on input

[24] The Hanoverian example is to be found in Scriven (1959, 452), and discussed in Skow (2016, 10).

[25] Note that my distinction between case and whole explanation is orthogonal to that between scientific and metaphysical explanation in the context of the Best System Analysis. See below, Chapter 13, section 2.

[26] One notable outlier here is Thomas Brown; see below, Chapter 13, section 2.

[27] I take this idea to be common between the deductive-nomological account of explanation and 'unification' accounts such as those of Kitcher (1989) and Friedman (1974), though I realize those accounts diverge in other ways.

conditions.[28] Whether nomic explanation involves counterfactual dependence of this kind is controversial, but it's worth keeping in mind as we go.

So we have our stipulative, 'thin' concept of laws: laws are general propositions that play the axiom role relative to a given science, where part of the axiom role involves explaining instances of the law at least in the sense of assimilating them to other cases. I think this thin concept of laws has a pretty good claim at being 'our' concept of laws, and also one we can sensibly look for in earlier philosophers without begging any questions. The thin concept says nothing at all about the ontology: it doesn't say that laws are regularities, or divine decrees, or powers, or anything of the sort. Nor does it even say that laws *entail* regularities. Consider Newton's first law: '[e]very body perseveres in its state of being at rest or of moving uniformly straight forward, except insofar as it is compelled to change its state by forces impressed.'[29] On its face, that law neither states nor entails any regularities at all, since every body is indeed being acted on by some force or other at every moment. Unless those forces are precisely balanced, no body will persevere in its current state. One virtue of our thin concept is its silence on the whole question of the laws' relationship to regularities.

Nor does the thin concept say that laws have to govern anything. Some—both skeptics and proponents of laws—wish to build governing into their concept. Stephen Mumford, for example, argues that denying laws a governing role is tantamount to denying laws *tout court*.[30] This claim is of course controversial. So for our purposes, it's much clearer to have a thin concept that sidesteps governing and doesn't take a position on the top-down/bottom-up distinction.

To help make the case that the thin concept fits what we typically call 'laws,' I propose to look at a few more examples. Some of these will also be with us later in the book, so it helps to have them on the table. I largely prescind from mathematical detail in their statement; those wanting quantitative statements can easily find them elsewhere. For our purposes, it's the ideas behind the equations that count.

Most of the examples we'll deal with in the book come from the world of classical mechanics, so we'll start there.[31] It may be that these laws are not fundamental, if it turns out they can be reduced to the laws of quantum

---

[28] As James Woodward puts it, 'explanation is a matter of exhibiting systematic patterns of counterfactual dependence' (2003, 191).

[29] Newton (1999, 416).

[30] Mumford (2004, 143–59); see also the symposium on Mumford's book in Bird et al. (2006).

[31] Mark Wilson (2013) and (2017) has persuasively argued that the very notion of 'classical mechanics' is suspect. Wilson argues that there is no single, unproblematic set of coherent claims

mechanics.[32] I take no position on that claim, nor on whether or not quantum mechanics will really turn out to have laws at all.[33] And there are of course contexts in which classical mechanics breaks down. Still, the laws of classical mechanics are at least the *sort* of thing we would want a philosophical theory to cover. We'll be returning to quantum mechanics, especially the question of entanglement, below; but for now, let's make things easy on ourselves.

- Energy conservation. When you toss a ball into the air, it leaves your hand with a certain amount of kinetic energy. It will return to your hand with exactly that same amount of kinetic energy, which entails that the speed at which it leaves your hand is exactly the same as when it returns. What happens as it rises from the surface of the Earth? On the way up, kinetic energy is getting converted into gravitational potential energy; at the top of its arc, the conversion has been completed: there is no kinetic energy at all. On its return, the ball has to exchange its gravitational potential energy for kinetic. So the amount of kinetic energy on its return must equal the kinetic energy on its departure.

Where do all the 'must's and 'has to's come from? Emmy Noether's theorem says that if time is homogeneous, then the total energy is a constant of the motion. This means that having put kinetic energy into the ball at the start, that energy is not fading away but merely getting converted into a different form.

In practice, of course, things do not work quite this way. In the real world, although energy is not annihilated, it is 'leaking' out of the system defined by the ball. Some amount of energy is being dissipated into the environment through friction with the surrounding air. If we switch examples—to a car traveling down the road at constant speed—this is even more clear. Energy is

---

that fall under that heading. Instead, as he sees things, 'classical mechanics' is a hodgepodge of different methods and assumptions that need to be invoked in different contexts. For example, dimensionless point masses are subject only to 'action-at-a-distance forces'; when macro-level objects collide, we also need to invoke 'contact forces,' to explain, e.g., how the surfaces of colliding bodies are deformed (2013, 45), and it is far from obvious how the two kinds of force are supposed to act in concert. I take Wilson's points but still find the category of 'classical mechanics' useful.

[32] For example, it may be that the Schrödinger equation entails Newton's second law. For doubts about such a reduction, see Healey (2017, 240–1). Other approaches simply reject classical mechanics, treating it as only a useful approximation of the behavior of a wide, though far from exhaustive, range of bodies, and then only in certain circumstances.

[33] Tim Maudlin calls the Schrödinger equation 'the fundamental law of non-relativistic quantum mechanics' (2007, 11); Healey, by contrast, claims that the equation is not a law at all, but 'a constraint on models of non-relativistic quantum mechanics' (2017, 240). From my point of view, what counts is whether quantum mechanics puts the Schrödinger equation in the axiom role.

continuously being transferred from the car engine to the air and the road, an exchange that is described by the first law of thermodynamics: $\Delta E = W + Q$. Here, the energy of our system changes if work (W) is done on it by the environment or if energy (Q) is transferred across the system boundary due to a temperature difference between the system and the environment. Although energy can come and go from real-world systems (as none of them is isolated), the point is that it must come from or go to somewhere, rather than simply popping up on the scene.[34] Still, the law of the conservation of energy has the same features as the inverse square law: it makes no essential reference to any points in space and time and it permits the inferences we need for prediction and explanation. Nor is it limited to Newtonian mechanics: Noether's theorem retains energy conservation in the context of general relativity.

We find quantities being conserved at every level of analysis, from the gross macroscopic level to the sub-atomic. A list of conserved quantities would have to include 'energy, momentum, angular momentum, charge, lepton number, [and] baryon number,' along with the 'spin' of quantum mechanics.[35] So conservation laws will be important test cases for any philosophical analysis.

- Coulomb's inverse square law. In addition to gravitation, there seems to be another force at work, one that governs charges. Particles with like charges repel each other; with unlike charges, they attract. The force of attraction or repulsion is inversely proportionate to the square of the distance between the particles.

These examples fit our thin concept of laws. They are general statements with content that suits them to the role they do, or once did, play: axioms of a theory that allow us to derive conclusions irrespective of the quantities taken by their variables.[36] I am less sure that other so-called laws meet even the meager criteria of the thin concept.[37] Let's consider some other candidates:

---

[34] My thanks to Doug Juers for setting me straight on the details, here and throughout this section (any mistakes that remain are mine).

[35] Bigelow, Ellis, and Lierse (2004, 154).

[36] Another commonly proposed feature is the claim that law statements introduce opaque contexts; see Dretske (1977). I'm unsure about this, so I don't want to build it into the thin concept.

[37] I leave aside here the fascinating case of laws in biology. One might think that the Hardy-Weinberg equation is roughly analogous to a law of inertia, in that it tells us that the degree of genetic variation in a population will remain the same unless interfering forces are introduced. And yet, Marjorie Grene and David Depew (2004, 264) argue, the equation is not spatio-temporally unrestricted, unlike its Newtonian counterpart: the equation 'describes the working of an admittedly large, but finite number of historical entities that come into existence at a certain time and go out at another.' Ernst

- The ideal gas law.[38] One aspect of the ideal gas law ('Charles's law'), says that the ratio of volume to temperature will remain constant. As temperature increases, so does volume. (When we add that density is inversely proportional to volume, we can see why planes have a hard time taking off in extreme heat: the medium on whose resistance they depend has been thinned.) I have some worries about counting this one as a law. It assumes the gas is ideal in the sense that its particles have no volume, which of course is false. Whether we want to count this a law in the thin sense or just a close cousin seems to call for decision rather than discovery.
- Price and demand. As prices rise, demand falls. This is only one formulation of a so-called 'law of supply and demand.' There are many cases where this one fails: in so-called 'Giffen goods,' demand and price rise together. Indeed, Apple's entire business model seems predicated on the falsity of the 'law' of price and demand. We have strayed so far from our initial paradigmatic laws that the term seems applied in this case merely metaphorically, a glorified 'rule of thumb.'

No doubt that there are many marginal cases over which we might want argue. But I think the thin concept is sufficiently clear and uncontroversial to be getting on with. Let's turn now to the origin of the thin concept before turning to the moderns' ways of thickening it up.

4. Starting over

I don't aim to give a complete history of the thin concept. That would be a substantial project in the history of ideas. I do think we can sketch enough of the background for our purposes here.

The thin concept requires only that laws be general statements that play the role of an axiom in a theory. We can trace its origins at least as far back as

Mayr (1988) explicitly rejects the idea that biology has, or even should aspire to have, laws at all, understood on the model of physics; for further discussion, see Grene and Depew (2004, 265–7). On the other side of the debate, Peter Turchin (2001, 18) argues that population ecology has plenty of laws of its own, particularly the law of exponential growth: 'a population will grow (or decline) exponentially as long as the environment experienced by all individuals in the population remains constant.'

[38] One might argue that this law is no longer fundamental, since Rudolf Clausius showed in 1857 how to derive it from kinetic theory. As an anonymous referee pointed out to me, however, there's no need to insist on absolute fundamentality: arguably, any science other than physics will fail to produce laws that are fundamental full stop. So on my view, the ideal gas law meets the criterion of relative fundamentality written into the thin concept of laws.

Aristotle's principles in *Posterior Analytics* I.10. Aristotle claims that any chain of demonstrations must depend on some propositions that do not admit of further proof, on pain of regress. Some principles are proper to a given science, but others are common, such as the fact that 'if equals are taken from equals, the remainders are equal.'[39] Whether or not the line of influence is direct, this classification also appears in Euclid. His *aitemata* or 'postulates' are particular to geometry, but his 'axioms' or 'common notions' (*koinai enniai*) are not; he lists Aristotle's principle about equals as his third axiom.[40] So right from the start, the basic framework is one of a hierarchy of propositions, in which principles are singled out as fundamental and not susceptible of independent proof, at least within the domain at issue. A wide array of propositions might fall under this umbrella. Throughout the seventeenth century and later, such metaphysical dicta as 'every effect has a cause' and 'every body occupies space' are explicitly called 'laws,' 'principles,' and 'axioms.'[41]

There is nothing distinctively modern or contemporary in the thin concept: Roger Bacon's *leges* of reflection and refraction and Ptolemy's law of planetary motion seem to meet the criteria for the thin concept of law. Nothing in the thin concept mentions regularities; nor does it tell us whether laws are divine commands, whether they govern events, or indeed anything at all about the ontology. To say that the thin concept goes back much further than the modern period is not to say much. After all, the thin concept is what contemporary Humeans, top-down theorists, and proponents of the powers view have in common. And by its very thinness, it flattens out any differences between the seventeenth century and earlier ones.

Our search, then, is for competing thick conceptions of laws; to be conceptions *of* laws, they have, trivially, to satisfy the thin concept—otherwise they fail to be in competition. What, then, really is new in seventeenth century concepts of law, not in relation to us, whomever 'we' might be, but in relation to the scholastics? Even those writers who date the invention of the concept well before the modern period acknowledge that Newton's and Descartes's

---

[39] *Posterior Analytics* 76a41–42, in Aristotle (1984, 1: 124).

[40] For a helpful discussion of the relationship between Aristotle and Euclid, see Katz (1998, 55).

[41] In the 1649 *Syntagma*, Gassendi tells us that it is a '*lege naturae*' that bodies occupy places (1658, 1: 381); for discussion, see LoLordo (2006, 222), who notes that this is one of the very few occurrences of the phrase in Gassendi's corpus. Walter Charleton (1654, 263) lists the same principle, and adds a catalog of 'the *General Laws* of Nature,' including: '(1) That every Effect must have its Cause; (2) That no Cause can Act but by Motion; (3) That Nothing can act upon a Distant subject' (1654, 343). These broadly Epicurean philosophers are joined by Descartes's successor Jacques Rouhault, who lists a set of '*axiomes*' that includes the principle that every effect has a cause (1671, 28). As late as the early nineteenth century, Mary Shepherd adduces the following law: 'that similar qualities in union necessarily include similar results' ('That human testimony is of sufficient force to establish the credibility of miracles' (1827), in Shepherd (2020, 167)).

laws have far greater scope than Bacon's or Ptolemy's: they are not intended to apply only to optics, or only to the movements of the planets, but to all bodies. I think the difference runs deeper.

To get at this difference, we first need a keener sense of the thick conception of laws-cum-principles in the pre-Cartesian world. Any laws or principles to be found outside of mathematics and geometry are going to need a foundation in the nature of things. Scipion Dupleix's *Physique*, first published in 1603 but re-published some twenty times through 1645, illustrates the point. For Dupleix, the *principes* of natural philosophy are form, matter, and privation: these are the fundamental notions that provide us with our starting points for demonstrations.[42] To start where Dupleix starts is to block from the beginning any attempt to provide a mathematical treatment of nature. Put differently: the chief development of the seventeenth century is the extension of *mathematical* principles from the realm of geometry and even astronomy to the whole of the natural world.

Consider what Newton claims sets 'the moderns' apart from their ancestors:

> Since the ancients (according to Pappus) considered *mechanics* to be of the greatest importance in the investigation of nature and science and since the moderns—rejecting substantial forms and occult qualities—have undertaken to reduce the phenomena of nature to mathematical laws, it has seemed best in this treatise to concentrate on *mathematics* as it relates to natural philosophy.[43]

Newton contrasts the substantial forms and occult qualities of the scholastics with the laws he and his fellow moderns propose. He is of course keen to emphasize that his *principia* will be mathematical: quantitative, not qualitative. But this is tied up with—and made possible by—the shift from explanation in terms of the forms and qualities of things to explanation in terms of laws.

For the scholastics, 'laws' play a much more limited role in natural philosophy simply because the fundamental unit of explanation is qualitative, not quantitative: it's the nature or essence of the objects involved in the event. The gold standard for Aristotelian natural philosophy is demonstrative explanation built on principles that capture the essences of things.[44] The basic idea

---

[42] Dupleix (1640, 85–6).   [43] Author's Preface to the 1687 *Principia* in Newton (1999, 381).

[44] I think J. R. Milton gets it right when he says, 'The fundamental reason why no clear well-defined notion of a law of nature had emerged by the end of the sixteenth century is that there was no room for any such idea within the inherited and still intellectually dominant systems of Aristotelian physics and epicyclic astronomy . . . What was still lacking was a new kind of natural philosophy, which could serve

from the *Posterior Analytics* is carried right through the early modern period in the works of such authors as John Sergeant, whose *Method to Science*, as late as 1696, still prescribes the syllogism as the basic tool for natural philosophy. Like Aristotle, Sergeant takes the premises of such demonstrations to be necessary truths that reflect the essences of things.[45] While one might formulate many such propositions that satisfy the thin concept of laws, they will not be conducive to mathematical formulation. Seeing the world as populated by a panoply of kinds, each conferring its own powers, is, if not an insuperable barrier, at least hardly conducive to formulating specific laws that apply to all bodies as such, such as the Galilean laws of free fall.

An interesting case here is that of Francis Bacon, whose *Novum Organum* appears in 1620. Bacon was well aware of at least some of Galileo's writings, and engaged in a dispute with him over the explanation of the tides.[46] Bacon was also aware of Kepler's work. But when Bacon crafts his own notion of laws of nature, he casts it in a thoroughly Aristotelian mode: a law, for him, is a law *of* a given nature, such as heat. There is no suggestion in Bacon, as far as I can tell, that there might be laws that are true propositions that both apply to all bodies as such and are capable of being formulated in mathematical terms.[47]

By contrast, the Cartesian accounts we'll focus on leave behind the natures of created substances and are independent of the bodies they regulate. Their laws are top-down, not bottom-up.[48] Here's one way to put the contrast. Consider the Humean mosaic, the instantiation of properties at points in space time. A top-down picture insists that the laws are not fixed by the

---

as a satisfactory replacement for scholastic Aristotelianism' (1998, 684). Against Milton, Daryn Lehoux argues that 'there is nothing about Aristotelian philosophy that has been shown to preclude such laws'; that is, there is no inconsistency between the claims that things have essences and that '[n]atural regularities can be described by laws' (2006, 540). But the Aristotelian approach does block the top-down picture. Since the scholastics take the powers of created beings to be responsible for their behavior, they are committed a bottom-up view.

[45] See Sergeant (1696). Sergeant offers four kinds of demonstrations, from each of the four causes. '[T]is evident that we can Demonstrate Proper Effects from Proper Efficient Causes, which we call Demonstrating *a priori*; and Proper Efficient Causes from Proper Effects, which is call'd Demonstrating *a posteriori*. For, since a *Cause* and a *Reason* do only differ in this, that the word (Cause) speaks the thing as it is in *Nature*, and (Reason) the same thing as 'tis in our *understanding*; and Proper Causes and Effects in Nature are necessarily *connected* to one another, and, consequently, do *Infer* one another naturally; it follows, that those Causes (and, for the same reason, Effects) as they are *in our Understanding* must be the *Reason why one infers the other* in our Understanding: Whence follows, that those Causes and Effects can be used as *Proper* Middle Terms to *Infer* or *Conclude* one another' (1696, 276–7). For example, Sergeant thinks we can demonstrate, from the 'notion' or essence of fire, that it '*Desegregates* the *Heterogeneous* parts' of what it touches (1696, 282).

[46] As Lisa Jardine and Michael Silverthorne note in their introduction to Bacon (2000, x).

[47] I cover Bacon in more detail below (Chapter 7, section 2).

[48] Milton also makes this sort of observation, though without using the same terminology (1998, 684). I draw the distinction in my (2009, 5–6). More recent authors to emphasize this feature of the moderns' conception include Harrison (2019, 62).

mosaic. It's just the other way around: laws fix the mosaic. To sharpen the point, we can ask the following question: suppose the intrinsic properties of bodies and their locations were fixed to the last detail. Would subsequent events (human and divine intervention aside) necessarily follow a single course? The bottom-up picture answers 'yes,' for it derives whatever predictions it can muster from the natures of bodies. The top-down picture answers 'no,' for it locates the explanation and governance of events in laws imposed by a divine being.

If this account is broadly right, then to hunt for an early modern concept of laws as regularities is to miss the crucial point. A regularity theory is resolutely bottom-up: it takes the regularities in nature to be fundamental and casts their statement in science as their shadow. By taking a top-down view of laws, the Cartesians are able to wrest the framework of laws as axioms free from its geometrical and Aristotelian roots.[49] The moderns don't conjure a thin concept de novo, but they do fill it out in ways incompatible with any that might be found in their scholastic predecessors.

Another question remains: why should the transformation of laws in the seventeenth century have needed divine backing? Our case studies of Descartes and Newton below provide detailed answers, but in broad strokes, I think the story is this. The new science of Galileo, Kepler, Descartes, and the others hijacks the axiomatic structure of Euclidean geometry. But in order to cast its basic principles in mathematical terms and apply them to natural philosophy, it needs to yank this hierarchical structure from free from the Aristotelian framework of principles in natural philosophy such as form, essence, matter, and the rest. But we then have an edifice whose buttresses have been removed. If the laws or principles can no longer be taken for granted, as in Euclidean geometry, or founded on familiar Aristotelian notions, how will they get the ontological support they need? This is where the moderns draw on the tradition of divine voluntarism that runs back to the thirteenth century: the requisite backing can be found in God's will.[50]

---

[49] I speak here of 'the Cartesians' rather than the moderns generally. As we'll see in later chapters, there are plenty of figures on the other side, who resist the new top-down conception and strive to preserve a roughly Aristotelian picture of metaphysics and natural philosophy. Here one might list not just Aristotelian die-hards like John Sergeant but Pierre-Sylvain Régis, Baruch Spinoza, John Locke, and G. W. von Leibniz.

[50] I am indebted here to John Henry's perceptive discussion. As Henry sees things, this appeal to medieval voluntarism is the grain of truth in the work of Oakley (1961) and Funkenstein (1986). (For a more recent statement of Oakley's view, see Oakley (2019); for further criticism of the claim that divine voluntarism is the moving force in seventeenth century concepts of law, see Harrison (2002)). Henry writes, 'By claiming that this theology led to the concept of laws of nature, Oakley, Funkenstein and their fellow travelers were putting the cart before the horse: in fact, the theology was taken up in order

To sum up: as I see things, the thin concept antedates the moderns. Nevertheless, the occasional outlier aside, even the thin concept of laws does not really take off in the realm of natural philosophy until the modern period. Roger Bacon and Ptolemy, as we've seen, can lay claim to it, and I have no doubt others lurk in the history of science, waiting to be found. But only during and after the modern period do competing thick concepts begin to emerge, growing alongside a new crop of specific candidate propositions for lawhood. While the Cartesians can hardly be said to have invented the thin concept, it was their particular *thick* concept that allowed for the break from the increasingly moribund Aristotelian view. And in breaking from that view, they made possible the whole project of assembling mathematical formulae to play the axiom role.[51]

5. Descartes

Let me back up these programmatic claims with a reading of the first figure among the moderns to put laws front-and-center in his natural philosophy: Descartes. My central claim is that the new 'way of laws' leads to a version of occasionalism, the view that God is the only cause of body–body interaction.[52] So far from being an aberration of seventeenth-century France, occasionalism is the dominant metaphysics of the way of laws. It reaches through the work of Descartes, Malebranche, Berkeley, and even Newton himself, as we'll see.[53] It outlives Hume's mockery and resurfaces in 'common sense' philosophers

to make sense of, and to persuade contemporaries of the validity of, the concept of laws of nature' (2004, 97). On Henry's (very plausible) view, the appeal to medieval voluntarism is a kind of *post hoc* rationalization, not a moving force.

[51] It is worth pausing to compare this account to that of Wigner (1967), taken up and extended by Woodward (2018). Wigner writes, 'The surprising discovery of Newton's age is just the clear separation of laws of nature on the one hand and initial conditions on the other. The former are precise beyond anything reasonable; we know virtually nothing about the latter' (1967, 40), quoted in Woodward (2018, 162). Wigner treats laws of nature as regularities, so it is hard to see how he can think that the moderns discover the independence of laws from initial conditions. Indeed, as Woodward notes, the tight link between laws and initial conditions required by the regularity account is the source of some of the main objections to it. Nor is it clear just why the scholastics should be incapable of making this split between laws-cum-regularities and initial conditions. If laws-cum-regularities are fixed by the powers and arrangements of bodies, then the same set of regularities is capable of being produced by an indefinitely wide array of initial conditions (though those conditions would have to include the same set of powers).

[52] This definition of 'occasionalism' brackets whatever causal contributions human minds and angels might make.

[53] I give a fuller defense of the claim that Descartes's account of laws commits him to occasionalism in my (2009).

such as Thomas Reid.[54] The otherwise surprising persistence of occasionalism is no accident; it is driven by the top-down picture of laws in the modern period. And when occasionalism begins to wane, it is only because the top-down view is supplanted by quite different thick conceptions of laws.

I begin by making the case that the top-down view of laws leads to occasionalism. To see how, we need to supply a premise articulated by some of the moderns but, I think, held by all: that top-down laws cannot operate on their own. I then move on to Descartes's laws themselves. Since I am concerned with the form and structure of Descartes's views, rather than with their precise content, I bring out Descartes's position by seeing how it responds to a contemporary concern, namely, the problem of *ceteris paribus* clauses. By testing Descartes's view against this problem, we get some insight into one of its essential features that will be important in evaluating his successors: the claim that the laws of nature work together as a web.

5.1 Laws and Ontology

In his *Principles of Philosophy*, Descartes announces that '[f]rom God's immutability, we can ... know certain rules or laws of nature [*regulae quaedam sive leges naturae*], which are the secondary and particular causes of the various motions we see in particular bodies.'[55] The laws flow from God's nature, not that of bodies. And while God is the primary cause of motion, the laws are secondary causes. Before we worry about the details, we should note how startling such claims would be to Descartes's predecessors and contemporaries.

Descartes himself is well aware of the novelty of his position. In a letter from 1630, he exhorts Mersenne, '[D]o not hesitate to assert and proclaim everywhere that it is God who has laid down these laws in nature [*ces lois en la nature*] just as a king lays down laws in his kingdom.'[56] Why should he

---

[54] In his *Essays on the Active Powers of Man*, Reid writes, '[t]he physical laws of nature are the rules according to which the Deity commonly acts in his natural government of the world; and whatever is done according to them, is not done by man, but by God, either immediately or by instruments under his direction' (1788, 344–5). Here and elsewhere, Reid is agnostic on whether God acts through instruments or not. What is clear is that the laws of nature are rules according to which God acts.

[55] *Principles* II.37 (AT VIIIA 62/CSM I 240). References to Descartes are to the Cambridge translation (Descartes 1984), abbreviated 'CSM,' and to the Vrin original language edition (Descartes 1996), abbreviated 'AT.'

[56] Letter to Mersenne, 15 April 1630 (CSMK III 23/AT I 145). As John Henry notes, the passage begins by discussing mathematical truths and then switches to discussing 'laws in nature.' So one might try to read the whole passage as concerned with mathematics. But as Henry points out, Descartes had

instruct Mersenne to go around proclaiming this? I've argued that the top-down view of laws adopts the hierarchical structure of mathematics and geometry. But in pulling this structure free of its Aristotelian roots so that it can be applied to natural philosophy generally, the moderns have to provide it with a new support. Unlike the axioms of geometry, which plausibly stand on their own, or Dupleix's *principes*, which are supported by the Aristotelian ontology, Descartes's laws stand in danger of coming loose from the world entirely. This is why he insists, almost from the beginning of his career, that the laws of nature are given directly by God. His exhortation to Mersenne is a directive to announce the new foundation for—and hence new thick conception of—the laws of nature.

Now, the scholastics were quite comfortable with God 'governing' the universe *indirectly*, by creating bodies with essences that provide them with powers. But to say that God simply decrees the laws is to court unintelligibility. How can a planet be said to 'obey' God's dictates? Indeed, a typical complaint about Descartes's talk of laws is that it is mere metaphor. As Robert Boyle puts it: 'to speak properly, a *law* being but a *notional rule of acting according to the declared will of a superior*, it is plain that nothing but an intellectual being can be properly capable of receiving and acting by a *law*.'[57] Boyle's remark suggests two alternatives: either bring God in to the picture, so that he moves bodies around in accordance with the laws he gives himself, or, even less plausibly, inflate the conception of bodies to the point that they can obey laws just as subjects do.

In this same vein, Ralph Cudworth argues that Descartes and his followers must 'either suppose these their laws of motion execute themselves, or else be forced perpetually to concern the Deity in the immediate motion of every atom of matter throughout the universe, in order to the execution and observation of them.'[58] Cudworth plainly regards the first option as absurd.

---

already formulated the laws in *Le Monde* by the time of his writing to Mersenne. Moreover, Henry writes, '[i]t seems hard to imagine why Descartes would think it was "especially" important to discuss the metaphysics underlying [mathematical or geometrical claims], much less why he would want Mersenne to continually proclaim this' (2004, 104). I am unsure whether Henry is right in thinking that Descartes turns to metaphysics precisely in order to provide a backing for his laws of nature, but it is certainly a claim worth taking seriously.

[57] 'A Free Inquiry into the Vulgarly Received Notion of Nature' (1991, 181). Note that Boyle's preferred alternative points to a bottom-up picture: '[i]t is intelligible to me that God should at the beginning impress determinate motions upon the parts of matter, and guide them as he thought requisite for the primordial constitution of things, and that, ever since, he should by his ordinary and general concourse maintain those powers which he gave the parts of matter to transmit their motion thus and thus to one another.' The issue is complex, however; for a fuller discussion of Boyle, see my (2009, 151–8).

[58] Cudworth (1837, 1: 214).

For his part, Leibniz writes, 'to say that, in creation, God gave bodies a law for acting means nothing, unless, at the same time, he gave them something by means of which it could happen that the law is followed; otherwise, he himself would always have to look after carrying out the law in an extraordinary way.'[59] Cudworth and Leibniz, for all their differences, are united on one point: if laws are 'top-down,' holding in virtue of divine decree alone, then God must move the bits of matter around himself. There is simply no way for the laws to 'execute themselves.'[60] While it's impossible to prove a negative, I can only report that I've been unable to find anyone in the modern period who thinks that laws can govern events without an enforcer to do the governing.

If God is to govern the world immediately and directly, then his action must be similarly immediate and direct. For Descartes, God is 'the universal and primary cause—the general cause of all motions in the world.'[61] So far, the scholastics would agree: they use the primary/secondary split to mark the distinct contributions of God as the primary cause and creatures as the secondary cause. Descartes's innovation comes when he describes the secondary causes. Just as Aquinas maintains, the secondary causes must be invoked to explain the diversity we find in the natural world. But what plays this role, for Descartes, are 'the rules or laws of nature': they alone are the *causae secundariae* 'of the various motions we see in particular bodies.'[62]

The laws of nature apply only because 'God preserves the world by the selfsame action and in accordance with the selfsame laws as when he created it' ('*ac cum iisdem legibus cum quibus creavit*').[63] God's activity of continually preserving the world is fundamental. To say that our world is governed by laws is to say that God re-creates the world in accordance with these laws throughout time. What sense, though, does it make to say that the laws themselves are *efficient* causes at all?[64] Surely God's non-nomic activity, creating motion and indeed the world at every moment, puts the laws out of any sort of causal business, primary or secondary. I think Descartes has an answer to this worry, and it's one that draws again on the parallel with the scholastics' secondary causes.

[59] 'On Body and Force, Against the Cartesians,' in Leibniz (1989, 253–4).
[60] Cudworth and Leibniz both, in effect, attribute what we'll see Malebranche calling 'little souls' to bodies: they think that there is in fact a mind (or a little one, anyway) in each bit of matter that governs its behavior. Descartes's extension of course has no room for any such mental agent.
[61] *Principles* II.36 (CSM I 240/AT VIII-A 61).
[62] *Principles* II.37 (CSM I 240/AT VIII-A 62). For a different reading of this passage, see Platt (2011, 865).
[63] *Principles* II.42 (CSM I 243/AT VIII-A 66). For a very different reading of Descartes on laws, see esp. Helen Hattab (2007) and (2018). For a fuller defense of my reading, see chapter 9 of my (2009).
[64] My thanks to Travis Tanner for pressing me on this point. Tanner's original and very interesting work, forthcoming in due course, defends a different position, according to which Cartesian nomological explanation is formal causal explanation.

A useful analogy here is that of a determinable, such as color, to one of its determinates, such as scarlet. A scarlet object doesn't have two properties, scarlet and color. Scarlet is simply the determinate way that object has of being colored. Volitions sometimes admit of a parallel structure. Suppose—to keep the parallel with Descartes's account—my nature is such that I will to do the dishes tonight. I might also will to do them according to a particular method (plates first, of course). But it's not as if I have thereby sprouted a new, distinct volition. My will to do the plates first stands to my will to do the dishes as determinate to determinable: it is the particular form in which my will is made specific enough to be efficacious. In the same way, the laws of nature are determinates of God's determinable creative volition: they make precise the form his creative volition is to take. They are not distinct volitions that compete with his general creative volition for the title of 'efficient cause.'[65]

Note that we might have asked the very same question of the scholastics' primary and secondary efficient causes: surely God alone is sufficient to bring about any effect; what need is there for creatures to contribute anything? Aquinas's answer points to the role secondary causes play. On its own, God's activity is undifferentiated. A secondary cause, such as fire and its power to burn, must be on the scene to 'particularize and determine the action of the first agent.'[66] Here again we see the parallel between Descartes's and the scholastics' secondary causes. *Qua* secondary efficient causes, Descartes's laws also 'particularize and determine' the actions of God. The volition to re-create the world at a given moment is made precise and determinate in the volition to follow the laws.

Whatever we make of the tenability of all this as divine psychology, the fact remains that Descartes's laws are divine volitions. The notion of laws as directives God gives himself is to be found throughout the inheritors of the Cartesian tradition. Nicolas Malebranche, for example, identifies laws with God's general volitions. But note how the general volitions actually work: 'when a ball strikes another, I say God moves the second ball by a general volition, because He moves it in consequence of the general and efficacious laws of the communication of motions.'[67] It's God, not the first ball, that moves the second one; and he does it because of the laws of nature he has willed. George Berkeley shares this picture, even if he excises the world of extension: 'Now the set rules or established methods wherein the mind we

---

[65] In a very different context below (Chapter 8, section 8), we'll make further use of the point that determinables and determinates are not causal competitors.

[66] *Summa Contra Gentiles* chapter 66, in Aquinas (1997, 2: 119).   [67] Malebranche (1992, 195).

depend on excites in us the ideas of sense, are called the laws of nature.'[68] In his last work, *Siris*, Berkeley is equally explicit, when he speaks of 'the laws and methods observed by the Author of nature.'[69] To be sure, neither Malebranche nor Berkeley (any more than Newton) sees the laws as necessary consequences of God's nature, as Descartes does. But we shouldn't allow that divergence to obscure the respect in which their converge: God's commands are in the first instance directed at himself, at his own future actions.

5.2 The Problem of *Ceteris Paribus* Clauses

Here, I need to interrupt the discussion of Descartes to present a philosophical problem. Although none of the moderns, as far as I can tell, explicitly poses or responds to this problem, testing Descartes's view against it gives us a better sense of its philosophical shape. In particular, doing so brings out a central feature all Cartesian views share: the claim that laws of nature work together, not singly, to produce and explain events.

In his 1638 *Discourse on Two New Sciences*, Galileo has Salviati put forth a basic principle: '[t]he speeds acquired by one and the same body moving down planes of different inclinations are equal when the heights of these planes are equal.'[70] His interlocutor, Sagredo, is quick to add: ' . . . provided of course that there are no accidental or external resistances, and that the planes are hard and smooth and the shape of the moving body is perfectly round, so that neither plane nor moving body is rough.'[71] Sagredo's list of provisos covers a lot of ground in a very indeterminate fashion. Those 'external resistances' might include wind, a magnetic force acting on the bodies, and on and on. The principle in question is simply false as it stands: it needs to be qualified, and spelling out the qualification is no easy task. Although the moderns are aware of the need for qualification, as Galileo's work shows, it was not until the twentieth century that the problem acquired a name: the 'problem of provisoes,' in Hempel's phrase, or 'the problem of *ceteris paribus* clauses.'

I propose to gather the three main threads of the contemporary version of the problem in a single trilemma. At the risk of anachronism (in both directions), let's use a simple version of Newton's law of gravitation: any two objects exert a force of attraction on each other that is inversely proportional

---

[68] PHK I 30. References to Berkeley's *Treatise concerning the Principles of Human Knowledge* (1710; rev. 1734) are to Part and section number in Berkeley (1975).

[69] Siris 243, in Berkeley (1901, 3: 238–9).

[70] Galilei (2008, 342–3); I have omitted the italics in the original.   [71] Galilei (2008, 343).

to the square of the distance between them. Given two otherwise isolated bodies, we can calculate the rate at which they will approach each other. But of course no two bodies ever act this way in the real world. For one thing, there are never just *two* bodies involved. We can call this the problem of token interference: what keeps the law from being true is the number of tokens of the same type involved in calculating its effects.

There is also what we might call type interference: what gets in the way of the law is an interferer of a different kind altogether. Any two bodies subject to gravity might also be subject to the electrical force described by Coulomb's law. As Cartwright puts it: '[n]o charged objects will behave just as the law of universal gravitation says; and any massive objects will constitute a counter-example to Coulomb's law.'[72] So we have our first horn: stated baldly, the law of gravitation is false. This is clearly a last resort, but it can seem forced on us: the exceptions, after all, might be construed simply as counterexamples.

We can insulate the law from such empirical refutations by adding a *ceteris paribus* clause. But if we just add a blanket 'other things being equal,' we court triviality. To say that the law of gravity obtains *ceteris paribus* is to say that it holds unless it doesn't. And of course that move is available to any general claim whatsoever: we might as well say, all Scotsmen eat oatmeal for breakfast, except when they don't.[73]

If we try to fill in for the blanket *ceteris paribus* clause, we run into our third horn. Suppose we wrote out all the qualifications needed to make the law of gravity, or the law of free fall, come out true. We would then have a single, extremely complex law that covered very few instances (perhaps only one). There is an epistemic problem on this horn as well: a fully qualified law could be known only by an omniscient being. Such a proposition would hardly be suited to playing the axiom role.

---

[72] Cartwright (1983b, 58). One possible response, which Cartwright considers, is vector addition: the truth of the law is saved if it predicts, not a resultant force or velocity, but a force that interacts with others to produce velocity. Cartwright argues that these alleged constituent forces are merely useful fictions. For a response to Cartwright on the composition of forces, see Lewis Creary (1981). Creary argues that Cartwright has it backwards: the component forces are real, while the resultant force is a mathematical fiction.

[73] Falsity and vacuity are the two horns of the dilemma Marc Lange (1993) presents. John Earman and John Roberts helpfully distinguish Lange's dilemma from Hempel's (1988) 'problem of provisoes' (*sic*). Roughly, Hempel's claim is that the needed provisos must be stated in the language of the theory modified, not in the antecedently accepted observation language. Hempel doesn't doubt that the provisos can be supplied, at least for fundamental physics; the point is that there is no stating them without using the language of fundamental physics itself. See Earman and Roberts (1999). The third horn of my trilemma comes from Nancy Cartwright's notion of a 'super law,' so fully detailed it comes out true but applies only to one or a handful of instances, and (most important for Cartwright) sacrifices explanatory power (1983a).

So those seem to be our options: believe in laws that we know to be false; qualify them so that they come out vacuously true; or qualify them substantively so that they come out true but apply only to a handful of instances and are all but unknowable. The trilemma, in various forms, has prompted a huge literature in the last half-century or so.[74]

5.3 Intentions, not Regularities

When we present Descartes with this problem, what answer can he make? Here we need to look at the content of the laws he presumes to deduce from God's immutability. Consider the first law:

Law 1.   *Each and every thing, in so far as it can, always continues in the same state; and thus what is once in motion always continues to move.*[75]

From the start, it's clear that Law 1 does not describe or summarize a regularity. The much-debated phrase 'in so far as it can' (*quantum in se est/autant ce qu'il peut*) indicates that Law 1, so far from setting out what happens, only purports to tell us what happens other things being equal. The claim is not that bodies always persist in the same state, but that they do so as far as they are able. It is thus a little surprising to see Descartes infer that what is in motion always continues to move. I take it that the qualification of the first clause is meant to apply to the second. As Descartes goes on to say, if a particular piece of matter is square, 'we can be sure without more ado that it will remain square for ever, unless something coming from outside changes its shape.'[76] The claim is not that motion persists— it obviously does not—but that its lack of persistence is due to some cause external to the body that is moving.

Law 2.   *All motion is in itself rectilinear; and hence any body moving in a circle always tends to move away from the center of the circle which it describes.*[77]

Here again we find a dispositional clause embedded in the law: it's only motion in itself that is rectilinear. 'Considered in itself,' each body, even one that is moving in a circle, 'tends to continue moving' in a straight line.

---

[74] See esp. Schrenk (2007b).   [75] *Principles* II.37 (AT VIIIA 62/CSM I 240).
[76] AT VIIIA 62/CSM I 241.   [77] *Principles* II.39 (AT VIIIA 53/CSM I 241).

Now, this is a surprising thing to read, coming hard on the heels of law 1. According to our first law, we should expect a body moving in a circle to continue moving in a circle. Law 2 says that a body that is moving in a circle instead tends to slingshot away from the center of the circle. How is this not a violation of law 1?

This is where it becomes necessary to mention a complication in Law 1. Law 1 doesn't apply to all things, but only to 'each thing, in so far as it is simple and undivided.'[78] Motion in itself, that is, motion taken as simple and undivided, is rectilinear; circular motion is complex. Law 2 is then a special case of law 1, in the sense that it specifies what kind of motion counts as simple and undivided.[79]

Why does rectilinear motion count as simple, and circular motion complex? After all, as Descartes is aware, any motion takes place over time and so in *that* sense is divisible.[80] In his proof of law 2, Descartes does not in fact call rectilinear motion simple; instead, he appeals to 'the immutability and simplicity of the operation by which God preserves motion in matter.' 'God always preserves... motion in the precise form in which it is occurring at the very moment when he preserves it, without taking any account of the motion which was occurring a little while earlier.'[81] So it is the simplicity of God's activity, driven, of course, by his nature, that is doing all the work.

Already from the first two laws, it's clear that Descartes's laws are not prefaced by any blanket *ceteris paribus* clause, but instead sometimes include qualifications in the form of dispositions.[82] Before evaluating Descartes's particular response to the problem of *ceteris paribus* clauses, we should ask how well the dispositional maneuver in general fares.

There is an obvious problem with appealing to dispositions: we have merely re-positioned the clause *inside* the law. To see this, let's begin with an epistemic issue. Suppose you know it's a law that x-particles are disposed to φ. That law by itself tells you nothing. It is not capable of disconfirmation, since any instance of an x-particle not φ-ing could simply be a result of the disposition's not being manifested. Nor does it help explain the x-particles that *do* φ, since the disposition to do so on the part of the x-particles might be interrupted, while another disposition of some surrounding bodies might

---

[78] AT VIIIA 62/CSM I 241.   [79] Daniel Garber (1992, 224) makes a similar suggestion.

[80] See *The Principles of Philosophy*, II.39 (AT VIIIA 64/CSM I 242): 'no motion takes place in a single instant of time . . .'.

[81] AT VIIIA 63/CSM I 242.

[82] For our purposes, the third law is less important ('if a body collides with another body that is stronger than itself, it loses none of its motion; but if it collides with a weaker body, it loses a quantity of motion equal to that which it imparts to the other body' (AT VIIIA 65/CSM I 242).

produce the same result.[83] So we would then need to specify the conditions under which the disposition is made manifest. And that *just is* the problem of *ceteris paribus* clauses, all over again. Peter Lipton calls this 'Hume's Revenge'; it has been developed most explicitly by Marcus Schrenk.[84] Just saying that the content of the laws includes a dispositional waiver, as it were, does not by itself protect them from the problem of *ceteris paribus* clauses.[85]

But Descartes's view has special features that make the dispositional maneuver more appealing in his context. A central property of Cartesian laws is that they work together. The manifestation conditions of God's dispositions are, in effect, specified by the contents of the laws together. Any apparent exception to one will be due to the others. It is only as a web that the laws can be used to predict any individual event. So Descartes is not simply waving his hands at any old potential source of interference: he can tell us (even if empirically he is quite wrong) just what those sources are. That's why no *one* law on its own makes any predictions whatsoever. To accuse such a law of vacuity simply because it is part of a web of laws that together governs events is to set the bar absurdly high.

To come at the point from the epistemic direction: someone might accuse the web approach of making laws immune from disconfirmation. If no law on its own makes predictions, then it is consistent with practically any state of affairs whatsoever. In fact, I take this to be a virtue, not a vice, since I think individual laws *are* immune from falsification. One doesn't need to embrace Quine/Duhem-style holism to think that the laws, at least, have function together to predict and explain events; only as a group can they meet the 'tribunal of experience,' in Quine's phrase.[86]

A further question looms over Cartesian laws with dispositional qualifications: exactly what is responsible for the dispositions? If they are reports on the natures of bodies, then we would have a bottom-up account after all: the laws are simply a convenient shorthand for the powers bodies have on their own.

[83] This is broadly similar to a point Earman and Roberts (1999) make.
[84] See Lipton (1999) and Schrenk (2007b). In fact, it is the recrudescence of an early twentieth-century problem: that there can apparently be no finite analysis of dispositions in terms of conditionals. This first came to light, I believe, in the debate over the phenomenalist project of exploiting conditionals to reduce statements about physical objects to statements about sensations. It was soon pointed out that no finite list of conditionals could capture even the most quotidian claim about a physical object. The same dialectic occurs in the context of behaviorism.
[85] The point recurs in debates over whether causes necessitate their effects, with Stephen Mumford and Rani Lill Anjum arguing that no list of potential 'completers' for a sufficient cause can necessitate its effect, since there is always the possibility of some additional feature interrupting the cause as it tries to go about its business. Mumford and Anjum refer to the strategy of explicitly excluding all possible interrupters in the description of a cause as 'strategy Sigma' (2011, 64–70).
[86] By 'holism' here, I simply mean the view that the basic unit of empirical testing has to be a theory as a whole, not a single sentence. On the Quine/Duhem kind of position, only a theory as a whole has any implications for experience.

Such a reading makes a hash of Descartes's epistemic and metaphysical derivation of the laws: as I've argued, they are reflections of God's nature and its consequences for his behavior in the realm of bodies.

Here we must draw a distinction we'll need later. Bodies might be said to have dispositions or powers in the 'thick' sense when their own intrinsic states just are, or are the sole grounds for, those dispositions. Even someone who denies this, however, can still accept that aspirins are disposed to cure headaches, for example. For the 'thin' sense of dispositions requires only that some conditionals and counterfactuals be true of the bodies said to be so disposed.[87]

So when the first law tells us that a body will continue in the same state *quantum in se est*, I take that as a shorthand for 'as long as no other law prompts God to do something else.' When Descartes speaks of the powers of bodies, he need mean nothing more than powers in the thin sense. Descartes tells us that the 'power' (*la force*, *vis*) of bodies 'consists simply in the fact that everything tends, so far as it can, to persist in the same state, as laid down by our first law.'[88] And the first law tells us what God intends to do with bodies.

The top-down picture helps to remove some of the mystery surrounding dispositions. If the dispositions of Cartesian laws are built in to divine intentions, then they are no more problematic than the dispositions in human intentions. We can see this in the way in which claims to intend to do something can and cannot be challenged. If I say in the morning that I intend to wash the dishes in the evening, my utterance will not be disproved if, during the day, I learn nuclear war is imminent and leave the dishes unwashed. My intention was always qualified, even if I cannot exhaustively spell out what those qualifications are.

Two points are salient here. First, the qualifications implicit in an intention are far-reaching, even if impossible to fully specify. For all that, there seems to be no problem in forming and understanding our own intentions. If anything, Descartes's divine nomic intentions are much more fully spelled out than those of any finite mind, if, as I've argued, the laws together tell you exactly what will be responsible for foiling any one of those intentions. Second, the ontology of the disposition is also not mysterious: it is part of a mental state. Obviously, we have to play along with Descartes's theological fiction here, but at least so far there is no problem in principle with the view.

---

[87] This important distinction is insisted upon, albeit in other terms, by Alexander Bird (2016). Bird rightly argues that a great many views reject thick dispositions (or 'powers,' in his sense), and yet happily avail themselves of thin dispositions ('dispositions,' in his idiom). Bird points to a number of arguments for powers that trade on this ambiguity.

[88] *Principles* II.43 (AT VIIIA 66/CSM I 243).

5.4 Epistemic Problems

Not everything looks rosy for Descartes, of course. To begin with, his laws, even taken as a web, are completely false. By the time of Voltaire, Descartes's laws had been empirically disproved by measurements of the poles; earlier, they'd been subjected to withering a priori critiques by Leibniz. For our purposes, what's much more interesting is the price he pays for solving the *ceteris paribus* problem as he does, within the context of the rest of his metaphysics. I've argued that he avoids the threat of vacuity and Hume's revenge, even though his laws conceal dispositional waivers. But that doesn't mean that we humans have any hope of actually applying them in the practice.

It's important to be careful here. We are now worried, not about the form of the laws themselves and the *ceteris paribus* problem as it affects them in principle. Instead, we are moving from the laws to their application. Descartes thinks laws not only govern how God will move things about; they also allow you, in practice, to make empirical predictions and explanations. Even if they play the first role, can they play the second?

The third law tells us that all changes happen through surface contact. The details of the 'impact contest' need not concern us here, except to note that the only collisions Descartes envisions are between two bodies. Indeed, all of the laws concern the behavior of at most two bodies. But Descartes's world is a plenum; there is never a case in which *only* two bodies are in contact with each other. Now, this is no problem for an omniscient being: he has no trouble taking account of the surrounding bodies and applying his laws to them, too. But it's a serious problem for us.

Descartes is aware of the gap between the laws and their application in scientific practice. After stating his three laws, he goes on to explain how we can 'determine, in the light of this, how individual bodies increase or diminish their motions or change direction as a result of the collision with other bodies':

> [A]ll that is necessary is to calculate the power of any given body to produce or resist motion...Our calculation would be easy if there were only two bodies colliding, and these were perfectly hard, and so isolated from all other bodies that no surrounding bodies impeded or augmented their motions.[89]

One is led to hope that he will provide some bridge principles that take us from the idealized cases to the real world. And he does indeed offer seven further

---

[89] *Principles* II.45 (AT VIIIA 67/CSM I 244).

rules, which take into account differences in size and motion.[90] Even these, however, deal with at most two bodies. As Descartes concedes, 'the application of these rules is difficult because each body is simultaneously in contact with many others.'[91] To calculate the behavior of any individual body, you would need to know the state of motion or rest and direction of all the moving bodies with which it is in contact.

Descartes himself tells us that 'experience may appear to conflict' with the rules.[92] How can we be sure that the conflict is merely an appearance, and not the real thing? Descartes here claims that his rules are 'self-evident,' adding that 'even if our experience seemed to show us the opposite, we should still be obliged to have more faith in our reason than in our senses.'[93] The only explanation for this bravado is Descartes's belief that he has deduced the behavior of bodies from the nature of God.

6. Virtues and Vices

I've argued that Descartes cannot be credited with the invention of the thin concept of law. Still, he provides a thick conception of laws that makes them central to the scientific project, to a degree that prior conceptions never were. This achievement is bound up with the particulars of his thick concept. By making laws divine volitions, Descartes jettisons the whole Aristotelian project of explaining events in terms of natures and powers. Along with the occasion-alist underpinning comes a re-orientation of the nature of scientific inquiry, explicitly articulated by Malebranche and Berkeley: laws supplant efficient causes as the source of explanation. To adduce the efficient cause, in the context of occasionalism, is metaphysically correct but scientifically uninter-esting, since every effect (human actions aside) has the same efficient cause. So the birth of this particular thick concept of laws in the modern period results in a split between different kinds of explanation: a scientific kind in terms of laws, and a metaphysical kind in terms of efficient causes.[94]

---

[90] *Principles* II.46–52 (AT VIIIA 68–70).   [91] *Principles* II.53 (AT VIIIA 70/CSM I 245).

[92] *Principles* II.53 (CSM I 245); the sentence appears only in the French version (AT IXB 93) but the substance of the claim is there in the Latin, on the prior page (AT VIIIA 70).

[93] French version (CSM I 245/AT IXB 93).

[94] Berkeley is very clear on this: 'It is not, however, in fact the business of physics or mechanics to establish efficient causes, but only the rules of impulsions or attractions, and in, a word, the laws of motions, and from the established laws to assign the solution, not the efficient cause, of particular phenomena' (*De Motu* 35 in Berkeley (1975, 218)). In this, he anticipates Bertrand Russell's (1953)

There are some virtues of the Cartesian conception we would do well to preserve in our own theories, if we can. Chief among these is the implicit interconnectedness of laws. No one of Descartes's laws on its own is intended to be used for the purposes of explanation and prediction. Only as a group can they serve that function. Later on, we'll see how one might try to leverage this web metaphor into a more satisfying answer to the problem of *ceteris paribus* clauses than Descartes can muster, given his reliance on God. If laws have to function together, then there might be no need to add a *ceteris paribus* clause. Put differently, the laws might be implicitly 'web-qualified': they can only be measured against evidence as a battalion, not single spies.

Note that, in terms of our taxonomy, Descartes's picture is one-to-one, in the sense that each law statement corresponds to a single truthmaker. In Descartes's case, the truthmaker is a divine volition. Fulling spelling this out would require more detail about the nature of volitions generally and some complicated theology. But Descartes's intentions at least are clear. And there's a lesson we'll need to apply later on: even views that pair law statements and truthmakers in this one-to-one fashion can, at least in principle, help themselves to the web approach.

There are two more virtues worth bringing out, although we need some anachronistic terminology to do so. The first is the ability to support counterfactuals. If it's a law that Fs are followed by Gs, then we should be able to infer that, had this particular x been an F, it would be have been followed by a G. How to cash out this notion of support in terms of possible worlds is a vexed issue we'll return to, as we'll see even anti-realist views scrambling to purchase this kind of support. But in the context of Descartes's view, it should be fairly straightforward: since the laws are divine intentions, and God is omnipotent, there is no sense to be made of a world in which God wills that Fs be followed by Gs and yet this particular F is not so followed. Given this, it's hard to see how the laws could fail to support counterfactuals: had this x been an F, God would have seen to it that it was followed by a G.

The robust support of counterfactuals provided by Descartes's laws figures in their explanatory role as well. Above, I suggested that the thin concept at least requires laws to unify their instances: the law of free fall lets you see what a falling piece of limestone and a meteor have in common. A more controversial element was, for that very reason, barred from the thin concept:

view. In the contemporary debate, the distinction between scientific and metaphysical explanation resurfaces in Loewer (2012, 131) and the ensuing debate with Lange (2013) and others. See below, Chapter 13, section 2.

explanations in terms of laws seem to involve counterfactuals in a straightforward way. If I'm adducing the law of thermal expansion as part of an explanation, I'm also saying that the event to be explained might've gone differently, had the initial temperature been different. Descartes's laws get us exactly this kind of modal, contrastive explanation.

Counterfactuals also enter into the question in a different way. What we'll call 'nomic stability' is the flip-side of counterfactual support. Rather than asking what would happen in a world with the same laws but different local matters of fact, we're now leaving open which laws obtain and asking whether changing one of those local matters of fact would change the laws. It's natural to think that the laws can remain constant, regardless of (at least a huge set of) variations in whatever we are taking as our initial conditions.[95] A single butterfly flapping its wings might cause a typhoon, but it shouldn't knock the law of gravity from its throne.

In his early work Le Monde, Descartes insists that even a completely chaotic state of affairs would resolve into perfect order, given the laws of nature.[96] That goes well beyond, but certainly entails, that the laws are unchanged by some imagined variations in the conditions of the world they govern. Some, but not all. If we go to a world without a god, or a god with a different nature, then Descartes's laws cannot hold in that world. Of course, Descartes denies that there are such worlds, given God's necessary existence and nature; so for him, the laws are stable under any possible variation in worlds they govern. The important point is just that nomic stability shouldn't require that every possible world has the same laws, even if that is in fact Descartes's position.

So we have three features that are prima facie worth preserving from Descartes's view: that the laws function together as a web, even if each has a distinct truthmaker; that laws support counterfactuals; and that laws are stable under (at least a wide array of) variations in the conditions we take as their inputs. Given all this, it's somewhat startling to realize how wide of the mark Descartes's specific views are. From his appeal to divine volitions to his

[95] Note that 'initial conditions' need not refer to the first state of a given world, if there is one. Typically, what counts as the 'initial conditions' will vary with the context, and by stipulation. So even a world with an infinite history can have states that count as initial conditions.

[96] 'From the first instant of their [bodies'] creation, he causes some to start moving in one direction and others in another, some faster and others slower (or even, if you wish, not at all); and he causes them to continue moving thereafter in accordance with the ordinary laws of nature. For God has established these laws in such a marvelous way that even if we suppose he creates nothing beyond what I have mentioned, and sets up no order or proportion within it but composes from it a chaos as confused and muddled as any the poets could describe, the laws of nature are sufficient to cause the parts of this chaos to disentangle themselves and arrange themselves in such good order that they will have the form of a quite perfect world...' (AT XI 34–5/CSM I 91). Cf. Woodward's (2018, 162) discussion of Wigner and von Weizsacker on the derivation of Bode's law.

particular claims about bodies and motion, he is quite wrong. Oddly enough, it took a figure mistaken on almost every empirical and metaphysical point to make conceptual space for the development of the indisputable advance of the era: Newton's laws. To see how that happened, we need to turn next to Descartes's successors.

3

Descartes's Legacy

1. Malebranche: Intervention or Autonomy?

I have been arguing that the Cartesians' distinctive, thick concept of law requires ontological support in the form of the volitions and activities of a divine being. The view that God is the only cause of physical events is sometimes regarded as a way of dodging the problem of mind–body interaction: if one of Descartes's fellow-travelers can't explain how a mental state causes a physical one, he can always wheel in God. We would then have to see occasionalism as an anomaly, a twig on the branch of Cartesian dualism. I think that picture is quite wrong. So far from being an ad hoc solution to the problem of mind–body interaction, occasionalism is largely a response—ad hoc or not—to the problem of laws. It's a necessary concomitant of the governing conception. Seen in this light, it's no wonder that occasionalism thrives outside of its original French territory, and well beyond the borders of Cartesian dualism.

Contemporary authors have suggested that laws might be irreducible features of the universe they govern. Such a picture is not seriously entertained by the moderns; if laws are to govern at all, it must be in virtue of God's ubiquitous activity. But there is one potential counterexample to that claim: the view of Nicolas Malebranche.[1] Exploring Malebranche's view, and the critique Leibniz levels against it, will help us put Isaac Newton's position in context. Moreover, the issues of this chapter will recur in the coming ones, when we turn to contemporary top-down accounts.

Unlike Descartes, Malebranche does not take the laws of nature to follow straightforwardly from God's essence or immutability. Nevertheless, the laws are still divine volitions. And like Descartes, Malebranche calls the laws 'causes,' although he drops the primary/secondary distinction so important in understanding Descartes's position. Treating the laws as unqualified causes might suggest that they somehow operate independently of God: that they are autonomous actors on the stage. Malebranche sometimes talks this way:

---

[1] Malebranche's central work, *The Search After Truth*, appears in 1674–5, thirty years after Descartes's *Principles* and thirteen years before Newton's *Principia*.

The Metaphysics of Laws of Nature: The Rules of the Game. Walter Ott, Oxford University Press. © Walter Ott 2022.
DOI: 10.1093/oso/9780192859235.003.0003

MALEBRANCHE: INTERVENTION OR AUTONOMY? 43

All natural forces are therefore nothing but the will of God, which is always efficacious. God created the world because He willed it: "Dixit, & facta sunt" [Ps. 32:9]; and He moves all things, and thus produces all the effects that we see happening, because He also willed certain laws according to which motion is communicated upon the collision of bodies; and because these laws are efficacious, they act, whereas bodies cannot act.[2]

Even here, of course, Malebranche insists that it is God who 'moves all things.' But this passage also says that the laws themselves act. If so, Malebranche might be able to parry the objection—usually attributed to Leibniz—that occasionalism requires a 'busybody' God, who runs around doing every little thing at every moment. On what we might call the 'Autonomy' reading of Malebranche's laws, God wills the initial conditions and the laws of nature; together, those two factors suffice to produce all that follows.[3] There is no need for God to go around adjusting the states and positions of every physical object.

I shall argue that the Autonomy interpretation is wrong. Instead, I'll argue for the 'Intervention' reading: God does indeed have to will every physical state of affairs as such.[4] When the details of Autonomy are spelled out, it collapses into my reading. Any version of the Autonomy interpretation robust enough to distinguish itself from its competitor must sin against the core doctrines of occasionalism. I'll also argue that Autonomy misconstrues Leibniz's critique. The core of Leibniz's objection has nothing to do with how many acts of will the divine being has to engage in; instead, he's objecting precisely to the top-down nature of Malebranche's laws. As we'll see, Newton and Clarke are both sensitive to the criticism Leibniz raises.

1.1 Efficacious Laws

Note that the passage above—some of the best evidence for Autonomy—begins by reducing all 'natural' forces to the will of God. What work, then, is left for the laws to do? The pressure to absorb the laws into God's volitions is

---

[2] SAT 6.2.3, OC 2:314 in Malebranche (1997b, 449). Travis Tanner pointed out to me that Malebranche's citation of the Psalms matches the numbering of the Latin Vulgate of St. Jerome and not that of modern Bibles.
[3] Such a reading is suggested by Nicholas Jolley (2002). Jolley's reading is actually more nuanced than this bare statement allows, as we'll see.
[4] For a further statement and defense of Intervention, see esp. Nadler (1993). I go into greater historical detail, especially with regard to the role Arnauld plays in the debate, in chapter 11 of my (2009).

hard to resist. The same pressure comes from a different quarter: the divine concursus argument. Malebranche argues that

> Creation does not pass, because the conservation of creatures is—on God's part—simply a continuous creation, a single volition subsisting and operating continuously. Now, God can neither conceive nor consequently will that a body exist nowhere, nor that it does not stand in certain relations of distance to other bodies. Thus, God cannot will that this armchair exist, and by this volition create or conserve it, without situating it here, there, or elsewhere. It is a contradiction, therefore, for one body to be able to move another.[5]

To create and conserve creatures is just for God to have a single, eternal volition that includes these creatures. And that volition must be fully determinate in that it spells out the precise location of the creatures conserved. God's continual creation drives out any competitors to the title of 'cause' of bodies or their states. Talk of laws as causes must be reduced to talk of God as the cause; and this, of course, is what one expects, given Malebranche's constant insistence that, where bodies are concerned, God is the only true cause.

There is another way to come at the same point. If Autonomy were correct, Malebranche's God would have to will the initial state and the laws of nature *without* willing the determinate states of every body throughout all time. But the divine concursus argument has just told us that this is impossible. The defender of Autonomy might reply that God can still be said to will all the determinate states, as the divine concursus argument requires, in the sense that he wills all logically necessary consequents of his volitions. In willing the laws and the initial conditions, he is also willing all the states that result from them.[6]

The problem now is that, developed in this way, Autonomy collapses into Intervention. If the two agree that God wills every bodily state of affairs as such, the appearance of the laws' autonomy vanishes. The laws can only be the rules that God follows when willing each of these states of affairs. In fact, Nicholas Jolley, the ablest proponent of Autonomy, has found a striking passage where Malebranche says just this: 'Order requires that [God] follows the laws which he has prescribed to himself so that his conduct may bear the

---

[5] Malebranche (1997a, 115). Berkeley runs this argument in *De Motu* §34, in Berkeley (1975, 218).
[6] This appears to be Jolley's reply to my (2009); see Jolley (2019).

mark of his attributes.'[7] This divine self-legislation is the same idea s'Grave-sande later chooses to enshrine in the awful lines of poetry that preface his popularization of the *Principia*.

If Intervention is correct, we face the same problem we did with Descartes: how can the laws be efficient causes? Malebranche's God must will every state of affairs as such; what causal role is left for the laws to play? To start, we can note that there is room for a relation of logical, though not temporal, priority, among God's volitions. He might 'first' choose a set of laws, and 'then' will the states of affairs that accord with them. Next, we should see that we shouldn't split God's volitions so finely. His willing of a particular state of affairs just is the precise way in which he wills the laws of nature. Suppose I decide to play a game of chess against myself. Once I sit down to play, I have to make particular moves. But my willing those particular moves need not be an extra volition, alongside my willing to play the game. In the same way, God's volition that a particular determinate state of affairs obtain is simply the way his will to obey the laws is actualized.[8] The two volitions are not causal competitors.

1.2 Leibniz's Critique

Part of the motivation for the Autonomy reading is the desire to parry Leibniz's critique. What, exactly, is Leibniz's worry? Here, some historical context is necessary. Although Leibniz repeats his criticism in a number of venues, the clearest is his exchange with Antoine Arnauld. The issue arises when Arnauld accuses Leibniz's pre-established harmony of being occasionalism by another name:

> [I]t seems to me that this [pre-established harmony] is to say the same thing
> in other terms as those people who maintain that my volition is the

---

[7] From the reply to Arnauld, regarding the *Treatise on Nature and Grace* in Malebranche (1958, 8: 651), trans. Jolley (2019, 130).

[8] This story differs only superficially from Descartes's own view. For Descartes, the laws are determinates of God's determinable volition to create. For Malebranche, volitions aimed at particular states of affairs are determinates of his nomic volitions. But the two accounts might be compatible: there's no reason not to think of Malebranchean nomic volitions as themselves determinates of God's determinable volition to create. One might then ask why Malebranche doesn't use the scholastic framework of primary and secondary causes. I don't think there's a deep philosophical reason here. Descartes wants his *Principles* to persuade the late scholastics and make converts of them, so it's natural for him to use their own terminology in developing his own very different position. A generation later, Malebranche has no such aspirations.

occasional cause of the movement of my arm, and that God is its real cause. For they do not maintain that God does this in time by a new volition he has each time I will to raise my arm, but by that single act of the eternal volition by which he willed to do everything he foresaw it would be necessary for him to do in order that the universe be such as he judged it ought to be.[9]

Arnauld's interpretation of Malebranche is spot-on: Malebranche does not think God has to formulate a unique volition for each and every state of a body. Instead, God, being omniscient, simply wills a single 'super-volition,' which includes everything. Nor does Malebranche's God do this in time, as if he were watching what happens and making decisions on the spot. The single volition is willed from eternity. The temporal indexing has to be inside the volition: it's not as if God, at time *t*, wills that S; he wills, for all eternity, that S obtain at *t*.

In reply, Leibniz does not dispute this characterization of occasionalism, as he would have to if he wanted to level the charge of the 'busybody' God. Instead, Leibniz argues that

[the occasionalists] introduce a miracle that is no less one for being continual. For it seems to me that the concept of a miracle does not consist in its rarity.... For example, if God had made the resolution to give his grace immediately, or to do some other action of this nature, every time a certain thing came to pass, this action would still be a miracle, although commonplace. [A miracle as Leibniz uses the term] differs internally and by the substance of the act from a common action, and not by the external accident of frequent repetition; and...properly speaking, God performs a miracle when he does something that surpasses the forces he has given to creatures, and preserves in them.[10]

The charge of invoking miracles has nothing to do with God's need to multiply his efforts to absurdity in moving bodies around. Leibniz's objection is that

---

[9] Leibniz and Arnauld (2016, 167).

[10] Leibniz and Arnauld (2016, 195). Leibniz makes the same point in the *New System of Nature*: 'It is quite true that, speaking with metaphysical rigor, there is no real influence of one created substance on another, and that all things, with all their reality, are continually produced by the power of God. But in solving problems it is not sufficient to make use of the general cause and to invoke what is called a *Deus ex machina*. For when one does that without giving any other explanation derived from the order of secondary causes, it is, properly speaking, having recourse to a miracle. In philosophy we must try to give reasons by showing how things are brought about by divine wisdom, but in conformity with the notion of the subject in question' (1989, 143).

God is bringing about events that have nothing to do with the bodies involved: their natures and powers are totally irrelevant. This is a resolutely bottom-up figure pushing back against the new way of laws.

Why have we bothered going into this level of detail with Malebranche? I think two results are worth emphasizing, and will be with us as we go. The first is the well-worn path from Cartesian ideas about laws to occasionalism. It is an open question whether these Cartesian ideas can be preserved without going in for the implausible theology with which they walk hand in hand.[11] Second is the initially surprising absence of anything like Autonomy as a live option in the moderns: whether they are treading the Cartesian way of laws or not, they all insist that laws on their own cannot govern anything.

2. Newton

Newton's 1687 *Principia* presents three 'axioms or laws of motion' (*axiomata sive leges motus*), the first two of which he credits to Galileo. It seems clear that he modeled this part of his *Principia* on Descartes's own. Since the first law is similar in spirit if not detail to Descartes's, Newton in all likelihood intends his own laws as a replacement for Descartes's.[12] However that may be, over the next century or so, they certainly did supplant Descartes's laws, first in England, then on the continent, where their spread was perhaps retarded by the attacks of Leibniz and others.

Our concern is with what Newton can tell us about the top-down approach. I'll argue that the central features of Cartesianism about laws—the severing of any dependence on the powers of bodies; the web-like function, with its dispositional *ceteris paribus* qualification and the ensuing failure of laws to state regularities—are very much in evidence in Newton. And even though he is famously reluctant to engage in speculation about the metaphysical under-pinnings of his laws, especially gravity, when he does tip his hand, Newton reveals an unswerving allegiance to God as the source and continual enforcer of the laws. Such remarks as he does make confirm the tight connection in the modern period between the top-down conception of the nomic and

---

[11] Although explicitly theological derivations of the laws are rare these days, John Foster's (2004) is a notable outlier.

[12] Newton calls them 'laws of motion' in the *Principia*; in other places, he calls them 'laws of nature.' For this and more on Newton's debt to Descartes, see esp. I. Bernard Cohen's introduction to the *Principia* in Newton (1999, 43; 109).

occasionalism: only by depriving bodies of causal efficacy could they be governed from without.[13]

2.1 Methods and Varieties of Explanation

Something must be said first about Newton's famous criticism of the Cartesians, and the method with which he proposes to replace theirs.[14] As Newton presents it in his preface, his strategy is to 'discover the forces of nature from the phenomena of motions and then to demonstrate the other phenomena from these forces.'[15] One first allows nature to teach him the forces at play, and then uses these forces to demonstrate other phenomena. In contrast, the Cartesians, as Newton's spokesman Roger Cotes puts it in his own preface to the work, 'are merely putting together a romance.'[16] Rather than basing their philosophy 'on experiment,' as Newton does, they merely 'contrive hypotheses.'[17] There is something disingenuous about this way of drawing the contrast: Newton, no more than Descartes, could pretend to have derived his first law from experiment: '[e]very body perseveres in its state of being at rest or of moving uniformly straight forward, except insofar as it is compelled to change its state by forces impressed.'[18] If anything, experience would tend to confirm the Aristotelian position that motion decays of its own nature.[19]

Where the contrast does have bite is its characterization of the Cartesians' particular way of deriving the laws from God. Descartes himself, as we've seen, thinks the laws flow necessarily from God's nature. His immediate followers,

---

[13] A very useful discussion of the literature on Newton is to be found in Janiak (2008). Janiak distinguishes three main lines: the anti-metaphysical reading, on which Newton dismisses or ignores metaphysical questions; the radical empiricist reading, on which Newton transforms what had been metaphysical questions into questions that can be answered empirically; and Janiak's own, on which Newton's metaphysical view is 'bifurcated into two intimately related but distinct elements: the first element expresses his conception of God's active role within the natural world, forming a fundamental metaphysical framework that is immune to revision from physics; the second element, concerning the material world we inhabit, lies within this framework and fully reflects the results of physical theory' (2008, 13). I think Janiak has it right, and find convincing his criticisms of both the anti-metaphysical and radical empiricist readings.

[14] A helpful source here is Domski (2018).   [15] Newton (2004, 41).

[16] Newton (2004, 43).   [17] Newton (2004, 43).

[18] Newton (1999, 416). For further argument that Newton's pretense at having discovered his laws empirically is just that, see esp. Westfall (1971, 386f.). A helpful discussion of Newton's empirical evidence for the laws, especially Law 3, is Harper (2012). But note that Newton's own *reductio* argument for Law 3 (1999, 428) presupposes the first law, and to that extent, the *reductio* does not provide independent support.

[19] Westfall's *The Construction of Modern Science* (1978) ably challenges the common idea that early modern science is distinguished from its predecessors by superior attention to experience generally and experiments in particular.

having given this up, have to rely on more general features of God's will and activities, such as simplicity and perfection. By declining to 'feign hypotheses,' Newton rejects their manner of proceeding.

But Newton also, at least officially, abandons part of their project. From their point of view, he is shirking the duties of a natural philosopher when he brackets all discussion of 'physical' qualities in favor of 'mathematical' ones (in Newton's idiolect). In the scholium to Proposition 69, Newton warns that, throughout his treatise, he will consider 'not the species of forces and their physical qualities but their quantities and mathematical proportions.'[20] It's important not to identify the physical/mathematical distinction with the real/ideal or real/imaginary one.[21] Both sides of Newton's distinction refer to things that are equally real. To see what Newton has in mind, we need to look at how the distinction functions as a bulwark against the charge that gravity is an occult quality.

Leibniz famously argues that Newtonian attraction represents a step backward into the scholastics' muddle of antipathies and hidden, inexplicable powers. Leibniz doesn't deny that the phenomena are as Newton describes them; even attraction is a perfectly respectable notion, so long as it is kept in its place. The mistake comes when one thinks that no mechanical explanation of the phenomena is needed:

> [I]f certain people, abusing this beautiful discovery [the mutual gravitation of the planets], think that the explanation [*ratio*] given is so satisfactory that there is nothing left to explain, and if they think that gravity is a thing essential to matter, then they slip back into *barbarism in physics* and into the *occult qualities of the Scholastics*.[22]

Newton himself, as we'll see, is far from thinking that 'there is nothing left to explain,' even if he thinks the explanation cannot be a mechanical one. Instead, his official position is one of agnosticism. But the physical/mathematical distinction is not primarily an epistemic one, even if Newton imbues it with epistemic force.

For Newton, gravity is not an occult quality for the simple reason that it is not a quality at all: it is a perfectly manifest phenomenon. What is occult or hidden is the *cause* of gravity, that is, the metaphysical underpinning of the observable phenomenon. In the General Scholium following Rule 4, Newton

---

[20] Newton (1999, 588).   [21] I owe this point to Janiak (2008, 53–65).
[22] 'Against Barbaric Physics' in Leibniz (1989, 314); italics in original.

claims he has 'explained the phenomena of the heavens and of our sea by the force of gravity, but I have not yet assigned a cause to gravity itself.' 'I have not yet been able to deduce from phenomena the reason for these properties of gravity [i.e., those he has established], and I do not feign hypotheses.'[23] Where Newton offers epistemic modesty and a refusal to speculate about the metaphysical explanation, Leibniz sees a positive, dogmatic thesis: there is no explanation necessary.

To see Newton's position clearly, we need to distinguish two things we might want from an explanation. One is metaphysical: to explain x is to show, either that x is primitive and fundamental, or that x is caused by, or supervenes on, something else. This kind of explanation is the one Newton officially eschews. But there is another legitimate sense of explanation, in which one shows that a phenomenon agrees with the laws of motion.

Despite his reputation as an innovator in philosophy of science, George Berkeley is just being a good Newtonian when he says that

> It is not...in fact the business of physics or mechanics to establish efficient causes, but only the rules of impulsions or attractions, and, in a word, the laws of motions, and from the established laws to assign the solution, not the efficient cause, of particular phenomena.[24]

In Newton and Berkeley, then, we have the first clear instance of a distinction that will be important in the contemporary debate: the contrast between explaining something scientifically, where it is reduced 'to those most simple and universal principles,' as Berkeley puts it, and explaining something metaphysically.[25] In Newton's terms, this is the distinction between mathematical explanation and physical explanation.

What consequences does this distinction have for Newton's ontology? It forces us to pull apart gravity as a mathematically describable phenomenon and as a physical force. Taken in the mathematical sense, 'gravity' does not refer to a quality or thing that might be had by bodies. 'Force,' 'impulse,' 'attraction,' and the like, in this sense, do not refer to ontologically basic categories. To see this in Newton's work, consider his Queries to the *Opticks*.

---

[23] Newton (2004, 92). See also his letter to Leibniz of 1712 in Newton (2004, 116).

[24] *De Motu* (1721), §35, in Berkeley (1975, 218). (References to Berkeley's works use the following abbreviations: *A Treatise concerning the Principles of Human Knowledge* (PHK), *De Motu* (DM), and *Siris* (S). In all cases, references are to a section number.) I realize many commentators read Berkeley as an instrumentalist, and so will balk at my claim that Berkeley follows Newton in the respect I've isolated. I argue below (this chapter, section 2.2) that Berkeley is not in fact an instrumentalist.

[25] *De Motu* §37 in Berkeley (1975, 218).

Here, he refers to gravity and cohesion as causes or active principles. But he goes on to clarify:

> These principles I consider, not as occult qualities, supposed to result from the specific forms of things, but as general laws of nature, by which the things themselves are formed; their truth appearing to us by phenomena, though their causes be not yet discovered.[26]

Here Newton is self-consciously *replacing* talk of causal principles or qualities with talk of laws. To say that gravity is the cause of the rotation of the Earth is not to say that there is some thing or quality pushing the Earth; it only reports that the Earth's movement obeys the law that constitutes gravity at the level of phenomena. Newton does not deny that there is some underlying cause: physical gravity is responsible for mathematical gravity. But it would be a mistake to suppose that there is anything more to Newton's mathematical gravity than what is given his laws.

At times, Newton talks as if there were two layers of causation happening: mathematical gravity causing events at the level of phenomena, and an underlying physical gravity causing its mathematical counterpart. This would be an awkward picture, to say the least; overdetermination threatens just out of sight. But if we take seriously the *Opticks'* move, it would be wrong to think of mathematical gravity as a thing or quality, and so as an efficient cause, at all.

Here again Berkeley is instructive. He claims a Newtonian pedigree for his own treatment of the key terms of physics:

> *Force, gravity, attraction*, and terms of this sort are useful for reasonings and reckonings about motion and bodies in motion, but not for understanding the simple nature of motion itself or for indicating so many distinct qualities. As for attraction, it was clearly introduced by Newton, not as a true, physical quality, but only as a mathematical hypothesis...[27]

Moving from qualities to laws disperses any air of mystery about mathematical gravity. Taken in that sense, 'gravity' does not refer to a distinct quality, alongside the others we observe.[28]

---

[26] Newton (2004, 137).   [27] DM §17, in Berkeley (1975, 214).

[28] I am not, of course, suggesting that Berkeley follows Newton in other matters such as ontology.

2.2 Ontology

Newton officially declines to comment on the 'physical' cause of gravity, contenting himself with gravity as a phenomenon. But we can learn much, even from his official pronouncements, if we look at how he describes the kind of thing he refuses to comment on.

In the anonymously published 'An Account of the Book Entitled *Commercium Epistolicum*' (1715), Newton points out that at the end of the *Principia* in the second edition, the author 'said that for want of a sufficient number of experiments, he forbore to describe the laws of the actions of the spirit or agent by which...attraction is performed.'[29]

Note that the laws are not the laws *of* bodies, or of motion: they are laws of the actions of the agent who 'performs' attraction. Despite his manifest departures from the Cartesians, in this respect he is at one with them: the laws govern the actions of the agent who brings about the phenomena we witness. Laws can be said to govern the motion of bodies only in a derivative sense.

The text is puzzling, however, insofar as it seems to draw a distinction between the laws we observe and the laws of the agent responsible for them. The analogy of a game can be used to make sense of this. We can think of the rules of chess plus the strategy a player observes as together making up the 'rules' the player follows.[30] An opponent witnesses only the moves she makes. So this opens an epistemic gap between the 'rules' the opponent formulates on the basis of the limited evidence available and those the player really has in mind. In an analogous way, we might say that the evidence available to us lets us formulate laws that enable the prediction and explanation of phenomena; if we get it right, the rules we come up with are the very rules God legislates for himself.

This picture is confirmed later in the text, when Newton sums up his defense against the charge of a perpetual miracle. Here, I think he is sensitive to the issue we've seen Leibniz pressing: that the top-down picture has to treat quotidian events as miraculous in the sense that they do not flow from the natures and powers of the bodies involved. In reply, Newton asks, 'must the constant and universal laws of nature, if derived from the power of God or the action of a cause not yet known to us, be called miracles and occult

---

[29] Newton (2004, 123). Newton is presumably referring to the 'General Scholium,' added to the second edition.
[30] I owe the analogy of a chess strategy to Tzuchien Tho.

qualities, that is to say, wonders and absurdities?'[31] From Newton's point of view, as long as the laws are constant and universal, they are not miraculous, regardless of their independence from the bodies that are governed by them.

So we have two levels, corresponding to the mathematical/physical distinction: the universal laws of nature on one hand and the action of God or some other agent on the other. Newton's rhetorical question asks us to consider the force of Leibniz's charge: there is nothing absurd in the two-tiered picture, even if it omits the powers and natures of bodies.[32]

Newton's advocate Samuel Clarke is equally strident in his response to Leibniz. Leibniz has been arguing that a perfect being would not create something he would have to continually interfere with. A world designed by such a being should itself be so perfect as to go on its way without the hand of the creator intervening at every moment, even if he did so by means of a single, eternal super-volition. As we've seen, Leibniz instead thinks that God has given his creatures the ability to act on their own, according to their essences. Now, Clarke might have replied that Newton's picture requires no intervention: he might have said that God creates the laws of nature and sets the initial conditions, with everything unfolding as it must thereafter. Such a view would be tantamount to the Autonomy reading of Malebranche we've examined. Just as one would expect, given the results so far, this is precisely what Clarke does *not* say. Instead, he vigorously defends God's perpetual involvement in his creation:

[God] not only composes or puts things together, but is himself the author and continual preserver of their original forces or moving powers: and consequently 'tis not a diminution, but the true glory of his workmanship, that nothing is done without his continual government and inspection.[33]

Clarke does not spell out just what God's involvement amounts to: does God conserve the powers he imbues bodies with? That would be a theological commonplace; indeed, it would be so common Leibniz himself would agree. That argues for a stronger reading, according to which it is God himself who is acting.

Indeed, when we turn to Newton's correspondence, we find more evidence of the stronger reading. Some twenty-three years before the 'Account,' Newton

[31] Newton (2004, 125).
[32] In the seventeenth century, 'absurdity' is often used as a technical term to refer to a contradiction. Minimally, Newton is claiming that there is no inconsistency in the two-tiered picture.
[33] Leibniz and Clarke (1956, 14).

wrote to his friend Richard Bentley. In that famous letter, Newton offers another important clue to what he conceives of as the 'physical' side of the two-tiered picture: 'Gravity must be caused by an agent acting constantly according to certain laws, but whether this agent be material or immaterial, I have left to the consideration of my readers.'[34] What is left to the reader is the status of the agent involved: is it an immaterial God, or a God who is somehow material and hence in space? What is *not* left to the reader's consideration is the fact that mathematical gravity is not a metaphysical *explanans*. Gravity at the level of observed phenomena is caused by an agent who acts according to laws. Again, we have the two-tiered picture: there is gravity *qua* manifest phenomenon, a phenomenon exhausted by the inverse square law, and the physical cause of this mathematical quality, what we would call the *metaphysical* explanation of the behavior captured by the inverse square law. This physical cause can only be an agent who acts according to laws. If we're getting the science right, then the agent will be formulating to himself precisely the same laws. Even Newton, then, does not take seriously the idea that the law of gravity could 'execute itself,' in Cudworth's phrase.

For this very reason, neither Newton nor Berkeley can be considered an instrumentalist about the laws of nature. Laws are useful in making predictions, of course; but it doesn't follow that that is *all* the laws do. If there is an agent observing these laws as he moves bodies about (Newton) or creates ideas (Berkeley), then human statements of laws have an objective truthmaker: the laws God formulates for himself.

My interpretation of Newton can be further buttressed by considering his rejection of action at a distance.[35] Just before insisting on gravity as the result of an agent acting according to laws, Newton tells Bentley:

> That gravity should be innate, inherent, and essential to matter, so that one body may act upon another at a distance through a vacuum without the mediation of anything else, by and through which their action and force may be conveyed from one to another, is to me so great an absurdity, that I believe no man who has in philosophical matters a competent faculty of thinking can ever fall into it.[36]

[34] Newton (2004, 103).
[35] There is of course a huge literature on this topic. See Henry (1994), Janiak (2008), and the references therein.
[36] Newton (2004, 102).

We know from the queries to the *Opticks* that Newton's God is omnipresent.[37] His rejection of action at a distance allows him to turn Leibniz's accusation of a miracle against him: a God who is outside of space and 'above the bounds of the world' 'cannot do anything within the bounds of the world, except by an incredible miracle.'[38] Where Leibniz takes any event that is not grounded in the natures and powers of the substances involved to be a miracle, Newton takes any action from outside the world of space and time to be a miracle. And if the physical cause (in Newton's sense) of gravity must act locally at every point in space, what better—indeed, what other—candidate is there, besides God?[39] I conclude, then, that Newton, just as much as Descartes and Malebranche, takes God to be the sole cause (human minds aside) of physical events. All three see laws as divine volitions.

There are other points of continuity. First is the hierarchy of propositions, derived from ancient mathematics. The laws of physics are fundamental in that they are the basic principles one uses in calculation. Newton calls his *leges* '*axiomata*' to signal their role in his system. Descartes does not call his laws 'axioms,' but I suspect that is because Descartes purports to derive the laws from God's nature, something Newton never attempts. And if the laws could be so derived, they would stop being axioms in the strict sense. Nevertheless, Newton and Descartes are both clear that their laws are to be understood as functioning in concert, as a source for the derivation of further truths and explanations.

The fact that the laws must function as a web brings Newton and Descartes together in another respect, namely, their implicit response to the problem of *ceteris paribus* clauses. Newton's first law, no less than Descartes's, explicitly includes such a qualification: every body perseveres in its state of motion or rest 'unless it is compelled to change its state by forces impressed.'[40] Nor is this a blank check, written on the good will of his readers; Newton thinks his remaining laws and rules can specify those forces, and thus make good the debt incurred by the qualification. So stated, the first law is not really an idealization at all; it relies for its completion on the rest of the system. A corollary of this web-of-laws approach is that none of them purports to state a regularity. The first law is a rule to be applied in tandem with others at the mathematical level of nature.

---

[37] Queries to the *Opticks*, quoted in Janiak (2008, 37).
[38] Newton (2004, 125). I owe this point to Janiak (2008, 38).
[39] For further evidence, including other unpublished writings and the account of Newton's contemporary David Gregory, see Janiak (2008, 38f.).
[40] Newton (1999, 416).

Both the place of laws in the hierarchy of scientific principles and the web of laws approach are underwritten by a shared—and to me at least, highly implausible—metaphysics. For Newton as for Descartes, the laws are laws a divine being gives himself. In the first instance, they govern God's behavior as he moves things about in the natural world.

3. Berkeley

To round out our sketch of the long shadow of Descartes, let's look in more detail at the work of George Berkeley. Although in other respects a benchmark empiricist, Berkeley takes a position on laws that puts him squarely in the Cartesian camp.[41] And unlike either Locke or Hume, Berkeley is well aware of the details of Newtonian science. Berkeley may well be the first modern philosopher who both is fully aware of, and self-consciously aims to accommodate, Newton's results.

3.1 Early Berkeley

Most readings of Berkeley cast the young philosopher as a regularity theorist about laws.[42] I shall argue that such readings ignore the difference between laws and the patterns they generate. The difference is easy to miss, however, because of the dialectical context in which Berkeley first introduces the laws of nature.

Bodies are ideas, and ideas are always passive. What then distinguishes reality from dreams or illusions? All of them are equally made up of ideas, and so on the same ontological level. What makes an idea part of reality is, in part, its connection to other ideas. God produces our ideas in an orderly way, and this provides us with 'a kind of foresight' that enables us to navigate the world. God's orderliness justifies induction and makes the difference between dreaming and seeing.[43]

---

[41] Berkeley's debt to Malebranche is well known. It's not hard to see why the first readers of the *Principles* dubbed Berkeley a '*Malbranchiste de bonne foi*' and named Malebranche 'his master.' (These are from early anonymous reviews of the *Principles* quoted by Charles McCracken (1983, 205–6)). For more on Berkeley and Malebranche, see also Kenneth Winkler (1989, 104–5) and Lisa Downing (2005).

[42] See esp. Brook (1973, 91), Downing (2005, 233), Airaksinen (2010), and Stoneham and Cei (2009, 76), the last of whom read Berkeley as beginning with a regularity theory and changing his mind during the course of the *Principles*.

[43] PHK §31.

It then becomes natural to think that when Berkeley speaks of 'laws of nature,' he means the patterns we cotton on to in everyday life: propositions such as 'food nourishes, sleep refreshes' (PHK 31).[44] It is striking, then, that Berkeley's first explicit mention of laws blocks this natural reading:

> The ideas of sense...have likewise a steadiness, order, and coherence, and are not excited at random, as those which are the effects of human wills often are, but in a regular train or series, the admirable connexion whereof sufficiently testifies the wisdom and benevolence of its Author. Now *the set rules or established methods wherein the mind we depend on excites in us the ideas of sense*, are called *the Laws of Nature*; and these we learn by experience, which teaches us that such and such ideas are attended with such and such other ideas, in the ordinary course of things.[45]

Experience shows us patterns, from which we infer the laws or rules God follows in producing them.[46] This notion of laws as divine rules, and patterns in experience as our evidence for them, runs right through Berkeley's last work, *Siris*. There, Berkeley describes the laws of motion as 'rules or methods observed in the productions of natural effects.'[47] A bit later, he speaks of 'the laws and methods observed by the Author of nature' and calls the patterns we observe in nature 'a foundation for general rules.'[48]

What, then, accounts for Berkeley's tendency to speak as if laws were simply patterns? For in the very next section of the *Principles*, Berkeley tells us that we know 'that food nourishes, sleep refreshes, and fire warms us; that to sow in the seed-time is the way to reap in the harvest...only by the observation of the settled Laws of Nature.'[49] If laws are aspects of the divine will, there is no such thing as observing them. Berkeley must mean we observe the patterns that result from God's will. If so, the laws just are the patterns.

It seems to me that Berkeley is exploiting a usually harmless ambiguity. Ordinary language permits us to slide between speaking of a pattern and of the rule followed by someone producing the pattern. The same answer could serve to enlighten someone about either. If I say that the series '2, 3, 5, 8, 13' is part of

---

[44] As Downing puts it, 'Berkeley holds [in PHK] that laws of nature are regularities in the phenomena...[A]ny simple inductive generalization describes a law of nature for Berkeley' (2005, 233).

[45] PHK 30, first emphasis mine, in Berkeley (1975, 85–6).

[46] See PHK 150: 'by Nature is meant only the visible series of effects or sensations imprinted on our minds, *according to certain fixed and general laws*...' (my emphasis; see also PHK 62, 107, 108, and 153).

[47] S 231.   [48] S 243 and S 245, respectively.   [49] PHK 31.

the Fibonacci sequence (each number is the sum of the two preceding numbers), I am equally describing the rule followed in producing the series. Nevertheless, rules and patterns are not the same thing. Laws or rules license inferences mere patterns do not; these inferences need to be underwritten by claims about the agent producing the pattern. If I came across the series written above, I could not conclude that the next number would be 21 unless I knew the series was deliberately produced by someone intending to go on in the same way.[50]

This last point helps explain why Berkeley does not think even an ideal epistemic agent could predict the future with Laplacean certainty. Knowing the laws allows us to 'deduce the other *phenomena*, I do not say *demonstrate*; for all deductions of that kind depend on a supposition that the Author of Nature always operates uniformly, and in a constant observance of those rules we take for principles: which we cannot evidently know' (PHK 107).[51] Nothing stops God from making exceptions if he chooses, as when he transforms Moses' rod into a serpent, or more broadly changing course as he wishes.[52]

Here we have a divergence from Descartes. Descartes thinks that the laws of nature follow from God's essence, particularly his immutability. If he were right, one could indeed demonstrate the laws of nature a priori.[53] Berkeley rejects any such project, for he denies that God's nature logically entails that he choose the laws he does.

In the 1710 *Principles*, Berkeley has yet to grapple with the details of Newton's views. One sign of this is how he treats gravity. Somewhat embarrassingly, Berkeley presents the fixed stars and the upward growth of plants as

[50] For this reason, I'm skeptical of instrumentalist readings of Berkeley's mature view, which take him as depriving law statements of a truth value; see Downing (1995) and (2005). On my reading, a true law statement reports on a divine volition.

[51] Berkeley is here using some Aristotelian terminology. An Aristotelian demonstration is more than a deduction because it proceeds from first principles, which are necessary truths (see *Posterior Analytics* 70b9f., in Aristotle (1984, 1: 115)). To use a tired example: the definition of human being (a necessary truth) allows us to infer that human beings are risible. One might be forgiven for thinking that Berkeley is making two distinct points at PHK 107: (a) that we might be mistaken in thinking the general principles of Newtonian science really are the rules God follows (since we merely 'take them for principles') and (b) that even if they are the rules God follows, we cannot rule out the possibility of miracles, since God's other attributes (such as benevolence) might require him to violate his own rules. Given the Aristotelian pedigree of 'principles,' however, it seems clear that Berkeley is making only point (b). To say that we take the rules we've discovered for principles is to say that we assume (wrongly) that they are necessary truths that can support demonstrations, rather than mere deductions.

[52] PHK 84.

[53] Some qualification is needed here, since Descartes doesn't deny that God can perform miracles. (How to square the possibility of divine miracles with divine immutability is a thorny question.) Still, I think there is a substantive contrast between Descartes and Berkeley here: in performing a miracle, Descartes's God must suspend the laws of nature, which follow from his essence, while Berkeley's God would not be suspending the laws at all when he departs from the ordinary course of things. (Thanks to Travis Tanner for pressing me on this point.)

counterexamples to the law of gravity.[54] If gravity's scope were universal, Berkeley seems to think, the stars would move toward each other and eventually collapse into one; plants would be strangled down by its force. I suspect that Berkeley, at that stage, had yet to take on board the lessons of Descartes's web of laws approach. Nor does he seem to have a grip on the details of Newton's work. Those developments had to wait another decade, when Berkeley produces his essay on motion.

3.2 The Web of Laws in *De Motu*

Berkeley's essay 'On Motion' makes a very important refinement to this basic picture.[55] For the first time, Berkeley sees that the laws have to function as a web. God doesn't simply follow the inverse square law in producing his effects; he follows all the laws, and takes account of both type-interference and token-interference. Neither plants nor planets can be counterexamples. Berkeley gets to this result by elaborating on Newton's own work.

We've already seen passages in which Berkeley is sensitive to Newton's distinction between mathematical and physical forces. The key passage comes when Berkeley applies this distinction to the composition of forces:

> A similar account must be given of the composition and resolution of any direct forces into oblique ones by means of the diagonal sides of the parallelogram. They serve the purpose of mechanical science and reckoning; but to be of service to reckoning and mathematical demonstration is one thing, to set for the nature of things is another. (*De Motu* §18)

In his notes on *De Motu*, Douglas Jesseph directs us to Corollary I, Book I, of the *Principia*.[56] Newton has just stated part of the second law of motion: 'to any action there is always an opposite and equal reaction.' The corollary to this law is meant to tell us what happens when a body is acted on by two forces at the same time. Newton writes, '[a] body acted on by [two] forces acting jointly describes the diagonal of a parallelogram in the same time in which it would describe the sides if the forces were acting separately.'[57]

---

[54] PHK 106.

[55] First published in 1721, Berkeley apparently submitted *De Motu* to a competition for the best essay on motion held by the Paris Academy of Sciences. Here I rely on Downing (2005), who cites Berkeley's first biographer, Joseph Stock, as the original source of this claim.

[56] In Berkeley (1991, 80).   [57] Newton (1999, 417).

The passage from *De Motu* in effect anticipates and replies to an objection based on Newton's text. How, one might wonder, can we reject (or at least bracket) realism about forces, if we need to mention competing forces to explain events, as Newton's corollary shows? But for Berkeley, as for Newton, the truthmakers for law statements are the rules God follows, not the natures or powers of bodies. Statements of the rules mentioning forces can be true, full stop, even when there are no forces, and even when no single event conforms to them. Casting the laws themselves as divine rules is precisely what allows Berkeley to accept laws about gravity and force without according these 'things' any place in his ontology.

The key point for our purposes is Berkeley's clear recognition, both in *De Motu* and his final work, *Siris*, that the laws of nature have to function in tandem. Any apparent counterexample such as the fixed stars must be due to the operation of other forces besides the attraction of the stars one to another. If we applied the inverse square law just to two bodies in the universe, it would give us the wrong prediction of their behavior. And if we *only* applied the inverse square law to generate a set of predictions about highly charged bodies, we would get the wrong results.[58] None of this means that the inverse square law is not a rule that God follows. It just means that God is following other rules besides.[59] Sadly, some of the epistemic problems that afflict Descartes are still present: the problem of token interference takes the form of the three (or *n*) body problem in classical orbital mechanics. Still, Berkeley's and Newton's implicit solution to the problem of *ceteris paribus* clauses is the Cartesian web of laws.

I don't mean to exaggerate the uniformity among Descartes, Malebranche, Newton, and Berkeley. Descartes's successors are skeptical of his claim to demonstrate the laws on the basis of God's nature; this of course affects how each thinks we can come to know what the laws really are. And their ontologies could hardly be more different: Berkeley is an immaterialist, while the others are not. Nevertheless, the continuities are more important. All treat laws of nature as, in the first instance, laws that govern God's actions. This makes possible the web treatment of laws: no law on its own states, or prescribes, a regularity. Only when we take laws together can we calculate how a given system is going to evolve.

---

[58] Berkeley develops these points in S 230–4.
[59] Here we have another advance over Malebranche: I can find nothing in the latter's work to suggest the 'web of laws' approach Berkeley develops.

4. Summing up: Top-down Laws in the Modern Period

Through much of the literature on the origins on the concept of law, the procedure seems to have been:

(i)   isolate the going thick concept of laws of nature;
(ii) declare our concept *the* concept of laws of nature;
(iii) assume that at least some of the early moderns share this concept;
(iv) hunt for it in the historical texts.

In practice, step (i) has picked out the regularity theory, plus or minus various accoutrements. Once it is enshrined in step (ii), it guides the hunt away from its real target. For step (iii) embodies a mistake: none of the paradigmatic treatments of laws in the early modern period identifies laws with regularities. The surface evidence is the way in which *ceteris paribus* clauses are built in to at least some of the laws, including Descartes's and Newton's first: a law that says something happens just in case some other conditions aren't on the scene can hardly be stating a regularity. The deeper reason for rejecting the regularity requirement is ontological. Regularities are bottom-up in the sense that they supervene on the tiles of the mosaic. But what's distinctive about the Cartesians is precisely their rejection of the bottom-up approach. That remains true even though the going version of the bottom-up picture, against which they are reacting, is a powers, rather than a regularity, view.

Instead, I've argued we should follow a different procedure:

(i)' isolate the necessary ingredients in the thin concept of laws of nature;
(ii)' look to the texts to see how philosophers of the relevant period add to the thin concept to develop their own thick concept.

Neither step requires the myopia of step (ii); nor is this procedure subject to the vagaries of fashion, as is step (i). Instead, we can sensibly require that any concept of laws of nature worthy of the name have at least two features: laws are general propositions that play the axiom role in a given science.

When we look at the thick concept in Descartes and Newton in particular, we find they add a trio of closely allied features to thin concept. First, the laws govern or fix the course of nature, not the other way around. One way to think of this is to say that God creates the laws and the initial conditions; the laws determine what takes place from there on. But the moderns are united in thinking that one cannot leave it at that. The laws *themselves* have no power, any more than the emoluments clause of the United States Constitution can stop a sitting President from enriching himself through foreign powers.[60] Instead, we get our second feature: the laws are edicts or decrees an agent gives to himself, directing his own actions. The powers and essences of bodies are now all but irrelevant.[61] These two features make possible a third: the laws function as a web, freeing each law from any duty to report a single, independent fact about the world. I'll argue below that this web approach is an important virtue of top-down views that we ought to try to preserve if we can.

One consequence of these three features is the ease with which top-down views deal with conservation laws. That motion is conserved is just a reflection of God's ubiquitous activity in moving things about. Here it pays to compare the Cartesians with one of the moderns who resists the top-down picture of laws, Thomas Hobbes.[62] One difference is terminological: Hobbes does not call his rules of motion 'laws.' That signals a deeper one: Hobbes does not think his rules need a foundation in the mind and actions of God. They are instead founded on the properties of bodies themselves, with an appeal to the principle of sufficient reason. We can see this in action when we look at Hobbes's version of the principle of inertia. Hobbes splits it in two, with one rule for rest and another for motion, but the argument for each is much the same. Here is his treatment of motion, from the 1655 *De Corpore*:

> [W]hatsoever is moved, will always be moved, except there be some other besides it, which causeth it to rest. For if we suppose nothing to be without it, there will be no reason why it should rest now, rather than at another time; wherefore its motion would cease in every particle of time alike; which is not intelligible.[63]

Without interference from another body, a moving body will continue in motion. If it stopped on its own, there would be no way to explain why it

---

[60] As Robert Boyle puts the point in his 1686 'Free Inquiry into the Vulgarly Received Notion of Nature': '[I]t is plain that the *law*, which being in itself a dead letter is but a *notional* rule, cannot in a physical sense be said to perform these things [e.g., punishing a murderer with death]; but they are really performed by judges, officers, executioners, and other men, acting according to that rule' (1991, 184).

[61] I say 'all but' simply because bodies must have *some* nature or other. As both Malebranche and Newton note, they must, minimally, exclude other bodies from existing at the same place and time.

[62] Stathis Psillos and Eirini Goudarouli (2019) provide an excellent treatment of Hobbes's position on laws of nature; what follows is much indebted to their paper, as well as the works referenced therein.

[63] Hobbes (1839, 1: 115–16). It's interesting to compare Hobbes's use of the principle of sufficient reason in generating his rules with Emilie Du Châtelet's versions of Newton's laws of motion in her *Institutions de physique* (1740). Du Châtelet is equally explicit in deriving her versions of the laws from the principle of sufficient reason; see her (1740, 222–8).

stopped at that particular time and no other. No God is needed to shore up the rule of inertia. Later, we'll consider another of the early moderns—Spinoza—who, like Hobbes, does not treat such principles as laws, and derives them from the nature of bodies and the principle of sufficient reason rather than the will of God.

Despite their resemblance to Descartes's laws, Hobbes's rules or principles have a striking gap: there is nothing that corresponds to Descartes's conservation of motion. Descartes appeals directly to God to justify that claim: it is God's activity of preserving the same quantity of motion that gets us the conservation law. As Stathis Psillos and Eirini Goudarouli argue, this lacuna in Hobbes's physics is no accident: there is no room in Hobbes's system for a principle like the conservation of motion.[64] There seems to be no way to derive any conservation law from Hobbes's geometrical principles.

So there seem to be some important and interesting advantages the Cartesian view can claim: it allows the laws to function as a web (and hence provides a glimmer of hope for dealing with the problem of *ceteris paribus* clauses), and it gives us a clear way to account for conservation laws. And as we saw in the conclusion to the previous chapter, Cartesian laws support counterfactuals and provide nomic stability in straightforward ways. Whether other views can claim these four virtues, or argue them away, remains to be seen. In Part II, we'll look at contemporary top-down positions and ask whether they can get God out of the picture while keeping what's valuable in the Cartesian approach.

For some contemporary philosophers, the theological origin of the moderns' thick concept of laws is cause for suspicion: perhaps, being born in the image of God, no top-down picture is acceptable. Less stridently, others maintain that the intuitions that drive the top-down picture are suspect. It is only the theistic background that encourages us to think of laws being fixed independently of the mosaic.

Whether the top-down family of views is stained by its original sin will be one of our chief questions in the second part of the book. Many of the arguments for top-down views take the form of thought experiments, and any such experiment must rely on some assumptions or intuitions. Are these arguments vitiated by their historical source, or is levelling that charge tantamount to committing the genetic fallacy? More broadly, without begging any questions, what can be said to vindicate or scuttle the top-down picture? That is the question to which we now turn.

---

[64] Psillos and Goudarouli (2019, 94).

PART II

CONTEMPORARY
TOP-DOWN VIEWS

4

Primitivism

1. Governing without God

To a contemporary reader, what is bound to stand out among all the features of the Cartesian view is its denial of supervenience. Neither Descartes nor Newton thinks that laws are fixed by powers, their instantiations over time and space, or the tiles of the Humean mosaic.[1] There is no doubt that the top-down view allows the Cartesians to wrench the axiomatic structure of mathematics and natural philosophy free from its Aristotelian roots. But was the top-down move necessary? Would any of the going views of laws today have done just as well?

My own position is that, although the top-down view has immediate and obvious advantages that make it natural for Descartes and his followers to adopt it, we should resist the temptation to think that *only* the top-down view affords those advantages. Nevertheless, I suspect that the top-down picture was a stage through which thinking about laws had to progress. Only after its virtues were clear could rivals aspire to claim them for themselves.

By contrast, those persuaded of the top-down picture are liable to turn Whiggish: only their view allowed for the decisive break with the nearly defunct scholastic system.[2] But I want to leave open the possibility that the Cartesians over-correct. For it's far from obvious that the Cartesian way is the only path out of the forest of Aristotelianism. As we'll see throughout the rest of the book, those opposed to the Cartesian way of laws almost instantly set about generating their own thick conceptions that would accommodate the axiomatic structure of Descartes's and Newton's laws without endorsing the underlying metaphysics. Some even wish to return to the forest. Whether any of these other thick conceptions can preserve the important virtues we've isolated will be among our chief questions. In this part of the book, however, we'll be concerned with contemporary versions of the top-down view, which

---

[1] In Chapter 12, we'll worry about the details of the Humean mosaic: what kinds of properties make it up, and how must they be instantiated? For now, all we need is the claim that the mosaic is made up of instantiations of properties across space-time.

[2] James Woodward (2018) provides an eloquent statement of this picture.

The Metaphysics of Laws of Nature: The Rules of the Game. Walter Ott, Oxford University Press. © Walter Ott 2022.
DOI: 10.1093/oso/9780192859235.003.0004

retain the governing role of laws. Our first candidate is primitivism, which simply rejects the demand for a metaphysical analysis of the laws.

If some moderns can be said to have given birth to this particular version of the top-down concept, they might well regard their offspring with horror. Primitivism treats laws as basic, unanalyzable facts about the universes they govern. I have argued that none of the moderns holds this view. Nor have I found any philosopher in the intervening centuries who entertains primitivism: whether one looks to Braithwaite, Mill, Mach, Comte, Whewell, Helmholtz, Lady Mary Shepherd, Dugald Stewart, or any of the other philosophers who interpose themselves between Newton and, say, D. M. Armstrong, one will look in vain for a philosopher who commits herself to the governing concept of laws without in the same breath adding that it is really God who does the governing.[3] Whether the absence of such a view in the historical texts reflects the benighted state of their authors or their tacit recognition of its unworkability remains to be seen.

As with many appeals to primitiveness, the best argument is negative: the primitive conception offers itself as a last resort in the face of the failure of its competitors. Not until the end of the book will we be in a position to weigh up the merits and demerits of all competitors, so in one respect, the main argument for primitivism cannot be evaluated till the end. Nevertheless, we can develop the view right away by looking at some of the points in its favor and examining the costs to be paid for endorsing it.

To anticipate: I'll be arguing that primitivism's virtues are ones we should hold on to, if we can. At the same time, primitivism labors under some burdens we would do better to lay down. In the coming chapters, we'll turn to another top-down conception, one that adds some metaphysical machinery: the Dretske–Tooley–Armstrong (or 'DTA') position, on which laws are relations among universals. Further winnowing will take us to what I believe is the best version of a top-down picture.

2. Motivations

In the plus column, we have primitivism's claim to four key virtues of the Cartesian approach. First, nothing stops the primitivist's laws from

---

[3] One exception might William Whewell (1794–1866), who in some respects anticipates the universals view. I suspect that the appearance is deceiving, since Whewell gives the laws an explicitly theological basis: the laws 'must exist as Acts of that Intelligence—as Laws caused by the thoughts of the Supreme Mind—as Ideas in the Mind of God' (1860, 359). For excellent treatments of Whewell, see esp. Ducheyne (2009) and Snyder (2019).

functioning together as a web, such that only in concert do they determine the course of events. Although this feature is rarely touted by primitivists themselves, it seems to me an important one. Since the laws are unanalyzable facts about the universe, there's no need to match them with any further fact or 'special entity.' Just as the rules of grammar, or of common practice musical composition, function together, and *only* together, so might the primitivist's laws. And were our own grasp of laws complete, we might well find that none of them requires a *ceteris paribus* clause: putative exceptions to one law are to be explained by the operation of others.

I've argued that even those views that pair law statements and truthmakers in a one-to-one relation can take the web approach; Descartes is a case in point. But nothing stops primitivism from holding that the relation is one-to-many. There might be many different primitive facts that are needed to serve as truthmakers for a single law statement. Although it's an open position in logical space, it doesn't seem especially attractive, if only because there seems to be no independent reason why the primitivist should prefer it to the simpler one-to-one pairing. Alternatively, a primitivist might claim there's only one super-law that serves as truthmaker for *all* law statements, so that the relation is many-to-one. That view, too, while open to the primitivist, doesn't seem to have independent motivation. So for now, I'll take the primitivist as holding a one-to-one view.

The primitivist can retain the Cartesians' second virtue and maintain that laws support counterfactuals. If it's a law that objects in motion will stay in motion unless acted on by another force, then we can say that had this lawn dart, which was moving at a given velocity before being pulled to Earth, not been subjected to any outside forces, it would have gone on just as it was. Now, the scope of worlds in which the laws themselves hold is up for grabs. Some primitivists might say the laws hold in all possible worlds; others, only in some. If the laws are contingent, then there are worlds in which the counterfactual fails. But all we can reasonably require of counterfactual support is that the counterfactual hold in worlds with the same laws as the world that is being taken as actual.

Our third Cartesian virtue—the ability to support conservation laws— merits a bit more elaboration, now that we are moving to the contemporary debate. Many of the laws we'll look at seem, at least at first glance, to be susceptible to all three of the families of analysis: top-down, powers, and anti-realism. But, as we've seen, there is a special subset of laws that poses serious problems for all realist (that is, non-Humean) competitors to primitivism: the conservation laws. I'll argue that, occasionalism aside, only the primitivist version of the top-down view neatly fits such laws.

Conservation laws, such as the first law of thermodynamics we looked at in Chapter 2, tell us that some quantity is conserved throughout a given system. Energy can be re-distributed, but it can never simply vanish or appear. If we couldn't assume that the total amount of energy remained constant, there would be no way to predict how fast the ball you just tossed up in the air will be going when it returns to your hand. But, of course, there is, and we can. The philosophical question is: what does the conserving? What is it that guarantees that non-negligible amounts of energy are not leaking out of, or coming into, the system? On its face, it's not *just* a matter of the gravitational force, alongside the force of your throw. Something appears to be making sure that kinetic and gravitational potential energy remain in proportion to each other. If there was ever a case where laws 'govern,' this is it. Somebody—or something—has to be in charge.

Philosophers defending competing views are of course aware of the problem conservation laws pose for their views, and we'll canvass their treatments in due course. But it is worth noting some prima facie problems with the alternatives. It's unclear how the next main competitor in the top-down category, the universals view, will handle the conservation laws. On that view, laws are relations among universals. But the conservation laws seem to be a kind of *meta*-laws, laws that govern the relations among universals.[4]

Nor does the powers view have any straightforward answer. If the laws are as they are simply in virtue of the powers of things, then those powers must somehow conspire to keep the total level of energy (and all the rest) constant. As we'll see, in the modern period, powers were routinely mocked as 'little souls' who know just what to do in every circumstance. If powers are to secure conservation laws, they must be very sophisticated souls indeed. It's not enough that salt should 'know' to dissolve when it meets water; salt must also calibrate its transformation precisely, to ensure that none of the conserved properties goes out of balance.[5]

By contrast, the primitivist—assuming her view is otherwise intelligible—is home and dry. All laws of nature are brute facts about the universe, and the conservation laws fit comfortably into their ranks. From the primitivist's perspective, the reason other realist views find conservation laws so intractable is that they are trying to give an analysis of the unanalyzable: there are no

---

[4] John Bigelow, Brian Ellis, and Caroline Lierse (2004, 156–7) use the conservation laws to argue against the universals view, a point we'll return to in due course.

[5] N.B. I am not presenting this as a knock-down argument against either the universals or the powers view; my point is only that there is a prima facie problem.

'special entities,' to use James Woodward's term, that could make the conservation laws hold.[6]

Our fourth virtue, nomic stability, can equally be claimed by primitivism. The laws can remain constant, regardless of (at least a huge set of) variations in whatever we are taking as our initial conditions. In fact, the most prominent arguments for primitivism advertise this fourth feature. Working through each argument's individual weaknesses will reveal a single, powerful charge: that no competing view can get us the stability laws require.

These arguments—spin, mirror, and the argument from science—can be used both to motivate primitivism and as weapons against the Humean position. The dialectic, then, is more complicated than one might wish, for some of the very same arguments that primitivists deploy are equally to be found in other opponents of anti-realism.[7] So each of the three arguments will need to be re-evaluated when we turn to the other views they have been used to target and support. They have generated an enormous literature in their own right, and will be with us again when we look at responses on behalf of the bottom-up pictures. Even without a full map of the debate, however, there is much to learn from the basic structure of these three arguments. As we progress through the arguments, we'll get closer to the core issue: nomic stability.

3. Mirrors and Spin

All three arguments stand out clearest against the background of a broadly Humean view. For such views, the laws of nature are fixed by the tiles of the mosaic. On the Lewisian version, the laws are statements of regularities that function as axioms of the best system, where the 'best' system is the one that achieves the best balance of simplicity and strength. However the details are spelled out, what is crucial is that the laws supervene on the mosaic: there is no way to change the laws of a world without re-arranging its tiles. Any two worlds that are mosaic-indistinguishable—worlds that have all the same tiles arranged in the same way—are also nomically indistinguishable.

To challenge the supervenience claim, all one needs to do is find a pair of worlds that are alike in all their non-nomic respects but differ with respect to their laws. We can begin with the simplest such argument, the

---

[6] See Woodward (2018, 176).

[7] Recall that I am using 'anti-realist' in my own, purely stipulative sense, to refer to the view that the laws of nature supervene on the tiles of the mosaic *and* that there are no 'enforcers' for the laws. Until we have other anti-realist views on the table, I'll use 'anti-realist' and 'Humean' interchangeably.

'spin' argument.[8] Suppose there is a possible world w_1 with X-particles and Y-fields. In this world, no X-particle ever enters a Y-field. And yet, we are invited to suppose, it might be a law, L1, that all X-particles entering Y-fields acquire spin-up.[9] Now suppose there is another possible world, w_2, just like w_1 in the non-nomic facts: no X-particle ever enters a Y-field. In w_2, L1 is not a law; instead, L2 obtains: any X-particle entering a Y-field acquires spin-down.[10] Anyone accepting all of these invitations will find himself committed to the top-down picture. Our two worlds are identical in their mosaics but different in their laws. So laws cannot supervene on the mosaic.

At first sight, these invitations seem easy to decline. I'm tempted to put the argument in reverse. The picture it presents is deeply *counter*-intuitive: in what could the supposed nomic difference between w_1 and w_2 consist? What is the further fact that nomically differentiates them, and what is it a fact *about*? By definition, primitivism insists these questions have no answer: that w_1 has law L1 and w_2, L2, are brute facts. We seem to quickly reach a standoff. Another way to put this point is to say that the spin argument has not done enough to persuade us that the worlds it posits are possible. As we'll see, many putative counterexamples to anti-realist views have crashed against the walls of modal skepticism. If the worlds are possible only on a non-Humean view, why should the Humean countenance them?

Our next argument tries to make good on this defect. John Carroll's mirror argument shares a structure with the spin argument but tries to give independent support to its modal claims. Like the spin argument, the mirror argument will tag along with us throughout the rest of our investigation. We will deal with its full complexities in due course; for now, our concern is with the structure of the argument and its presuppositions.[11]

---

[8] This is sometimes known as the 'impoverished worlds' objection to the Best System Analysis, since the worlds involved are given minimal descriptions. I don't see that anything turns on the impoverishment: we're free to embed the situation described in a fully worked-out world, even our own, as long as the details of the thought experiment aren't compromised. For the original objection, see Tooley (1977); for replies and discussion, Loewer (1996) and Demarest (2017).

[9] Nothing in the argument turns on the objects at issue being particles, or on the property of spin, but it might be worth adding a quick explanation. The behavior of some particles can be usefully predicted by attributing spin to them: they behave as if they had angular momentum, as Peter J. Lewis points out (2016, 7). In classical mechanics, angular momentum is continuous, but in quantum mechanics, spin is discrete: an electron, for example, can take one of only two values, spin-up or spin-down. Both Peter Lewis and Richard Healey warn against taking the metaphor of 'spin' too literally; Healey (2017, 24–5) suggests the polarization of light as a closer approximation.

[10] The original thought experiment is due to Michael Tooley (1977). Other uses of it can be found in Menzies (1993) and Carroll (1994); for discussion, see esp. Roberts (1998) and Beebee (2000).

[11] Carroll's argument comes in two forms, formal and informal. I discuss the formal version below (Chapter 13, section 1). But I agree with Susan Schneider (2007) that the informal version is the more powerful.

We begin, as with the spin argument, with two, seemingly impoverished, universes.[12] U_1 has two particles, a and b, and two Y-fields. They enter the Y-fields and acquire spin-up.[13] Next to particle a is a mirror on a swivel, such that, had it been turned slightly, it would have deflected particle a so it never entered a Y-field. In U_1, it's a law that

L1:   all particles entering Y-fields acquire spin-up

Now consider a different universe, U_2, which is just like U_1 except that particle a, on entering the Y-field, does *not* acquire spin-up. So L1 is not a law in U_2, because L1 is not even true in U_2. (The requirement that laws state truths is of course itself subject to dispute, but it's certainly part of orthodox Humeanism.) So far, the thought experiment does not pose a threat to any particular conception of laws. U_1 and U_2 are designed to be acceptable even to the Humean: we have a set of facts about what the particles do, and L1's truth or falsity turns on those facts.

What if we swivel the mirror so that it prevents particle a from ever entering a Y-field? Now we generate two more possible worlds, U_1^* and U_2^*. In the starred worlds, the mirror deflects particle a; it never gets to face the test of entering a Y-field. What are the laws in the starred worlds?

U_1^* is the mate of U_1; it's just like U_1, except that the mirror has been turned. In U_1^*, Carroll argues, L1 is still a law: just swiveling the mirror shouldn't change the nomic facts of the world. For the same reason, in U_2^*, L1 is still *not* a law: it's only the position of the mirror that blocks a from taking the Y-field test, but it's natural to think that, since U_2^* is a close relative of U_2, had a entered the Y-field in U_2^*, it would not have acquired spin-up. The problem is that U_1^* and U_2^* are alike in all their non-nomic facts: in each one, particle a is deflected from its course, while particle b happily goes about its business and acquires spin-up. So we have two worlds indistinguishable in non-nomic facts but with different laws. Thus laws fail to supervene on the non-nomic. There is

---

[12] As with the spin argument, the under-description of the worlds involved in the mirror argument is a matter of convenience. As Carroll (1994, 65) points out, the scenarios described 'could all take place in an isolated portion of an ordinary universe, one much like ours'; Roberts goes further, and allows that the scenarios could be embedded in any world, as long as the facts stipulated by the thought experiment still obtain (1998, 430). Beebee (2000, 586) claims that both thought experiments involve 'a dull, barren, and very distant possible world,' but I don't think this is necessarily so. If the rest of the universe is just like ours except for a small region where the drama of X-particles and Y-fields plays out, then it's not so distant at all. Nor do I think the spin argument has to take place in a world with only one law, as Beebee suggests; it can have as many others as you like, as long as they don't contradict the details of the thought experiment.
[13] In this presentation, I follow the statement of the mirror argument in Carroll (2018, 124). His original version (1994) has five particles, not two, though I can't see that that makes a difference.

of course much more to be said about the mirror argument, and when we consider the responses rival theories have proposed, we will say it. How, exactly, is the space of possible worlds taken to be ordered in the thought experiment? Is U_1^* sufficiently close to U_1 to suggest that the laws of the latter carry over to the former?

The most common response, however, runs something like this: the mirror argument depends on the intuition that the laws govern events. Carroll's challenge to his Humean opponent is to explain how U_2^* and U_1^*, agreeing as they do in all their non-nomic facts, can nevertheless differ nomically. To this, Helen Beebee retorts that, as a good Humean, she feels no compulsion at all to offer such an explanation.[14] She has 'no desire to find a way of grounding the "fact" that [L1] is a law in U_1^* but not in U_2^*,' since she thinks L1 'is a law in U_2^* and not an accident.'[15] As Beebee acknowledges, this response entails that the position of the mirror *can* change the laws. But that's just the sort of thing the anti-realist is claiming: the laws of nature are fixed by what happens, not the other way around. Presenting a case in which two worlds share a mosaic and yet seem to differ nomically is a useful way to draw out the Humean's commitments, but it falls short of compelling her to recant. Only by begging the question, the reply runs, can the mirror argument persuade us that there is indeed a nomic difference between the mosaic-identical worlds.

This debate takes an unusual shape.[16] Typically, accommodating pre-theoretical intuitions is taken to be an advantage of a view: one aims to reach a kind of reflective equilibrium, where the laurel is accorded to the position that can preserve the greatest number of our intuitions. Of course, the case is rarely decided by counting up numbers of intuitions: one has to weigh them, and sacrificing a number of peripheral beliefs might well be the price one pays for preserving the central ones.

And yet, in the mirror and spin arguments, the opponents typically reject the intuitions full stop. They claim that such assumptions are a hold-over from the theological picture, or else that they flow from an illegitimate analogy with moral and legal rules. I am sympathetic to this critique, and yet, as Susan Schneider points out, the historical origin of our intuitions is not automatically relevant to their status in the space of reasons: it may well be that, as a matter of historical fact, the governing picture has theological origins (as I have argued it does), but for all that, those intuitions deserve to count when totting

---

[14] Beebee (2000).   [15] Beebee (2000, 590).
[16] For a useful discussion of the role and status of intuitions in the debate, see esp. Susan Schneider (2007).

up the scores of the competing views.[17] The theological accusation, as we might call it, is in danger of committing the genetic fallacy.

True, a belief's origin is typically irrelevant to its justification. But when we are dealing with a claim that has no further justification—in this case, that merely changing the mirror's position cannot alter the laws of nature—a genealogy is in order. To trace such an otherwise ungrounded belief back to a questionable worldview is automatically to put that belief in doubt. If we really are asked to accept a claim without further justification, then pointing to its conceptual origins can amount to an error theory: the mirror arguments' opponents can say that the claim only seems irresistible from within a worldview they reject.[18]

If the mirror argument rested solely on the governing intuition, it would be question-begging. But I think the real action lies elsewhere. As Schneider argues, the core issue raised by the mirror argument is the stability of laws.[19] So we need not let the governing intuition and its provenance decide the matter. As Carroll himself puts it: '[i]t is natural to think that L_1's status as a law in U_1 does not depend on what position the mirror is in.'[20]

Why think that laws are stable? The real reason has nothing to do with the governing intuition; instead, it grows naturally out of the roles we expect laws to play in our theorizing and scientific practice.[21] Suppose I want to build a robot capable of throwing a baseball one hundred yards, into a neighbor's window. I expect the equations I'm using—and hence whatever laws they are, or depend on—to remain stable when I vary the input conditions. Imagine I work out a way to calculate the relationship among the relevant variables,

---

[17] Schneider (2007).

[18] Another option for the Humean is to take the blow from the mirror argument and argue that Humeanism should command assent based on its other virtues. As Roberts puts it, the defenders of Humeanism maintain that their view 'has theoretical advantages that trump the intuitions it offends against' (1998, 428). Schneider (2007) calls this the 'On balance' reply, and those responses that reject the intuitions the 'Negotiability Reply.'

[19] See Schneider (2007, 315).   [20] Carroll (2018, 124).

[21] Marc Lange (2009) makes stability the defining feature of laws. On his view, the ontological primitives are the subjunctive facts. They are the 'lawmakers' and no special entities are needed to back up these subjunctives. So Lange's view differs from Maudlin-style primitivism in that the laws do receive an ontological grounding, but only in terms of these subjunctive conditionals. Like orthodox primitivism, Lange's view benefits chiefly from the misfortunes of its realist rivals, who all attempt to shore up the counterfactuals with special entities. I am not sure how to understand the proposal that the subjunctives themselves are (or report on) brute facts: at least in everyday cases, there seems to be some further fact that underwrites the conditional. For example, if the window would have shattered had it been hit with the baseball bat, it's natural to ask after the further fact (whether the window's fragility, the bat's solidity, or the micro-structures of each) that underwrites the counterfactual. Such appeals to what's 'natural' in everyday contexts hardly suffice as a refutation, of course. Other philosophers have raised more technical difficulties for Lange's proposal; see esp. Demarest (2012) and Hall (2011), as well as the symposium on Lange's book with objections from Carroll, Loewer, and Woodward, as well as Lange's replies (Lange et al. (2011)).

including wind-speed, distance, and the rest. I don't expect to get a different answer—and hence different baseball-behavior—if I change some variable I've judged to be irrelevant. Nor do I want the equations to change every time I change the variables. This, I think, is the real point behind the mirror case: to do the jobs we want them to do, laws should remain invariant under a wide array of conditions.[22]

I'm not claiming that *only* primitivism, or indeed top-down views generally, is entitled to nomic stability. My claim here is that nomic stability is a reasonable desideratum, and we should award points to views that get it for us. As we go on, we'll see other views trying to lay claim to it. Only at the end will we be in a position to make a judgment.

4. The Underdetermination Argument

On the surface, this argument seems merely to ring the changes on the mirror argument. Both point us to worlds indistinguishable in non-nomic terms that nevertheless are characterized by different sets of laws. But the method of argument, and the real gravamen of each, is quite different.

Suppose possible worlds function as models of laws. If, as Tim Maudlin asserts, scientists routinely (or at least sometimes indispensably) avail themselves of such models to consider how a world governed by those laws might be, then scientific practice itself is committed to the underdetermination of nomic facts by the mosaic. All we need to ask is whether two worlds could be indistinguishable in their mosaics while differing nomically.[23] The answer, Maudlin thinks, is obviously 'yes'; so the top-down conception has to be correct.

It's helpful to keep in mind where the goal posts are set in this game. To defeat the Humean, the primitivist (and her fellow-travelers) needs to find two worlds that differ nomically but in no other way. Such worlds are easy to conjure by stipulation; the question is, should we accept them as possible? The mirror argument gives us a pair of worlds that, by virtue of their association with the worlds on which they are variants, end up being nomically different

---

[22] By 'wide array,' I don't mean 'all.' Below, I'll argue that the powers view can get us a version of nomic stability, and it would be absurd to require that the laws such a view can fund remain invariant regardless of the nature and number of powers instantiated in a world.

[23] See Maudlin (2007, 67). John Carroll presents Maudlin's argument as 'the argument from scientific practice,' alongside his own mirror argument; see Carroll (2018, 124). In work that antedates either of these sources, Barry Ward (2002, 195) presents the underdetermination argument as part of his case against Humean supervenience. Unlike Maudlin and Carroll, Ward goes on to defend projectivism. See below, Chapter 14, section 3.

but otherwise identical. The appeal to scientific practice aims to do exactly the same thing by other means. As Maudlin puts it, the question is, 'can two different sets of laws have models with the same physical state?'[24] If so, then the non-nomic facts cannot determine the laws. Humeanism is false.

The formula for generating such non-nomically indistinguishable models is simple: just subtract any feature that would differentiate them. If our chosen worlds remain silent on the points on which the competing sets of laws differ, they will by definition serve equally well as models of either one. This is just what the spin and mirror arguments above try to accomplish. But instead of trying to motivate the nomic difference of something like U_1^* and U_2^* by appeal to their relationship with their parents, U_1 and U_2, the argument from scientific practice tries to justify their nomic difference by appeal to their role in theorizing. To use Maudlin's example: two worlds, U_3 and U_4, each consisting only of empty Minkowski space-time, could instantiate the laws of General Relativity and of a rival theory, respectively.

This argument strikes me as interesting but less persuasive than the appeal to stability implicit in the mirror argument. I'm not sure why the scientist, *qua* scientist, needs to take a position on the metaphysical question, or how, *qua* scientist, she would justify it.[25] A natural reply is that there's no fact of the matter about whether General Relativity or the rival theory is true in U_3 and U_4, or indeed whether U_3 and U_4 are distinct possible worlds. We're free to take either or both as models of General Relativity or of any rival consistent with the existence of Minkowski space-time; nothing compels us to choose one over the other. (This case, perhaps, is where the accusation that the realist's possible worlds are 'impoverished' has bite.)

An analogy might help here. It's plausible to think that which opening is played in a chess game supervenes on the positions of the pieces in the first few moves. Imagine a chess board at the start of the game. White moves the king's pawn two places. Is this part of the Ruy López opening, or of one of the many other openings with that first move? The chess board with White's king's pawn moved two spaces can serve as a 'model' of any of these openings. Imagine no

---

[24] Maudlin (2007, 67).

[25] I take it this is the kind of point Richard Healey is making, in his (2008) review of Maudlin's book: '[A] *scientific* interest in physical possibility is limited to applications of laws to the actual world. The Schwarzschild solution represents a scientifically interesting possible General Relativistic world because it can be used approximately to model a system like a planet, star, or other local feature of the actual world. In such employment, of course the system's behavior will be "governed" by the laws of general relativity, insofar as these are assumed to hold in the actual world. If asked whether an infinite, empty Minkowski space-time is "governed" by the laws of Special or General Relativity (or perhaps some other theory), the practicing scientist should decline to answer, on pain of turning metaphysician.'

more moves are made. If pressed for an answer to our question, we should simply decline to answer. It's perfectly consistent to assert both that one move is not enough to fix the opening *and* that the opening supervenes on the first few moves. The underdetermination argument seems, on its face, to conjure nomic non-supervenience from under-description. And that's a sleight of hand we should at least be wary of; we'll encounter it again as we go on.[26]

5. What is Governing?

Although I happily join critics such as Loewer and Beebee in claiming that the governing intuition is fruit of a poisoned theological tree, we've seen that the real point of the mirror argument is nomic stability. So rendering the governing intuition powerless by presenting its pedigree is not enough to block the argument. But there is another way in which the history of this particular top-down view matters.

Taking the long view of the debate will display a central weakness in primitivism. The problem is not that primitivism derives its appeal from outdated or questionable intuitions. The worry instead is that the view is in danger of creating and then ignoring two substantial holes in its parent view. The first is the ontology of the laws themselves. What kinds of things are they? The moderns insist on locating them in divine volitions; without some such move, they are a substantial and inexplicable addition to our ontology. The second hole—and the one I focus on in this section—is what the laws *do*. Many of us, including my former self, have grown so accustomed to governing locutions we may well think we understand something that we do not. We can get so used to talking about laws governing events that we stop feeling the need for any further explanation. But to the moderns, it's simply obvious that the laws cannot do anything: to say that a law governs some event must be shorthand for some fuller description in terms of divine action. Now, here again, we will be told we're asking for an account where none can be given. But notice the price to be paid: primitivism no longer simply rejects the demand for an account of the ontology of the laws; it rejects any demand for a substantive story about how the laws carry out their characteristic function.

I do not claim to be able to show that primitive laws cannot govern. I think I can show that, when it comes to *what* governing is, the primitivist's pockets are empty. Of course, this may seem to belabor the point: the primitivist can

---

[26] See Chapter 12, section 5 below for a fuller statement of this argument.

always say that governing, like the laws themselves, admits of no conceptual analysis and is *sui generis*. But one can then be forgiven for simply not knowing what the primitivist is talking about. This 'dual' primitivism is an important position in logical space; but by abjuring the demand for an elucidation of *both* laws and how they govern, it is one we will be justified in passing by, unless further developments should make it our last resort.

5.1 The Governing Dilemma

It is as not as if the primitivists have said nothing at all about how governing is supposed to work.[27] My point is that the typical formulations of the governing thesis is no real advance over simply insisting there is nothing to be said about how laws govern.

Schneider offers a way of understanding governing, completely innocent of theology: 'once the initial conditions and the laws are set at a world, the subsequent history will evolve under the direction of the laws.'[28] Schneider argues that proponents of the top-down view are no more helping themselves to theological aid than is the Humean who equally (and just as dispensably) uses God as an illustration.[29] Fair enough. But such a claim hardly makes for an improvement: in place of a bizarre universe where Boyle's laws run around adjusting temperature and pressure, we are told that events unfold 'under the direction' of the laws. That seems no more or less mysterious to me. Note that this proposal for cashing out 'governing' embodies the Autonomy reading some contemporary interpreters project on to Malebranche. To show that the

---

[27] Roberts defends the governing conception by arguing that '[t]he proposition we call the law is not the agent of the governing, but the content of the governing' (2008, 46). In his review of Roberts's book, John Carroll (2012, 897) endorses this idea, implying that it removes the air of mystery surrounding governing. For my part, I don't see how this helps. Roberts likens the laws of nature to those of a polity: the laws themselves don't do the enforcing. But then what does, and how? Roberts seems to think the laws might be propositions that govern without need of enforcement. He appeals to another analogy, this time with the way in which the rules of English grammar 'govern' the speech of anglophones. 'So it is at least coherent to suppose that there could be contents of governing without a concrete governing agent' (2008, 47). But the rules of grammar are rules we can follow (or not); the governing agent is the speaker himself. All this becomes irrelevant once we see how Roberts proposes to understand 'governing.' On his view, to say that the laws govern is just to say that they are 'preserved under every counterfactual supposition that is logically consistent' with them (2008, 191–2; see 197). In my terms, this 'nomological preservation' is nomic stability, not governing. And if it's all that governing amounts to, then other parties to the debate, Humean or not, can try to claim it as well.

[28] Schneider (2007, 317).

[29] For example, in presenting the Best System Analysis, Beebee (2000) imagines God trying to summarize his 'Big Book of Facts.'

Autonomy view lives on in the current debate, consider this revealing passage by Tim Maudlin:

> Speaking picturesquely, all God did was to fix the physical laws and the initial physical state of the universe, and the rest of the state of the universe has evolved (either deterministically or stochastically) from that. Once the total physical state of the universe and the laws of physics are fixed, every other fact, such as may be, supervenes.[30]

Modulo the opening caveat, this is precisely the Autonomy reading of Malebranche. Above, I argued that Autonomy fails as a reading of Malebranche; I think it fails *simpliciter*. The accusation is not that primitivism relies on outdated intuitions, but that the view cannot be understood without God in the frame. Let me develop the case against primitivism by means of a dilemma.[31] The primitivist's claim is:

(the laws and the initial conditions) → the course of events[32]

The challenge is to solve for '→.' The governing dilemma is this: either the arrow is causation, or it's not. If it is, then the view is vulnerable to charges of overdetermination and, what is more important, effectively bars laws from their governing role. Alternatively, if governing is not causation but some weaker relation, then the view collapses into Humeanism.

Let's take the first horn. Suppose that the laws determine outcomes by producing them, somehow taking account of the prior states of the universe. (If this is what laws do, note that that limits the view to laws of temporal evolution, which presuppose a direction of time.[33]) How exactly the laws would by themselves calculate the right result to produce given the total state of the universe is the sort of thing that would worry the moderns, who wheel in God at that point. It's equally obscure how the laws could be the right *kinds* of things to cause anything. I can see how events, or property

[30]   Maudlin (2007, 158).

[31]   The problem I develop for primitivism is a cousin of the inference problem faced by the universals view (see below, Chapter 6, section 1). In its broadest form, the inference problem asks: how do we infer from the laws to the behavior of the things they allegedly govern?

[32]   For the sake of simplicity, I'm bracketing the whole question whether the laws of quantum mechanics will turn out to be deterministic or not, and whether there might not be probabilistic laws at higher levels.

[33]   I'm not sure if this is in fact an objection to governing as causing. But Chen and Goldstein (forthcoming, 17) argue that laws should not be limited to propositions that govern temporal evolution.

instantiations, or even objects could be causes; as for the laws themselves, I draw a blank. Now, that may well be just another point at which the primitivist would claim I am asking for elucidation where none is to be had.

Even if we waive both of those objections, another immediately presents itself. A metal bar is subjected to a blowtorch and then expands at a given rate. We have our cause (the flame of the blowtorch being applied to the bar) and our background conditions (the presence of enough oxygen to allow the flame to burn, the absence of interferers of every sort, and so on). Where, in this picture, is there room for yet another cause, namely, the law of thermal expansion? To add it to our list of causes-cum-conditions would be to have one thought too many. The events I described strike me as sufficient to produce the expansion of the metal bar; if so, making the law a cause, too, results in overdetermination.

A more persuasive point appeals to what place laws are meant to take in the governing picture. It's natural, on such a view, to think that causation itself is law governed. But if the law is just another cause, or part of the total cause, then we have in fact deprived the law of its governing role. To use a spatial metaphor: the law is now inside the causal relation, rather than directing it from above. We would now need a brand new law to govern the relation between the blowtorch, the background conditions, and the law of thermal expansion on one hand, and the expansion of the metal bar on the other. This last observation suggests that the laws describe what happens; they don't make it happen.

I suspect most primitivists don't want to think of governing as causing. So now we can turn to the second horn of the dilemma. If there's no 'oomph' to the laws, we still have two options. On the one hand, it might simply be that the laws entail the future states of the system. But entailment isn't governing: it's an abstract relation among propositions. What's more important, the Humean can equally say that, given the laws of nature and a set of initial conditions (assuming determinism), you can deduce other states of the relevant system. Indeed, the deductive-nomological model has traditionally been one of the Humeans' standard models of explanation.[34]

A different option is more appealing: we might try to identify the laws' activity with constraining.[35] There's an intuitive sense in which the laws constrain modal space: in worlds like ours, gases expand in proportion to

---

[34] Perhaps the classic discussion of D-N and its limitations is Salmon (1984). Nothing for me here turns on the truth or falsity of the D-N model.

[35] The most developed version of governing as constraining can be found in Chen and Goldstein (forthcoming).

temperature and pressure, given the laws. Perhaps the laws govern by fixing regions of modal space. Set the laws, and you thereby draw a Venn diagram that includes nomically identical worlds.[36] But that can't be all there is to 'constraining,' or else the notion is so weak as to be compatible with anti-realism. The Humean, too, can say that there's a region of modal space defined by the laws of the actual world.

Another potential issue with the governing-as-constraining proposal is that, by abstracting the direction of time, it doesn't yet account for '→,' which is supposed to take us from initial conditions as inputs to a further state of the universe. How would laws of temporal evolution fit in? Here, I take it the idea would be that in a deterministic universe, the laws would constrain by narrowing the possible outcomes for any complete set of initial conditions to exactly one.

But now the notion of constraint is too weak. It merely says that *given* the laws of the actual world, this is the outcome that must result. The arrow then represents entailment. But again, that doesn't capture governing, since it's consistent with Humeanism. Now, maybe something more is meant by constraining the development of a system over time. Perhaps the idea is that the laws *force* events to come into existence, given prior conditions. At that point, we're back to the first horn, on which governing is causing.

The upshot of the governing dilemma is this: there is no substantive way to understand what the primitivist's laws are supposed to do. Either governing is causation—which it isn't—or it's something weaker. But that weaker something had better not turn out to be compatible with Humeanism, or else primitivism loses its distinctive flavor. If there's any middle ground between accounts of governing that are too strong or too weak, I haven't been able to find it.

6. The Virtues of Primitivism

Contemporary primitivism has the merit of reminding us of what was both novel and worth preserving in Descartes's thick concept of laws. Our thin concept says only that laws are general propositions with a content that allows

---

[36] As Chen and Goldstein put it: 'We suggest that laws govern by constraining the physical possibilities. More precisely, laws govern by limiting the physical possibilities and constraining the actual world (history) to be one of them. In other words, the actual world (history) is constrained to be compatible with the laws.... $F = ma$ governs by constraining the physical possibilities to exactly those that are compatible with $F = ma$' (forthcoming, 19).

them to play the role of axioms in a given science. Primitivism adds the top-down elements of Descartes's view, without the theology. And yet it can retain the distinctive virtues of Cartesianism, especially the web approach and the ability to account for conservation laws.

Most of the debate has centered on a third virtue, stability. The real lesson of the spin and mirror arguments is that we need laws to come out stable under some range of counterfactual variance. I've argued that that feature is latent in the Cartesian conception, and that it's a reasonable desideratum. In this section, I want to say a bit more about why nomic stability is an advantage, if we can get it.

Above, I suggested that stability appears to be a feature of scientific practice. The laws let us consider, not just what would happen with different initial conditions, but what would happen to any system we choose to isolate within the actual world, treating whatever arbitrary point in space-time we select as the initial conditions. One thing knowing the laws of nature ought to do for you is allow you to select from a number of possible scenarios. The engineer relying on Newton's laws in entertaining and then rejecting a certain design of buttresses for a cathedral is treating that design, and the other relevant factors, as a set of initial conditions and then calculating how those materials would evolve over time. If the putative laws don't let you do *that*, then they have correspondingly less claim to be laws in the first place.

This desideratum is controversial. Proponents of the Humean Best System Analysis in particular might reject it; but they might also, I'll suggest, come up with ways to accommodate it. Let me explain a bit more about what we can and cannot sensibly require of a view, if we adopt nomic stability as a desideratum.

Counterfactuals like the ones involved in imagining different initial conditions are notoriously context-dependent. What would the world be like, if clouds routinely played hopscotch in the sky? There's just no single, right way to answer that question, for we have to know what *other* features of the world we're supposed to hold stable, and only a context of utterance (or brute stipulation) can tell us that. Another way to put the same point would be to say that precisely what proposition a counterfactual expresses is indeterminate absent a context.

But I think we can bypass these thorny issues. All we need is the claim that, if L is a law, L might still have been a law if some range of initial conditions were different. What fixes that range depends on the view of laws in question. Suppose you think laws just are powers. A state of affairs where the initial conditions added or omitted a power would by definition not be a world in which the laws are preserved. Similarly, suppose you identify laws with relations among universals: vary the universals, and you get different laws. So we shouldn't require that the actual laws be compatible with just *any* initial conditions; that rules out too many views in a purely ad hoc way.

What we *can* say is that the laws ought to be compatible with any arbitrary re-distribution of whatever the view takes to be its nomically relevant entities. A scenario in which a single power or universal is instantiated in Battambang instead of Baltimore shouldn't, by itself, generate a world with a different set of laws. Of course, how different such a world is depends on the laws in question, and also on what features the context of the counterfactual is sensitive to: do we hold steady the rest of the history of that world and assume indeterminism? Or do we conceive of that world as having a sufficiently different history to put the nomically relevant entity at some other point in space-time? Either way, the point is the same: the laws ought to allow us to generate models of worlds that are different from our own but nevertheless nomically identical.

The stability of the primitivist's laws clearly belongs in the plus column. But what of the view as a whole? Here, it pays to return to the wider historical context. I have argued that no one in the modern period entertains the Autonomy thesis, which is precisely the primitivist position. The philosopher who has been read as articulating it—Nicolas Malebranche—in fact rejects it, for he insists that it is really God who is doing the governing, by framing laws that he himself will follow. Those who insist that the contemporary governing conception relies on a theological picture are on the right track. But the problem is not that the governing conception depends on theological meta-phors or intuitions. It depends instead on a God to do the governing. The primitivist view helps itself to the power of laws to fix events without including what the moderns see as the necessary metaphysical support.

My argument against primitivism is necessarily incomplete at this stage. For it may well be that primitivism is the best anyone can do. Put more strongly: if indeed scientific practice and philosophy of science both insist that laws are stable under variations of contingent matters of fact, *and* if primitivism is the only way to accommodate that fact, we had better not dispense with it.

The primary objection we have considered is whether primitivism can, and needs to, explain how governing works. Now, we'll turn to the final top-down view, which offers to flesh out the governing picture. Where primitivism sees the laws as simply 'determining' how events play out, the universals view purports to tell us *how* they do it.

5

Universals (I)

1. Motivations

The signal challenge the primitivist faces is to provide a substantive account of '→' in

Laws plus initial conditions → the course of events.

By definition, no card-carrying primitivist would feel the force of this demand. But for those as yet uncommitted, I hope to have shown that it would be desirable to explicate '→' if we can. For the primitivist not only resists the demand to analyze laws; she also resists the demand to tell us what laws do. The governing dilemma suggests that there is no way to meet that demand. Without a grip on what laws are, or even what they do, we might well conclude that primitivism is not yet intelligible. Looked at differently, primitivism may seem more like a last resort, its sole virtue the weakness of its competitors.

The universals view aims to preserve the governing nature of laws while providing a substantial ontology to account for it. After developing the universals view and defending it from some attacks I believe can be parried, I turn to those I think cannot. These latter objections recommend a revision of the universals view. Once revised, the universals view can muster even stronger replies to the initial attacks. But let us begin with the patient in its unaltered form: the original view, apparently formulated independently in the late seventies by Fred Dretske, D. M. Armstrong, and Michael Tooley.[1] I focus on Armstrong's view, since it is the most developed.

2. Armstrong's View

The universals theory identifies laws with necessitation relations between universals. The view is both realist and top-down. Armstrong's laws are

---

[1] Armstrong (1983) and (1997), Dretske (1977), and Tooley (1977).

The Metaphysics of Laws of Nature: The Rules of the Game. Walter Ott, Oxford University Press. © Walter Ott 2022.
DOI: 10.1093/oso/9780192859235.003.0005

contingent, since there are possible worlds with precisely the same first-order properties as our own but with no necessitation relations to bind them. So, like our other top-down picture, this one rejects the supervenience of laws on the mosaic. In terms of our third axis, the universals view counts as a 'one-to-one' view, because it pairs each true law statement with a single fact that makes it true. 'Force equals mass times acceleration' is true in virtue of the necessitation relationship among the universals named in that proposition. If a given science is not issuing laws that involve genuine universals, it is not really dealing in laws at all. For example, if no universals correspond to 'price' and 'demand,' then, while economics might treat some relationships among those variables as laws, statements involving them will not count as laws. So whether the special sciences are investigating and formulating law statements will turn on whether there are genuine universals to be found at their level of inquiry.

The 'N-relation' is contingent; there are worlds where it connects F and G, and worlds where it does not. An immediate result of this picture is quidditism: the claim that properties are categorical rather than dispositional. To call a property categorical is to say that, on its own, it has no essential causal profile. Whatever dispositions there are must be conferred by the laws that govern the world. Suppose, for example, that shape is a categorical property. By itself, being spherical confers no power of rolling on the object that has it. There are worlds like ours where spherical objects tend to roll downhill near the surface of the Earth; there are other worlds where the laws dictate that they do not. This is just a reflection of the top-down nature of the laws: if the laws are not fixed by the denizens of the worlds they govern, those denizens had better not have any dispositions or powers written into them.

Armstrong argues that we can use an inference to the best explanation to establish the laws.[2] Confronted with a regularity such as $\forall x(Fx \rightarrow Gx)$, we may posit the nomic relation N, such that N(F,G), where that means F-ness necessitates or 'brings with it' G-ness. The N-relation itself is a second-order property: it is a property of the properties F and G, and only derivatively of the objects that instantiate F and G. We can now see how the universals view proposes to cash out the '$\rightarrow$' of governing: it is the fact that N(F,G), together with the initial conditions, that brings about the regularity $(\forall x)(Fx \rightarrow Gx)$. The mosaic of nature has an explanation, and must be as it is, given the universals that are instantiated and the N-relations connecting them.

---

[2] I am aware that many challenges have been issued to such abductive inferences in the past, especially by Bas van Fraassen; since I consider them adequately answered in the literature, I pass over them here. See especially Douven (2017).

Giving a metaphysics of governing is a substantial advance over primitivism. But can the universals view claim the same virtues as its sibling? At first sight, such laws do much of what we would want. They clearly satisfy the thin concept, which requires that statements of laws be sufficiently general and play the axiom role. And they go some way toward preserving the four virtues of Cartesian top-down views.

First, they provide some stability for the laws across a wide array of differing initial conditions. But the issue is complicated by Armstrong's immanentism: on his view, only instantiated universals exist.[3] A universal has no existence apart from its instances, and the same goes for the necessitation relation. Consider again the spin argument, which asks us to imagine a world with X-particles and Y-fields. No X-particle enters a Y-field, and yet it might be a law (L1) that all such particles would acquire spin-up should they enter the field. If that's all there is to the world, then the universals theory can't say that (L1) is a law. For L1 refers to a universal, namely spin-up, that is not instantiated, and so doesn't exist; trivially, it can't enter into any N-relations. The same considerations would make trouble for the universals theorist if we pressed the underdetermination argument.

One option is to reject immanentism and go for a Platonic theory of universals.[4] Merely not being instantiated would be no barrier to spin-up's existing in the impoverished world we envisioned. Armstrong would recoil in horror. For my part, I think the universals theorist should tough it out. L1 just isn't a law in the world in question. If we look at the more threatening mirror case, the universals theory gets us the right result, since spin-up is instantiated in all four worlds, U_1, U_2, $U_2{}^*$, and $U_1{}^*$. This leads us back to one of the rules of the game: if a given view claims that laws of nature are or depend on x, we cannot ask that it keep the laws stable even in non-x worlds. If laws are N-relations among universals that depend on their instances, then we can't demand that the laws remain stable even when those instantiations are absent. Admittedly, in the context of the universals theory, there are some counter-intuitive consequences: a law can pop into existence when either of the

---

[3] Given immanentism, one might wonder whether Armstrong's view isn't closer to the powers theory than it at first seems. McKitrick puts the point well: '[T]he Necessitation relation is located where x is. Hence, it seems that x is intrinsically necessitated to yield G, and this seems similar to having a power to produce G' (2021, 262). As we'll see with the key example below, however, there are other respects in which the universals and powers views differ.

[4] For discussion of Platonism as opposed to immanentism in the context of laws, see esp. Matthew Tugby (2016).

associated universals is first instantiated.[5] But I don't think the view can be dismissed on these grounds.

A second virtue of Cartesian laws, and primitivism too, is the ability to account for conservation laws. This is more difficult to evaluate in the case of the universals view. It seems that the conservation of a given property needs to be handled at a level above that of the N-relations; it seems to be a kind of meta-relation among the N-relations instantiated in a given world. This is an issue we'll need to look at in more detail.[6]

Finally, laws *qua* N-relations support counterfactuals. This feature has been implicit in our discussion of contemporary top-down views so far, but is worth bringing out now. I have resisted building it into the thin concept, since it is highly controversial. The controversy surrounds not just whether laws support counterfactuals but what it even means to do so. We'll get into the weeds of that debate when we turn to anti-realist proposals. For now, it's enough to see that if $N(F,G)$ obtains in a world, we can say that had x been F, it would also have been G. That falls out of the nature of the N-relation.

Note that the universals view can help itself to the language of powers without endorsing that ontology. Water has the power to dissolve salt, even if this disposition accrues to water only in virtue of the necessity relation linking the two universals. All parties can agree that there are dispositional predicates, in the anodyne sense that some objects are truly described as being such as to break easily when force is applied, and so on. But not all predicates refer to genuine properties, on pain of Russell's paradox.[7] So there is no straightforward inference from 'x is fragile' to 'x has the metaphysically weighty property *fragility*.'

In our terms, we can put this point by saying that the universals view casts all dispositions as dispositions in the thin sense. To bring out the difference in the context of universals, we should consider an intriguing argument against the universals view developed by Gabriele Contessa. He argues that the alleged

---

[5] I owe this point to Travis Tanner. One possible way to minimize the damage of the suddenly appearing law is to appeal to the counterfactual: had F been instantiated sooner in the world in question, $N(F,G)$ would have been true. This quickly gets tangled, however: in order to support that counterfactual, we would need to go to a nearby world in $N(F,G)$ holds. But since Armstrong's laws are contingent, not all worlds, even those that have instances of F and G, are such that $N(F,G)$. So I think the universals view is better off biting the bullet.

[6] Chapter 6, section 5 below.

[7] Take the predicate 'is not self-exemplifying.' Suppose it refers to a property. Now we have a problem: does the property of being non-self-exemplifying exemplify itself, or not? If it does, then it doesn't, and if it doesn't, then it does.

dispositions purchased by the universals theory are at best mimickings of dispositions.[8] Imagine two cases:

Key 1.   A key is disposed to open a given type of lock in virtue of its own intrinsic states.

Like many of us, I despair of finding a definition of 'intrinsic' impervious to all potential counterexamples. I think we can do well enough to get on with by saying that an intrinsic property F of an object *a* is one such that F*a*, regardless of what else goes on in the world in which *a* finds itself. Now, in Key 1, the key has a thick disposition: its shape is what gives it the power to open keys of a certain kind, and shape is an intrinsic property if anything is. But now consider

Key 2.   A powerful wizard casts a spell that enables hunks of tofu to open locks of a given kind.

Nothing about the tofu, not even its shape, gives it the ability to open locks of a given kind. We can put this by saying that the key only *mimics* the disposition to open the lock. The dispositions the universals view can get for us are *all* cases of mimicking, on Contessa's view. There is no relevant difference between Key 2 and the actual case: 'spells are virtually indistinguishable from laws' in the universals sense.[9] The disposition of Fs to bring about Gs is to be explained by the second-order relation N(F,G). So the presence of F is neither necessary nor sufficient to bring about, or to confer the disposition to bring about, G. Just as the wizard might have chosen pork chops instead of tofu as enchanted keys, the N-relation might have attached H rather than G to F. This is just a consequence of the universal view's insistence that the laws of nature are contingent. No object's being F confers on it the disposition to bring about G; that is due to the grace of the N-relation. From Contessa's point of view, this is disastrous. The universals theorist is stuck with what he calls 'neo-occasionalism,' on which property possession only appears to confer dispositions on the objects that instantiate the N-related properties.

I think this 'mimicking' argument nicely illustrates the universals view, and shows how it differs from a powers view. I doubt, however, that the universals theorist will be much bothered by it. Contessa might well be correct in thinking that 'the powers theory seems to be the only option currently available to those

---

[8] See Contessa (2015).   [9] Contessa (2015, 173).

who believe that (at least some) properties confer dispositions on their bearers.'[10] But the rejection of that view is built in to the universals position: it is only by virtue of instantiating universals that are themselves N-related that anything has a disposition. Nor is there anything deeply counter-intuitive about the universals view: a mushroom that merely mimics the disposition to bring about death is no less dangerous for that. Any appearance to the contrary is generated, I suspect, by conflating thick and thin dispositions. It might indeed be disastrous to say that nothing has so much as a thin disposition to do anything. But no one, as far as I can tell, is committed to that.[11]

3. Quiddities: Epistemic Worries

A 'quiddity' or 'categorical property' is a property that neither is nor essentially has a power. Lots of properties at least seem to be categorical. Being located in space-time, having a certain shape, being made up of carbon atoms: all of these might play some causal role, to be sure, but they are neither exhausted nor defined by that role.

Quiddities now occupy roughly the same intellectual position dispositions did in the modern period: despised as basic metaphysics, tolerated only when an energetic paraphrase has purged them of their mysterious elements. What are the reasons for this reversal? The chief arguments originate with Sydney Shoemaker's 1980 paper 'Causality and Properties.'[12] Subsequent opponents to quiddities ring the changes on his central theme.[13] It makes sense, then, to return to the source.

Shoemaker's basic argument is epistemic and savors of verificationism. Here it is in outline:

1. We know properties by their effects
2. Quiddities are not defined by their effects

---

[10] Contessa (2015, 162).

[11] As Bird (2016, 361) notes, Armstrong is perfectly happy with thin dispositions, in my sense. Here is Armstrong: 'Typical cases of dispositions are...solubility, elasticity, and brittleness. Associated with dispositions are certain truths...It is a plausible thesis that in every case of cause and effect the effect can be seen as the manifestation of some disposition or dispositions, and such a view would be a congenial one for a Dispositionalist. But the Dispositionalists go a great deal further than this. They wish to resurrect the old pre-Humean idea of powers. The Humean idea that there is no necessary connection between wholly distinct existences is completely rejected...' (2010, 48–9).

[12] In Shoemaker (2003, 206–33).

[13] See, e.g., Black (2000), Mumford (2004, 103), and Bird (2016, 347).

Therefore

3. Quiddities are unknowable.[14]

I can detect the presence of the rectangular shape of a table by light striking it and bouncing into my eyes. But if shape is not itself a causal power, it has no essential causal profile: as a quiddity, its identity does not consist in its potential or actual causal contributions. Given the contingency of the N-relation, there are worlds where rectangularity always causes the visual experience of rectangularity,[15] worlds where it causes the visual experience of a sphere, and on and on. The shape itself is then permanently beyond my epistemic reach.

To illustrate his argument, Shoemaker presents three unpalatable scenarios, each of which is allegedly permitted by the quiddities view. First is what we might call the 'ghost property' scenario: for all we can tell, there might be properties lurking in the world that are never in a position to make any difference to what happens. More common in the subsequent literature are the next two. The 'swapping' scenario envisions two properties that swap a causal role; again, we would have no way of knowing whether this is happening or not. Finally, we have the 'sharing' scenario: two quiddities might, for all we know, share the same causal profile. Whatever information we glean about the world, it will always leave these scenarios open; hence, quiddities are unknowable.

As it stands, the argument is in pretty bad shape. To begin with, the opponent of quiddities needs a much stronger premise than (1): he needs to claim that the *only* way we know properties is through their effects. Surely there are other justification-producing forms of inference than that from effect to cause. Even if we accept the strong version of (1), how is (3) meant to follow from (2)? The opponent of quiddities has left out a step:

2.5 Any property not defined by its causal powers makes no causal contribution at all.

And that seems pretty clearly false; even a powers theorist might reject it. To give just one example: Brian Ellis has argued that locations make a causal contribution in the sense that *where* a power is located determines where it

---

[14] Shoemaker (2003, 214).
[15] Of course our world is not one of them, as anyone who has tried to draw a table in perspective will know.

exercises its effects.[16] Now, locations might be particulars, rather than individual properties. But being located at a given region of space-time seems to be a property, and it seems to help determine which powers are activated.[17] A given chemical structure is not itself a power, but clearly makes a difference to where and when powers are instantiated and manifested. In short, 2.5 claims that no property not defined by its causal profile is ever causally relevant, and I see no reason yet to believe that.

Nor should the universals view accept 2.5. For Armstrong, a property F has its causal powers contingent on its being connected by the N-relation to another universal. How does it follow that F makes no causal contribution at all? Let's go back to our cases of Key 1 and Key 2. In the latter set-up, a wizard has endowed tofu with the ability open locks of a certain constitution. When I use the enchanted tofu to open a door, clearly the tofu is playing a causal role, even if it does so contingent on the wizard's spell.

We can construct a parody of Shoemaker's argument: the property of being tofu is not defined by its potential contributions. There are worlds where tofu is merely nourishing, and worlds, such as Key 2, in which it is an important aid to burglary. Therefore, the property of being tofu is unknowable. Even someone granting the possibility of Key 2 world should find this unpersuasive, precisely because 2.5 is false.

In short, the most prominent argument against quiddities is unpersuasive. We can grant that we know properties only through their causal contributions (1) and yet resist the premise that those contributions must be written into the essence of those properties, if they are to be causally relevant.

4. Quiddities: Metaphysical Worries

We have not yet taken the measure of the anti-quiddities argument. Some authors, such as Mumford, are quite clear that they regard the swapping scenario as a metaphysical, not epistemic, problem. It is not obvious that there *is* a metaphysical problem, even if such a scenario is both possible and counterintuitive. But let me try to bring out what is supposed to be so bad with the ghost, swapping, and sharing scenarios.[18] Those who insist on the individuating role of a

---

[16] Ellis (2010).   [17] See Ellis (2021, 280).

[18] Robert Black also offers a metaphysical argument, roughly to the effect that two possible worlds might share all their events and yet differ with respect to properties, since the quiddities of one world might be swapped (or presumably even inverted) in the other. See Black (2000); for discussion, Mumford (2004, 152).

property's effects might be thinking that a putative property that does not have those effects across all possible worlds fails to be individuated as a property at all.

Here is Mumford's statement of the upshot of the swapping scenario, in which F trades its causal powers with G:

> The telling question these cases produce is how, if the causal role of a property is altered, are we still talking of the same property? If something has the causal role of F, why are we not now talking of F? And if F now has the causal role that G had, why is F not G? The only available answer seems to be: if the property had a quiddity over and above its causal role. But this allows that F and G could swap their entire causal roles and yet still be the same properties they were. This apparent absurdity shows the general weakness of the categorical properties-plus-laws view.[19]

I think more needs to be done to make the absurdity apparent. Let me try to spell out an appealing line of argument:

1. A quiddity, by definition, can play a different causal role, or none at all, depending on the presence or absence of the N-relation
2. All properties are metaphysically individuated by their causal roles
3. Quiddities are not metaphysically individuated by their causal roles

/.: No properties are quiddities.

Just why the universals view should grant 2 escapes me. Whether that view has the resources to offer a different account of the individuation of properties, or must instead take it as a primitive, is a further question. One might argue that definitions are enough to achieve individuation: being spherical is not being cubical because to be a sphere is to have every point at your limits equidistant from the center, and that is not what being a cube is. True, cubes and spheres do different things in worlds governed by relevantly different laws. But when I talk about a world in which spheres don't roll downhill, no one doubts that I am still talking about spheres.

We can go further, since I think that the universals view can happily *accept* the causal individuation principle (2).[20] Causal roles will vary with the N-relation across worlds. Suppose, then, that we have a very simple array of modal space,

---

[19] Mumford (2004, 104).
[20] I feel sure my argument below must have occurred to others over the decades; I have been unable to locate it in the literature. My apologies to whichever philosophers have scooped me.

made up of only nine possible worlds. In worlds one through three, N(F,G) obtains; in four through six, in which G does not exist, N(F,H); in worlds seven through nine, in which F does not exist, N(G,H). Our example features a case of swapping: F brings about H in four through six, but in seven through nine, G brings about H. There is, then, no *one* effect that is associated with F in every possible world. And if having a causal role 'essentially' means that it has the same causal role in every possible world, then F fails to bring about G essentially.

Why, though, are we restricted to defining causal roles in this narrow way? On a broad—and to my mind, equally natural—construal, a property's causal role is defined as the world-indexed conjunction of *all* its causal contributions. That is, F's 'nomic profile' is really this: in w_1–w_3, N(F,G); in w_4–w_6, N(F,H). Similarly for G: in w_1–w_3, N(F,G), in w_7–w_9, N(G,H). Defined in this way, there is no swapping of causal role. The role, fully spelled out, tells us exactly what each property does in every world in which it exists. Call this fully elaborated description the 'wide nomic profile' of the property.

The initial swap, ghost, and share scenarios arise naturally out of the universals view: they are the plain consequence of making the laws of nature contingent. But once we properly and fully define the nomic profile of each property, the initial scenarios vanish. And since, by definition, a property has the same wide nomic profile across all possible worlds, we have met the criterion of essentiality: construed in this conjunctive way, each property *retains* its nomic profile across all possible worlds. Now, there is no further fact that explains why modal space exhibits just the structure it does and no other. That is, taking the laws as contingent means there is no further fact that makes it the case that in w_1–w_3, N(F,G). But that just is another way of putting the claim that the laws of nature are contingent.

For similar reasons, the universals theorist should not be worried by the accusation that wide causal profiles are cosmic accidents, or are distributed among properties by chance. They are essential to the properties that have them and there is no wider modal perspective from which to judge that their distribution is accidental or not. Even to speak of a 'distribution' of causal profiles is to court circularity, since, on this view, it is not as if the universals exist *and then* Zeus sprinkles them with causal roles. They come as a package.

Someone might try to reproduce the problem at this new level. Could there not be two properties that make the *same* causal contribution across all possible worlds, such that they share even this enriched nomic profile? Nothing in the universals view rules it out. But neither does such a possibility arise naturally out of the universal view's other commitments. This new argument is correspondingly less worrisome than the initial one.

To come at the same point from another angle, we can consider it in light of Armstrong's argument for the universals view. Recall that he takes regularities, and even singular cases of causation, for granted. We then are entitled to posit the N-relation by means of an inference to the best explanation. Now, the initial scenarios that feed the metaphysical argument arise when we combine this conclusion with the thesis that the laws of nature are contingent. The new scenario, where two universals enter into *exactly* the same nomic relations across all possible worlds and yet are distinct, does not rely on the contingency of the laws.

To make the universals theorist take this new scenario seriously, one would have to show that there could be a series of regularities that would justify the claim that two universals share a wide nomic profile. Anything one could offer in such a direction would in fact serve to make it more *plausible* that such a thing is possible. That is, anything one could do get the universals theorist to accept the possibility of wide profile sharing would at the same time render such a scenario toothless. Suppose, for instance, that we find reason to posit two universals—triangularity and trilaterality—that we distinguish by their definitions. They do not enter into causal connections with each other, but of course they do with regard to other universals. Now suppose, quite plausibly, that they necessarily go together: any world with triangularity is thereby a world with trilaterality. More strongly: anything that instantiates the one instantiates the other, across all possible worlds. I take it as obvious that they would then have the same wide nomic profile. But I also take it as obvious that we have no trouble individuating them, for all that. Such a scenario provides no reason to chuck out the universals view.

5. Epiphenomenalism?

Still, one might feel I have not done justice to the issues raised by the anti-quidditists. I think there is one last argument worth considering, though to my knowledge it has not appeared as such in the literature.

This argument sprouts from the mimicking objection we considered above. There, the worry was that the universals view cannot allow anything to have a thick disposition. I argued that there is no objection in that: if we require thick dispositions to be grounded entirely in the object or its intrinsic properties, then the universals view just denies that there any thick dispositions. But there is a deeper worry here. Has the universals view not thereby made the intrinsic properties of objects epiphenomenal? Quiddities become gears that turn nothing else in the machine. Here is one way to spell out the worry:

1. Whether property F endows an object O with disposition D to bring about G is determined by the laws of nature
2. Having F is not necessary for O to have D
3. Having F is not sufficient for O to have D
4. Any property that is neither necessary nor sufficient for having a disposition is epiphenomenal with respect to that disposition. Thus
5. For any property P, P is epiphenomenal with respect to any disposition D.

Steps 2 and 3 are simply the consequences of making the laws contingent: there are worlds in which the role of F is performed by some other property standing in the N-relation, as we have seen. Nor is instantiating F, or so the argument must go, sufficient for bringing about G: doing so relies on the N-relation.

So far, so bad for the universals theorist. For now having F is totally irrelevant to the disposition in question. Consider another example: being painted red is neither necessary nor sufficient for a bowling ball to (be disposed to) crash through a window. Being red is epiphenomenal with respect to that disposition. How are properties any different, on the universals view?

It will by now be apparent where I think the sleight of hand has taken place. The universals theorist would not, and should not, accept 2 and 3 as they are. Instead, she ought to insist they be indexed to worlds according to the N-relation:

2.′ In worlds such that N(F,G), having F is not necessary for O to have D.
3.′ In worlds such that N(F,G), having F is not sufficient for O to have D.

But on her view, these new premises are clearly false: her view is precisely that N(F,G)'s obtaining in a world *makes it the case* that Fs bring about Gs. At most, we have yet another illustration of the universals view, not a substantive objection to it. I conclude that there is nothing much wrong with quiddities, at least in the context of the universals view.

6. Laws and Explanation

I distinguished two questions one might have about the connection between laws and explanation:

(Case)   What is it for a law of nature to explain its instances?

(Whole)   Under what conditions does a theory of laws promise to explain the whole mosaic?

With regard to the universals view, answering case explanation must wait till we cover the inference problem below (Chapter 6, section 1). Here, I want to ask after its ability to achieve whole explanation.

As Tyler Hildebrand frames the issue, a crucial virtue of a whole explanation is beating competitors on the following metric: given the theory of laws, how likely is it that we should have ended up with a world that is uniform, in the sense of exhibiting predictable regularities?[21] Prima facie, the Humean view doesn't fare well on this criterion. If all we are given is that the world is a mosaic of property-instantiations in space-time, it seems fantastically improbable that we should have ended up with as regular and uniform a mosaic as we actually have.[22]

Surprisingly, the primitivist also faces a problem with whole explanation. The mere fact that the laws are primitives does not mean that they are such as to guarantee, or even raise the probability of, our world's being uniform. Primitivism on its own does not restrict the *kinds* of things can end up being law statements: nothing in the view as such rules out laws covering limited regions of space-time: nothing rules out laws that refer to 'gruesome' predicates. Suppose it is a law that all Fs are Gs, where F is 'is made of steel' and G is the gruesome predicate 'is observed before January 1, 2023, and conducts electricity or is observed after that date and does not.' With a little ingenuity, we might rig up any number of such disjunctive, time-indexed properties that would entail one pattern of events up to time t and just the opposite thereafter.

Now, we can always stipulate that laws have to range over natural properties, ruling out disjunctive, and hence gruesome, predicates by fiat. But as Hildebrand argues, this produces no gains in explanatory power: 'ungrounded restrictions do not allow primitive laws to explain natural uniformity, because the laws with probabilifying power over natural regularities are themselves very improbable—namely, they are just as improbable as that the regularities would occur by chance.'[23] If the restriction to natural properties is just an ad hoc stipulation, then we haven't really shown that primitive

---

[21] This is what Hildebrand (2013, 4) calls 'probabilifying power'; a different metric is 'initial plausibility.' For more details, see also Hildebrand and Metcalf (forthcoming).

[22] I am not claiming that the Humean view is seriously damaged by this argument; nor do I necessarily endorse Hildebrand's standard of 'uniformity.' See Chapter 13, section 3.

[23] Hildebrand (2013, 7).

laws raise the probability of a uniform world, for the chances of finding yourself in a world with primitive laws *that also produce uniformities* are just the same as finding yourself in such a world due to chance.

The same problems face the universals theorist. Hildebrand argues that we can use gruesome predicates to generate problems at two levels: the first-order universals themselves and the N-relation. At the first-order level, we should wonder what would happen if the universals themselves were gruesome. The same problem arises at the second-order level. Why not think that, even if the first-order universals themselves are not gruesome, the N-relation itself is? Suppose the 'N' of N(K,L) is such that, prior to 2050, it connects K and L, and thereafter connects K and M. So simply living in a world that is governed by N-relational laws fails to make our actual world, with its uniformities and patterns, any more probable than chance allows.

Unlike the primitivist, however, the universals view has a reasoned way to rule out gruesome properties, as Hildebrand points out.[24] That the universals are natural and not gruesome is 'baked in' to the view from the start: the whole point of introducing universals in the first place is to carve nature at its joints, something gruesome predicates fail to do. There is thus no need for the universals view to resort to ad hoc stipulation.

Even if that reply fails, another suggests itself. Once again, I think it pays to attend to the way in which the universals view is meant to be justified in the first place. Suppose we admit gruesome predicates into our ontology. Then, all the empirical evidence we have so far is consistent with both the hypothesis of stable N-relations among natural predicates on one hand and N-relations among gruesome predicates on the other. But if we are positing laws of the form N(F,G) on the grounds of an abductive inference, I am not sure we should be bothered by this. In short, this scenario looks very like the ghost, swap, and share scenarios above: states of affairs that cannot be ruled out a priori but nevertheless fail to pull their weight as competing explanations.

Even if we accept these replies, however, there is a closely related worry around the corner.

---

[24] Hildebrand suggests this response on behalf of the universals view: 'there just aren't (weaker: it's just *a priori* unlikely that there are) non-natural properties—not really, not fundamentally; by hypothesis, governing laws are objective, real/fundamental features of the world; therefore, since N is a real/fundamental relation (whereas N1 and N2 are not) relating real/fundamental properties, there is good independent reason to think that DTA laws will give rise to natural regularities' (2013, 10–11).

7. Intra-world Variance

The universals view casts the laws of nature as contingent: there are worlds where the N-relation stitches together F and G and worlds where it does not. So there are also worlds where the universal generalization $\forall x(Fx \rightarrow Gx)$ is true, and yet there is no corresponding law, because N fails to hold between the universals. The new problem is simply this: if the N-relation can vary like this *across* worlds, what is the principled reason for assuming that the N-relation cannot vary *within* worlds? Consider a scenario where the N-relation holds up until a given time t and then simply fails. All the evidence we have for thinking that N(F,G), right up until t, would be exactly as it is in our world. What entitles us to claim that that is not the case, in our own world?[25]

We can create the same difficulty with regard to space. Why not think there could be a region of space where the N-relation fails to hold? Perhaps the N-relation *itself* cannot include spatial or temporal information. But that is not the scenario under discussion. It is not that the N-relation itself stipulates that it holds only in New Jersey. Instead, it is a brute fact that it fails to hold everywhere else.

Note that neither scenario turns on making N a gruesome property. The claim is not that N itself is defined temporally or spatially. For that very reason, Hildebrand's move is not available as a solution: we can rule out gruesome properties and still be powerless in the face of this new objection. Again, if N can fail to connect F and G in other worlds, why not think it can fail in our world, too? Such a failure would cry out for an explanation. But since the laws themselves are contingent, there is, *ex hypothesi*, no explanation for their failure to hold in other regions of modal space; so the universals view is not entitled to rule out my scenarios by demanding an explanation for the failure of the N-relation. Again, that relation fails to hold across worlds; what is the metaphysically relevant difference between those failures and a failure *within* a given world?

Here's a reply I once found attractive. I have been arguing that some objections cannot be evaluated without looking to the mode of inference whereby the view is meant to be established in the first place. In Armstrong's case, this method is inference to the best explanation. Unlike deductive inference, abduction does not promise to secure its conclusion with complete certainty.

---

[25] Helen Beebee (2011, 511) first posed this problem. For discussion, and a helpful refinement of her objection, see Hildebrand (2016).

Based on the available evidence, the universals theorist might say, the best explanation is one that does not allow the N-relation to come and go.

I'm no longer sure this works. Abduction might be a legitimate and powerful form of inference, but it can't be a license to print money. What we're justified in positing depends on the evidential base we're working with. And since that base is limited to past cases (and observed ones at that), what we're justified in positing is confined to what serves as the best explanation *of those cases*.

Suppose that prior to the present moment, we have amassed evidence of a regularity: all Fs have been Gs. Here are three candidates for an *explanans*:

(1) There is an N-relation between F and G that has held in all cases of F and G up to the present moment within this world;

(2) There is an N-relation between F and G that has held in all such cases and will continue to hold throughout the rest of this world's future history;

(3) There is an N-relation between F and G that holds across all possible worlds.

It's true that both (2) and (3) buy us more *predictive* power. But they are tied for explanatory power with (1). So there is no reason to prefer (2) to (1), and of course Armstrong himself rejects (3).

Building on Armstrong's work, Hildebrand crafts a different response, one which doesn't rest on an appeal to abduction. Hildebrand's central move amounts to denying that (1) is possible within the framework of a theory of universals. The N-relation, being a relation among universals, is 'pure' in that it can *only* relate universals. Relations among particulars can come and go; I can be taller than Al Pacino one day and not the next. But since the relata at issue are themselves universals, the N-relation cannot come and go.[26] As Armstrong puts it, for two universals 'to be differently related at different times in the same world, different phases of the two universals would have to be present at different times. But that supposition is not a real possibility, if it is universals that we are dealing with.'[27]

My worry about this line of reply might simply reflect my own thick-headedness: I just can't see what *relevant* difference there is between holding

---

[26] As Hildebrand notes, the precise details of his reply depend on the theory of time one is working with; for our purposes, I prescind from such details.

[27] Armstrong (1983, 80).

at different times and holding at different worlds. If, as Armstrong thinks, nothing stops the N-relation from failing to unite F and G in other possible worlds, I can't see any principled reason to rule out N's failing to unite F and G within this one. If the variance of the N-relation within a world results in 'different phases' of the two universals flanking it, why shouldn't variance *across worlds* do the same thing? Whatever is wrong with fleeting temporal phases of universals is equally wrong with world-indexed phases.

The problem of intra-world variance, like other objections we'll examine below, suggests a revision to the universal theory. These problems all stem from making the N-relation hold contingently. By the end of the next chapter, I hope to establish that a more appealing view makes the N-relation hold necessarily.

6

Universals (II)

1. The Inference Problem

To base our inference to N(F,G) on an F–G regularity, we must be assuming that N(F,G) is somehow responsible for $\forall x(Fx \rightarrow Gx)$. Otherwise, the necessitation relation would fail to explain the regularity. But just how N functions to secure the regularity can seem mysterious. David Lewis might have been the first to be on to this issue. He writes,

> Whatever N may be, I cannot see how it could be absolutely impossible to have N(F, G) and Fa without Ga ... The mystery is somewhat hidden by Armstrong's terminology. He uses 'necessitates' as a name for the lawmaking universal N; and who would be surprised to hear that if F 'necessitates' G and a has F, then a must have G? But I say that N deserves the name of 'necessitation' only if, somehow, it really can enter into the requisite necessary connections. It can't enter into them just by bearing a name, any more than one can have mighty biceps just by being called 'Armstrong.'[1]

Bas van Fraassen calls this the 'inference problem': 'it simply does not seem that (irreducible higher-order) relations among universals can provide information about how particulars behave.'[2] Without a way to infer from the laws to the particulars, the universals view is just so much surplus metaphysical machinery.

At first sight, it can be hard to work out what the problem is. After all, there seem to be many relations among universals that are transferred down to particulars. 'Yellow is lighter than green' is true of the universals, and automatically true of the things that instantiate them. Might we not then subscribe to the general principle:

(Inheritance)   If any two universals stand in relation R, then so do their instances

[1] Lewis (1983, 366).   [2] van Fraassen (1989, 126).

The Metaphysics of Laws of Nature: The Rules of the Game. Walter Ott, Oxford University Press. © Walter Ott 2022.
DOI: 10.1093/oso/9780192859235.003.0006

Not so fast. The Inheritance Principle, as it stands, is false. To start with, there are all the relations that only universals can stand in, on pain of making a category mistake. 'Has more instances than' is certainly true when flanked by 'insect-ness' and 'dog-ness,' but the relation can't hold between Bobo the flea and the dog Odell on which he rides. So the simple inference from relations among universals to relations among their instances is blocked.

There are independent reasons to resist the Inheritance Principle in this case. It identifies the N-relation among universals with that among particulars, and there are costs to doing that. It does seem to be Armstrong's view, however. In replying to van Fraassen, Armstrong asks, '[I]f a certain type of state of affairs has certain causal effects, how can it not be that the tokens of this type cause tokens of that type of effect?'[3] In other words, the N-relation between F and G is exactly the same relation that holds between Fa and Gb: causation.

There are many prices to be paid for this move. By identifying the two relations, the universals view sacrifices the priority of the laws to their instances. And it is a short step to wondering why we need this higher-level N-relation at all. For his part, van Fraassen argues that we lose our grip on the N-relation as applied to universals.[4] It is just not clear what it could mean to say that the universals F and G are united by the very same causal connection that unites their instances.[5] Causation seems to unfold in time (even when causes and effects are simultaneous); but universals do not admit of temporal extension. Nor do they procreate: it is not as if the universal *fire* brings about the universal *heat*. Now, given that Armstrong is an immanentist about universals, he may not be bothered by these last points; to say that one universal brings about another one is not to say that they mate in Plato's heaven, but that some concrete particular x that is F must also be G. But immanentism makes the superfluity problem worse: why posit N as a separate, higher-order instance of causation to cover the relation between F and G in the first place? And we still seem stuck with the original worry: how can it be that the N-relation between F and G guarantees that any particular F is also a G?

A second attempt merits consideration, partly because it points us back to the moderns and some of the issues with top-down views we have already

---

[3] Armstrong (1993, 422); see also Armstrong (1997, 228–30).

[4] His case relies on an analogy with responsibility, into which I shall not delve here: see van Fraassen (1993, 436–7).

[5] This is similar to a point van Fraassen makes in a footnote. He points out that 'philosophical English' would need to be changed, for as it stood then, it was simply wrong to say that 'the universal *striking/being struck* causes (brings about) the universal *shattering*.' He takes this as no real objection, for 'our terminology can be regimented anew' (1993, 436).

canvassed. While admitting he has no proof for the validity of the inference from universals to particulars, Fred Dretske asks us to consider a political analogy. The laws passed by a government do not apply to individuals as such; they apply to individuals only *qua* office-holders.[6] Anyone familiar with the early modern debate, or U.S. politics in 2019, will see the problem here: the laws of a state can be said to bind individuals only in the normative sense. They do not actually bring it about that anything happens. Such efficacy as the laws can be said to have derives from the wills of individuals pledged to obey them.

1.1 Legitimacy

Before attempting a better defense on behalf of the universals view, I think we need to distinguish two questions:

Legitimacy: With what right does the universals view infer regularities from relations among universals?

Mechanism: What is the metaphysical account of the relation between laws and the particulars they govern?

The question of legitimacy asks: what entitles the universals theorist to say that there is a connection of the proper sort between what happens at the level of universals and that of particulars? The question of mechanism is logically independent: it asks, what precisely is the means by which such a relation among universals produces or brings about a relation between particulars?

The two are typically treated together, and it's easy to see why: an answer to one should, one might think, work for the other. In fact, I don't think this is quite right. I'll argue that we can indeed answer the legitimacy question, but that doing so falls short of answering the mechanism question.

Let's begin, then, with (1): what is the warrant for asserting the inferential connection between the universals-level and the particulars? Here it becomes appealing to run a version of my responses to some other objections to the universals view. Above, I suggested the view should appeal to the means by which it was initially justified, namely, inference to the best explanation. Could a similar move work here?

---

[6] Dretske (1977).

The basic idea is simple. If the universals view is indeed justified by an inference from regularities or patterns in nature, then it must include some means by which to infer those patterns from the universals and the relations among them. Developing a similar thought, Jonathan Schaffer argues that the universals theorist might add the axiom:

$$(Ax\text{-}1)N(F, G) \rightarrow \forall x(Fx \rightarrow Gx)$$

Moreover, Schaffer thinks the universals theorist is perfectly justified in so doing. On his view, whenever we make a posit, we need at the same time to stipulate axioms that will show how it is to be applied. Suppose we want to introduce the box to indicate necessity. We have to choose some modal axiom system, S4 or S5, for example, that will tell us what the box is to mean. Among those axioms will be something like

$$Box(p) \rightarrow p$$

From Schaffer's point of view, Lewis's refusal to see how you get from $N(F,G)$ to $\forall x(Fx \rightarrow Gx)$ is analogous to the refusal to see how we go from $Box(p)$ to (p). 'Such a person has simply not understood that [his opponent] has posited something whose work includes underwriting this very inference.'[7]

This maneuver does not license the positing of any old axiom one wants. The axiom earns its place in the theory by being necessary for the explanation to function *as* an explanation. Without (Ax-1), the universals view does not explain any regularities at all; so it can hardly be justified by serving as the *best* explanation. To take a different analogy: suppose one explains a patient's symptoms by appealing to dengue fever, but leaves out of the description that it causes just those symptoms. Without the causal link, that is no explanation at all. The right reaction is not to reject dengue fever as an explanation but to stipulate that part of its nature is to be such as to bring about just those symptoms. From Schaffer's point of view, the inference problem is based on a confusion.

Given its alignment with the responses I have pursued to other objections, Schaffer's reply seems to me both natural and plausible. I think it is perfectly satisfactory as a reply to the legitimacy problem. I'm dubious that it helps with the metaphysical problem. As Schaffer himself acknowledges, (Ax-1) states but does not explain the metaphysical connection we are worried about. The Box does not 'force' new states of affairs into the world; it is not as if

---

[7] Schaffer (2016, 580).

Box(*p*) goes about making it the case that (*p*). But the laws really are taken, in the universals view, as genuine producers of events.

Schaffer argues that one could, in principle, build in a new axiom, the 'productive' axiom:

Production: N(F,G) produces $\forall x(Fx\rightarrow Gx)$

The problem, as he acknowledges, would be meeting the epistemic burden: what would such an extra axiom get us by way of explanation that was not had by the laws plus A*x*-1 alone? The production axiom would be another thought too many. Schaffer, then, doubts that there is any real mystery to the inference problem.

Pace Schaffer, I think the metaphysical problem is still a real one.[8] Here it's instructive to compare Berkeley's arguments against materialism (here, the view that there is a material world lying behind our perceptions). He points out that we might have the very same evidence we do now, even in a world where matter does not exist. Fair enough; but if materialism is meant to be established by an abductive inference, that is precisely what we should expect.

This is why Berkeley immediately goes on to challenge materialism *as an explanation*.[9] Materialism, he thinks, has no good story to tell about how matter can act on immaterial minds. That genuinely *is* a problem for the materialism of Locke and Descartes: absent such an account, materialism can hardly claim to be an explanation at all, let alone the best one. Now, the materialist might add an axiom

(M-A*x*)   Physical states of kind K bring about mental states of kind M

That will hardly appease Berkeley. It is only centuries later, when we have some grip on the actual mechanism that produces our mental states, that this particular arrow is removed from the idealist's quiver.

I fear I'm being obtuse in insisting that the metaphysical issue is still pressing. There is yet another way to articulate what would be missing if we left matters where Schaffer does. In the context of this book's dialectic, at any rate, the move to the universals view was motivated by our misgivings about primitivism. Including the inference axiom (A*x*-1) would not dispel any of

---

[8] In this, I agree with Theodore Sider's (1992, 262) reaction to the 'solution by stipulation,' discussed in Schaffer (2016, 585–7).
[9] See PHK I §§18–19.

these misgivings; indeed, the universals view would seem superfluous since, as Schaffer himself shows, one could just as easily append an appropriately revised (Ax-1) to primitivism. That is, one could take the laws themselves as unanalyzable features of the universe and just append

(Ax-1′)   It is a law that $p \rightarrow p$.

Why bother with higher-order relations among universals at all? In short, then, I think we must face up to the mechanism question, even if we can declare the legitimacy issue resolved.

1.2 Mechanism

If the Inheritance Principle were true, the metaphysical inference problem would be solved. For any relation R, it would follow straightforwardly from R (F,G) that R(Fa,Gb). Sadly, as we've seen, there is just no reason to suppose that the Inheritance Principle is true. More carefully, we might say that it admits of so many exceptions that it can hardly be set up as support for the inference from N(F,G) to the F-G regularity. Let's look at a case where the inheritance does seem to go through, and ask what makes the case of laws disanalogous. For if we could suitably restrict the Inheritance Principle to those cases where it *does* work, and if we could show that the laws are among them, the universals view would be able to account for the mechanism.

Above, I suggested one prima facie case of inheritance: yellow is lighter than orange. Here it seems to be the case that a relation among universals is passed on to the particulars: any individual yellow object will be lighter than any orange object. Notice, though, that what does the work in the color case is an *internal* relation, that is, one that obtains solely in virtue of the intrinsic features of the relata. 'Is lighter than,' 'is similar to,' 'is taller than': these are all internal relations, in the sense that once the relata are fixed, the relation follows necessarily. (Note that this does not mean that once one relatum exists, the other exists, as well. The claim is conditional: an internal relation aRb is such that the existence of a and b is enough to make R hold between them.) As Armstrong puts it, internal relations are 'no addition to being': they come for free, riding in the train of the things they relate.[10] This is the reason why objects inherit the relations among the colors they instantiate: the universals in

[10] Armstrong (1997, 12).

question themselves, by their natures, stand in the relation 'is lighter than,' and cannot help but pass that on to the particulars that share in them. Let's assume that story about color relations works. No parallel story can be told where laws are concerned, for the N-relation is meant to hold only contingently. For that reason, it cannot be an internal relation: it does not supervene on the relata. That suggests that we ought to reject the contingency of the laws, and make N-relation internal instead.

It's not obvious that there can be any *external* relations at all where universals are concerned. A paradigm external relation such as distance cannot apply. Still, there are some external relations that can hold among universals. Consider the property of having n instances. Suppose there are nine billion human beings and a trillion insects on Earth at time t. It follows that, at t, the universals *human* and *insect* flank the relation '...has fewer instances than...' That seems to count as an external relation, since it is not fixed by the universals themselves.

But whenever an external relation obtains, there must be some further fact in virtue of which it obtains. The principle is hard to articulate with any precision, but it seems true for all that. Consider the insect and human case. That is relevantly similar to the case of Armstrong's N, since in both cases we have universals standing in an external relation. Being external, it can vary across worlds: no doubt there are some worlds where insects and humans switch sides on the '...has fewer instances than...' relation.

We know what further fact explains the external relation between human-ness and insect-ness: the populations of each group. What further fact could explain that our world is one where N(F,G) holds? The mere existence of instances of F and G is not enough. We appear to have reached the point where the universals view has to accept another primitive: it is simply the case that our world is an N(F,G) world.

More important, casting N as an external relation reverses the direction of explanation. We want something that will take us from a relation among universals to one among particulars. Whenever we have an external relation among universals, the order of explanation is just the reverse. It is the facts about the instances of *human* and *insect*, not those universals themselves, that grounds the fact that *human* has fewer instances than *insect*. The universals view wants precisely the opposite to be true. But so long as N must be an external relation, I cannot see how that could be.

If we have an external relation among universals, we must point to some further fact that grounds it. It cannot be a fact about the universals themselves; otherwise, the relation would be internal and apply to the universals necessarily, not contingently. The further fact, then, must be sought in the instances. But the universals view has to insist that there is nothing about the instances that makes N(F,G) obtain, not even the constant conjunction of Fs and Gs. So there is no explanation at all for why N(F,G) obtains when it does. At this point, we might as well go back to being primitivists about the laws.

2. A Revision: N as an Internal Relation

These last considerations suggest a substantive revision to the universals view: making the N-relation internal. I think this is a promising way out of the metaphysical inference problem; what is more, even if it fails to solve it, it provides a quite different means of resisting the anti-universals arguments we have already covered.

The basic idea is simple. If I'm right, at least part of the difficulty in providing a mechanism by which the universal view's laws can govern is its insistence that the N-relation is an external relation. This is just a consequence of the laws' being contingent. What if we reject both claims?

On such a view, the laws would be necessary. This need not mean that there are the same laws in every world.[11] We might say that the laws are hypothetically necessary, in the sense that they govern only worlds where the universals involved are instantiated. The N-relation is now an internal relation: the natures of F and G are such that they are N-related. There are further complications here, but I propose to postpone them so we can first see how such a view would fare against the chief objections to its contingent-laws counterpart. Let's begin with the mechanism version of the inference problem.

Above, we experimented with the Inheritance Principle to answer the problem. We have seen good reasons to reject the Inheritance Principle where external relations are concerned. If we restrict it to internal relations, however, we generate a more promising principle, albeit one with an unfortunate abbreviation:

Revised Inheritance Principle (RIP):   For any internal relation R, if R(F,G), then R(Fa,Gb).

In other words, an internal relation flanked by two universals must also obtain whenever those universals are instantiated. There is no further fact needed to

---

[11] Chris Swoyer (1982) makes a similar point.

knit together universals and their instances. To see how this might work, recall our color example: if yellow is lighter than orange, yellow objects will also be lighter than orange objects. We purchase this inference at the price of giving up contingency: internal relations cannot fail to obtain whenever their relata are instantiated. If N(F,G) is true at any individual world, it must be true at all worlds that have Fs and Gs.

Now, I don't claim to know how to mount an airtight argument for (RIP). A quick review of some thought experiments provides some minimal support. Suppose numbers are universals; the number 77 is, by its nature, larger than 2. It follows that a box with 72 apples contains more apples than one that has 2. That relation between the universals can even be said to 'govern' that between the particulars, although it is a non-causal governing.[12] When we consider the N-relation, by contrast, we can say that the N-relation between F and G causally governs the first-order relations between F and G: it explains and even brings about the fact that Fs cause Gs.

This mechanism for bringing particulars in line with the relations among the universals they instantiate leaves us with a set of brute facts: there is no explanation why F and G stand in the N relation, beyond the bare claim that it is simply their nature to do so. Still, these brute facts seems more acceptable than those of the unrevised universals view. If the laws are contingent, then the N-relation varies from world to world, with no explanation why it obtains here and not there. Even if the presence of the N-relation can't be called an accident, or said to obtain by chance, it still seems to need an explanation. The revised view gives us one: N depends on the natures of the universals themselves.

There are of course objections one might have to making the laws consist in internal relations among universals. Before getting to them, let's continue to look on the bright side: what else can the revision do for us?

3. Applications of the Revised View

Although I, naturally, find my replies to the remaining objections on behalf of the initial universals view pretty persuasive, I am sure others' opinions will

---

[12] Stathis Psillos writes, 'In any case, there is no conceptual difficulty in thinking that a supervenient relation, in some sense, determines (governs) its relata. I take it that being a solid wooden cube supervenes on having six wooden square sides of equal areas, but being a cube determines (in some clear sense) what arrangement must be in place among the elements of the subvenient basis' (Bird et al. 2006, 458).

differ. To such readers, I advertise the revised universals view as providing a totally independent means of answering some of them.

Recall the initial epistemic problem with quiddities: not being defined by their causal role, they are unknowable—or at least imperceptible—since they are not causal powers. The initial view makes properties independent of their nomic profile and hence of their contributions to events. Such a property then becomes a kind of mysterious Lockean substratum.

The present view faces no such worries. The nomic profile of F is given by the N-relations in which it necessarily stands. It can hardly then be said to be beyond the reach of knowledge or even sensory perception: when fire, say, burns your flesh, this is surely enlightening with respect to the nomic profile of the universal we call 'fire.' I am not sure what else anyone could want. Nor is there any worry that one property might swap its nomic profile with another, either within or across worlds: they have even their narrow nomic profiles essentially, given that N is now an internal relation.

Nothing about the revised universals view as such helps with the problem of ghost properties: there might be properties that have no nomic profile at all, that is, that stand in *no* N-relations whatever. Here I think the revised universals view has to fall back on my initial response: if the view is supported by an inference to the best explanation, it can discount, although not rule out a priori, the ghost scenario. By definition, there is no evidence that could support the postulation of such causally isolated properties. The same goes for the sharing scenario: nothing about either version of the universals view stops there being two properties with precisely the same nomic profile. And nothing suggests that we could ever be in an epistemic position to postulate them, either.

The revised view has ready replies to the metaphysical arguments against quiddities. It rejects the first premise of the opening metaphysical argument: it is no longer true that a property can play a different causal role, or none at all, depending on whether or not the N-relation is present. The second metaphysical argument also exploited the initial universal view's casting N as a contingent relation. It claimed that an object's instantiating a given universal is neither necessary nor sufficient to endow that object with a disposition. But that is no longer true, on the revised view: the N-relation does not come and go from world to world. What its instances can bring about is fixed by the universal itself, in collaboration with its N-related partners.

Finally, we now have a principled reply to the problem of intra-world variance. Since the N-relation cannot vary across possible worlds, there is no pressure to grant that it might vary within any given world. Our abductive inference to the N-relation, if it succeeds at all, automatically gets us an N-relation impervious to temporal variation.

On balance, then, I think the revised universals view is a significant improvement over the initial version. The revised view can at least go some distance in answering the metaphysical inference problem; it is entitled to my responses in earlier sections to other worries; and it affords new responses of its own to some.

4. Modal Inversion

There's an independent reason to think that the N-relation needs to be internal. To bring this out, we need to set up some of the dialectic between the universals and powers views, which will be useful below in any case.

Armstrong argues that powers views are committed to a bizarre, Meinongian ontology:

> Consider, then, the critical case where the disposition is not manifested. The object still has within itself, essentially, a reference to the manifestation that did not occur. It points to a thing that does not exist.... [H]ow can a state of affairs of a particular's having a property enfold within itself a relation (of any sort) to a further first-order state of affairs, the manifestation, which very often does not exist? We have here a Meinongian metaphysics, in which actual things are in some way related to non-existent things.[13]

Armstrong rejects any view that makes the nature of a property depend on its relations to unactualized possibilia. The problem cannot merely be that actual things are related to unreal things, as Armstrong puts it at the end; any view that aspires to ground counterfactuals will argue for just such relations. If N(F,G) governs a set of worlds, then in any of them, it's true that any x that is F is also G. So actual xs that are not-F are related in that way to merely possible xs that are F. Instead, Armstrong's objection seems to be that a thick disposition relies for its very nature on its relation to inexistents. Toby Handfield calls this the charge of 'modal inversion.'[14] How things are in the actual world should not depend on how things are in others. Dispositionalism has the ontology backwards.

For precisely this reason, Armstrong rejects any view that makes the laws of nature hypothetically necessary, as the internal relation revision proposes.

[13] Armstrong (1997, 79), quoted in Handfield (2005, 453). [14] See Handfield (2005, 453).

If that revision is accepted, then any worlds with Fs and Gs will be such that N (F,G). And even relative to worlds where no Gs exist, N(F,G) is a potential law, in the sense that were Gs present, N(F,G) would have to obtain. As Armstrong puts it, '[t]he universal F will be big with all its nomic potentialities, however impoverished the world in which it exists.'[15] Endorsing that view is tantamount to admitting irreducible powers: both are guilty of modal inversion, if sin it be.

But the proponent of the revised view shouldn't be bothered by this charge. As Handfield and Alexander Bird argue, Armstrong himself is committed to just such a modal inversion.[16] We can put the point in terms of a dilemma: is the N-relation in N(F,G) essentially such as to bring about the regularity $\forall x$ $(Fx \rightarrow Gx)$, or not? If not—if N in this occurrence is merely contingently related to the regularity—then we have a bizarre view, on which, as Handfield puts it, 'necessitation is not necessarily the relation of necessitation.'[17] There are worlds in which the very same N relation occurs, but instead of bringing about the regularity above, guarantees that no F is ever a G. That result, I assume, is intolerable.

So it must be that the N-relation in N(F,G) is, by its nature, such as to ensure that any worlds in which it holds are worlds where $\forall x(Fx \rightarrow Gx)$. But then what makes this N the relation it is, is its relationship to unactualized possibles. As Handfield says, 'it is essentially such as to bring about non-actual states of affairs under non-actual but possible conditions.'[18] If modal inversion is a problem, then it's Armstrong's problem, as well.

Our interest in the problem of modal inversion lies in whether the N-relation should be cast as internal or external. I think the argument we've just examined provides an independent and powerful reason to cast N as an internal relation. Only one horn of the dilemma is acceptable: the N-relation obtains in virtue of the universals it connects, and by its nature is such that what flanks it on the left brings about what flanks it on the right. That's what we need to solve the mechanism version of the inference problem, too. But then we have to be thinking of the N-relation as internal: in any world that has Fs and Gs, it must be the case that N(F,G).[19]

---

[15] Armstrong (1983, 168).   [16] See Handfield (2005) and Bird (2005).
[17] Handfield (2005, 461).   [18] Handfield (2005, 458).
[19] It may be that Bird and Handfield are right in thinking that the universals view collapses into a version of the powers theory (albeit a version with just one kind of power, namely, the second-order N-relations). I certainly think on this score the powers and universals theory are in the same boat: both are committed to modal inversion. Other convergences will emerge as we go.

5. Conservation Laws

There's a further problem we need to address, one that arises for either version of the universals view. One thing primitivism has going for it is its ability to handle the conservation laws. If laws are brute facts about the universe, one more brute fact—say, the first law of thermodynamics—makes the view no worse off than any of the others. But for the remaining non-Humean views, conservation laws will be a thorn that is maddeningly difficult to pluck out.

Here is how Bigelow, Ellis, and Lierse present the problem for the universals view:

> The claim that events and processes which are not X-conservative are impossible is not naturally construed as a relation between universals. For what are the universals which must be said to be related? Are energies at times, momenta at times, etc., to be counted as primitive universals?[20]

Although it's not incoherent to introduce a universal, 'the total amount of energy in the system at a given time,' it's hardly attractive, and raises the cost of adopting the universals view. Wouldn't the universals theorist then be committed to a different universal corresponding to each of the least moments of time, so that they can stand in the 'has the same amount of X' relation? Making that concession brings others in its wake: if there's a relation 'being conserved' that holds among the universals, how does it interact with, and perhaps limit, the other universals that are in play?

Here's a second approach. The universals view might want to distinguish levels of N-relations, such that first-order N-relations are governed (that word again!) by the second-order ones. Gravitational and kinetic energy, then, might be constrained by a single universal governing them both. Such a move further raises the price of the universals view by bloating its ontology.

A third possibility is worth considering. Our revised version of laws might be better poised to make sense of conservation laws than its predecessor. For now we have a set of laws that are necessary: there is no world with the same universals as ours that behaves differently. Is it absurd to simply 'bake in' the conservation laws at the level of the first-order N-relations? The universals would have to exist in a tightly woven net, with each prescribing behavior to its instances in accordance with the others. But since the N-relation is already a result of the intrinsic features of the universals it unites, we might just as well

---

[20] Bigelow, Ellis, and Lierse (2004, 156).

posit N-relations that of their nature secure the conservation laws. There is no need for a third-order N′ relation to do it for us. Of the three options, this is the most promising.

6. The Ontology of Relations

The revised view has it that the N-relation is itself necessary. If that is so, it must be an internal relation in the sense that the natures of the relata guarantee that the relation holds. And if the N-relation is, as Armstrong claims, just the production or causation relation at the level of universals, then it seems that any world with one member of a set of N-related universals will have to have all of them. Note that that doesn't follow just from the fact that the relation is internal. If you have a yellow fire truck, such that it would be lighter than a red fire truck, red fire trucks don't magically appear to complete the relation. All the internality of '...is lighter than...' gets us is the claim that any world that has both kinds of fire truck is such that the yellow one is lighter than the red one. It's the precise nature of the relation N between F and G that commits us to saying any F-world will also be a G-world, where that entails that those universals are instantiated. So the very fact that F necessitates G means that F and G stand in an internal relation of production. How could that be?

I see two exhaustive and mutually exclusive options here. One option is that universals have written into them their relations to all others. The relata here are partly constituted by their role *as* relata.[21] Such a view would amount to a kind of holism, the metaphysical analogue of semantic holism. Each universal would have to 'point to' the others, or at least the others with which its instantiations can have commerce. We'll encounter this kind of view below, in connection with powers below, so I defer discussion till then.[22] It in effect becomes a powers view, and so is better treated with them.

The second option is to claim that the N-relation is internal in the same way that 'is taller than' is internal: it is a free lunch, something that holds in virtue of the intrinsic, non-relational features of the universals. F and G would be such that N(F,G) holds; no extra glue is needed. By contrast, relations that require intrinsic directedness on the part of the relata, a kind of built-in intentionality, are not a free lunch but a five-course meal, with brandy to follow. So I think the revised

[21] Stephen Barker calls these 'Bradley relations,' and distinguishes them from 'Leibniz relations'; see Barker (2009, 247) and Tugby (2016, 1151). See below, Chapter 8, section 7.
[22] Armstrong himself is opposed to any such view; see Armstrong (1983, 168).

universals view needs the N-relation to be internal in the sense that the monadic properties of the relata bring N in their wake.

I have to admit I do not have much of a story about how such internal N-relations could hold among universals: what is it about electrostatic force, say, that makes it vary with respect to distance as Coulomb's law says it does? I think the best—and only—move here is to admit that this connection is a primitive. After all, the revised view refuses to admit that the universals in question could stand in any other relations than the ones in which they do. There is no pressure to give some further story about why electrostatic force and distance are related as they are in some worlds and not others. Many will find this move unattractive. But my goal with each view is to work out the best version possible, and leave its evaluation till later, when we have all its competitors on the table.

7. A Second Revision: Many-to-One

The original form of the universals view is one-to-one: it pairs each true law statement with its truthmaker, the N-relation holding among the universals involved. I think this feature of the universals view is detachable, and we ought to jettison it. For it makes answering the problem of *ceteris paribus* clauses all but impossible.

Let's start with how Armstrong himself deals with the problem. He distinguishes between iron laws, which issue in exceptionless uniformities, and oaken laws, which do not. Newton's first law—a body will remain at rest or in motion unless acted on by another force—is oaken, since it explicitly includes a *ceteris paribus* clause.[23] Suppose the law said that whenever there is a particle that is (F) at rest and not acted on by any external force, it will (G) remain at rest. So we have our N(F,G). The obvious problem is we have buried the *ceteris paribus* clause inside the alleged universal (F), by tacking on the qualification that the particle is not acted on by other forces. That would give you a pretty strange ontology of universals.

Instead, Armstrong puts the *ceteris paribus* clause within the quantifier of the regularity that is supposed to result from N-relation. So, with an oaken law, it's not true that N(F,G) entails $(\forall x)(Fx \rightarrow Gx)$. The scope of the quantifier has to be narrowed, from all Fs to 'all uninterfered with Fs,' where an interferer is anything that can prevent an F from being a G.[24]

---

[23] Armstrong (1983, 148).   [24] Armstrong (1983, 149).

I find this set-up bizarre. The law is not qualified at all; it's only the quantifier of the universal generalization it produces that is qualified. Why isn't that tantamount to saying that F and G are not, after all, N-related? If they were, how could anything interfere with it? We've seen Armstrong insist that the N-relation just is the relation of production or causation. How, then, can a universal F 'produce' or bring about G, when some of its instances don't? That seems to just be a contradiction. And it would mortgage any hope of solving the mechanism version of the inference problem.

The only route left is to insist that all laws are iron laws. We might have law approximations that are oaken, but the metaphysical machinery just doesn't support their being genuine laws at all. And with that, we are back to an unappealing horn of the Lange/Cartwright trilemma. If all laws have to produce regularities without interferers, then we will have to include all of those potential interferers in the statement of the law itself.[25] The law will be indefinitely long and complex, perhaps covering all of the universals instantiated in a given world. Indeed, there might only be *one* law for the entire world. Unfortunately, I think the prospects for getting such a maxi-law to play the axiom role are dim. No statement of it is available to us, and if it were, no human could understand it, much less use it as we expect to use laws. So I think we have yet another reason to reject the original universals view.

But the universals theorist can make a second revision: go for a many-to-one account. We can keep our N-relation as a truthmaker for laws without feeling compelled to identify the two. Suppose there is indeed just one instance of the N-relation that connects all the universals of a given world. On this revision, there would be many different laws, each of them made true by their relationship to the single fact N(F,G,H, . . .). Any law statements that include a *ceteris paribus* clause will depend, for their sense and truth, on the other universals tied up in the single N-relation.

Here is another way to come at this issue. Other top-down views have used the web of laws approach to deal with the problem of *ceteris paribus* clauses. On that approach, no single law entails a regularity. But once we identify laws with propositions involving the N-relation, we are stuck with each law entailing a regularity. The N-relation by its nature has to connect universals in such a way that their instances necessitate each other. That's what any solution to the inference problem tries to get us. The only way for the universals theory to

---

[25] As Armstrong (1983, 149) puts it: 'But cannot an oaken law always be represented, in principle at least, as an iron law by putting in all the negative qualifications? Yes, in a way it can, provided that we bear in mind how wide the qualifications may be which are implied by the phrase "in principle." It may even be that the statement of Newton's First Law as an iron law, would have to be of infinite length.'

construe laws as functioning together as a web is to split them off from their original truthmakers. Law statements will now be at best partial reflections of the one true maximally complete proposition that links *every* universal in a given world. Such law statements will be implicitly 'web-qualified,' in that any exceptions to them will need to be explained by appeal to other universals standing in the N-relation. This is a substantial revision to the universals view, and not one Armstrong would welcome. But I can't see any way to avoid it.

8. The Price of Revision

Let me try to bring all these points together by asking: what do the revisions get us, and what prices must we pay?

The revisions are twofold. Most recently, we've seen reason to give up any hope of one-to-one matching of true law statements and necessitation relations among universals. Instead, we need a single N-relation that knits together all the universals instantiated in a world. That's the only way to get the previously scattered N-relations to work together as a web. Put differently, once we recognize the need for each law to be sensitive to the others, we lose any right to individuate the N-relations that figure among those laws. So for the laws of fundamental physics, at least, there will only be one indefinitely long proposition, with an 'N' at the start, that governs all the universals that apply at that level. There might be many true law statements, but there will be only one lawmaker.

Now, it may be that at the level of the special sciences, there are different universals, not governed by this fundamental N-relation. I'm somewhat skeptical, but one's view here depends on how one sees the special sciences and their preferred vocabularies. If one thinks there can be a reduction of the core terms of say, psychology, to those of physics, then there would really be just the one 'maxi-N' proposition at the level of physics. But one might want to allow that other sciences are dealing with universals, and deny that those universals are those of fundamental physics. I take no position on that issue.

Second, I think the universals view is best served by making the N-relation an internal relation, in the sense that its presence is guaranteed by the natures of the universals it relates. This makes the laws necessary, in the sense that any two worlds with the same universals will have the same laws. Armstrong himself rejects this result, but the revision that leads to it more than pulls its own weight.

Making the N-relation internal gives us a more plausible way to deal with the conservation laws. We can see this by means of a quick *reductio*. Suppose the N-relation were external, in the sense that whether or not it obtains is not fixed by the universals that flank it. It would then be a mystery why some properties are conserved. We would need to solve it by positing a new third-order relation, N', that governed the N-relations. But that seems a step too far. Better to say that the nature of energy, charge, and all the rest is such that the conservation laws hold. This is precisely what making the N-relation internal does for us. Whatever N-relations conserved properties enter into must conserve them, by the nature of such properties. Our other revision is crucial here, as well: we need to pile all the relevant universals into a single proposition governed by N.

Just as important, the internal N-relation got us out of the inference problem. In its most threatening form, that problem asks for a mechanism by which the N-relation can entail the relevant universal generalization. We cannot bat the problem away as Armstrong does, by suggesting that universals cause one another and hence so do their instances. What we should say instead is that, given that N is an internal relation, it necessarily trickles down to its instances. If the natures of F and G guarantee that N(F,G), then it must be the case that any individual x that is F is also G. That's just what the Revised Inheritance Principle says.

Finally, the internal N-relation gives us additional resources in replying to the objections to quiddities. I've argued that these resources are not needed, and that the original view can give perfectly good replies to these objections. But anyone unpersuaded by my replies should welcome the extra help.

The revisions come with a price tag. From the primitivist's point of view, it introduces a lot of metaphysical machinery to no good effect. If the goal was to make governing less mysterious, we merely *seemed* to make progress when we said that laws govern by being necessitation relations among universals. The revised view makes such necessitation depend on the universals involved: it is the nature of force, mass, and acceleration to be related as they are. And that's where the spade turns, for the revised view.[26] The primitivist might well wonder how that improves on simply saying that $F=ma$ and leaving it at that. Nevertheless, I think the revised view is a significant improvement over the original. It faces fewer objections and does as well as, or better than, its predecessor in answering those that remain.

---

[26] My worry here is similar to Theodore Sider's (2020, 30) objection to nomic essentialism, the view that the essences or natures of properties determine the laws of nature. What's missing from such a view is 'some account of the internal nature of properties and laws that would give rise to the essentialist claim.' '[U]nless some more specific vision is articulated, it will remain unclear whether there is any sufficiently attractive specific vision of the natures of properties and laws that would underwrite the essentialist claim.' I don't think the revised universals view, as I've stated it, has the resources to provide the kind of account Sider's 'post-modal' metaphysician demands.

PART III
POWERS

7

Origins of the Powers View

1. The Moderns

Not everyone in the modern period is enamored of Descartes's top-down laws. Some try to preserve Aristotelian powers, even in the inhospitable climate of seventeenth-century mechanism.[1] Like Descartes's, their notion of laws involves a legal-cum-theological metaphor, but they choose a different aspect to exploit.

Descartes's *leges* emphasize one side of the legal analogy: the sense in which a god or king can lay down a set of rules his subjects must obey. Francis Bacon's 1620 *Novum Organum* and Baruch Spinoza's 1677 *Ethica*, which bookend Descartes's works, point to a different facet of the analogy: the sense in which laws describe what must happen in a variety of different circumstances. There are no laws of nature *tout court*; there are laws *of* individual natures, and these laws state the contributions those natures make to the events in which they figure. To fully grasp a nature like heat is to learn all of the conditionals that are true of it in virtue of its powers, and those of the natures it can encounter. Similarly, a legal statute might stipulate what is to happen under a variety of conditions. By pointing to this feature of the civil law, Bacon and Spinoza decline the seventeenth century's invitation to move beyond powers. Instead, they find a way to locate powers within the metaphorical space of laws.

This version of the legal metaphor results in a bottom-up picture. The dominance of the top-down reading of the metaphor has made this one all but invisible: both Bacon and Spinoza are routinely read by commentators as fellow-travelers, swept up in the burgeoning worldview of top-down mechanism. That they are in fact its opponents will, I hope, emerge in due course.[2] From a contemporary point of view, another feature of the history is more

---

[1] By 'mechanism,' I mean the view that the fundamental properties of bodies are size, shape, and motion.

[2] I discuss Stephen Gaukroger's reading of Bacon below. For the top-down reading of Spinoza, see esp. Curley (1969) and Curley (1988). I argue against the top-down readings of both more extensively in my (2018).

The Metaphysics of Laws of Nature: The Rules of the Game. Walter Ott, Oxford University Press. © Walter Ott 2022. DOI: 10.1093/oso/9780192859235.003.0007

interesting. As we go, we'll see a split emerging: Bacon identifies laws with powers, but later writers find this picture unattractive. During the eighteenth and most of the nineteenth centuries, the dominant powers view is one that relaxes Bacon's one-to-one pairing of powers and laws, and allows that any given law might obtain in virtue of a number of distinct powers.

2. Bacon

Unlike Descartes, Bacon presents an immanent or internal conception of laws; just as Spinoza puts it, laws are 'inscribed' in 'things as in their true codes, according to which all singular things come to be, and are ordered.'[3] But like Descartes, and anticipating him by decades, Bacon takes the discovery of laws to be essential to scientific progress. A very different view results if we push Bacon into the ranks of those who take explanation in terms of material structure to be fundamental. Since this is the orthodox way of reading Bacon, I need to clear it away before developing my own reading.

According to a widespread—and essentially correct—story, the scholastics' four causes (formal, final, efficient, and material) are gradually contracted into one, namely, the efficient cause. The other kinds of explanation are either epistemically out of reach (the final cause, since it is impossible to know the mind of God),[4] or part of the detested hylomorphism of the scholastics (the formal and material causes).[5] On one isotope of this sort of view, found in Robert Boyle and John Locke, the ultimate hope of natural philosophy is uncovering the micro-level structures that explain the efficient causal interactions among bodies. Bacon can seem an early harbinger of some moderns' insistence on mechanical explanations. This is how Stephen Gaukroger reads him: the form of something is 'its basic material structure.'[6] Gaukroger recognizes that Bacon talks a great deal about forms as laws, not structures, but takes such talk simply to be a way of 'emphasizing the causal/explanatory role of Forms,' that is, of micro-structures.[7] I think such a reading has things backwards. For Bacon, explanation in terms of law is primary; attending to the micro-structure is largely a waste of time.

---

[3] *Treatise on the Emendation of the Intellect*, §101 in Spinoza (1985, 41).

[4] As Descartes tells Gassendi, '[w]e cannot pretend that some of God's purposes are more out in the open than others; all are equally hidden in the inscrutable abyss of his wisdom' (AT VII 375/CSM II 258). See also Meditation Four (AT VII 55/CSM II 39).

[5] For the contraction of the four causes into the efficient cause, see esp. Carraud (2002).

[6] Gaukroger (2001, 140).   [7] Gaukroger (2001, 140).

In fact, Bacon opens Book II of the *Novum Organum* by rejecting the efficient and material causes: they are 'perfunctory, superficial things, of almost no value for true, active knowledge.'[8] Knowing what something is made of, or what produces such-and-such an effect, falls short of 'true' knowledge partly because it does not help one reproduce the effect, which is the aim of 'active' knowledge. Bacon is well aware this emphasis on forms will sound odd, since he has made fun of the scholastics' empty jargon in Part I. Here is his replacement doctrine of forms:

> For though nothing exists in nature except individual bodies which exhibit pure individual acts ('*actus puros*') in accordance with law, in philosophical doctrine, that law itself, and the investigation, discovery and explanation of it, are taken as the foundation both of knowing and doing. It is this *law* and its *clauses* which we understand by the term Forms, especially as this word has become established and is in common use.[9]

Here we have the central move: a body acts according to its laws. The clauses of the law, like the clauses of a statute of common law, spell out what is to happen under certain circumstances.[10] In place of the obscure scholastic notion of a form—a 'figment of the human mind'—Bacon has installed the laws 'of action or motion.'[11] Although Bacon rarely speaks in terms of *vis* ('power') in this part of the *Organum*, that is what his laws-cum-forms amount to.

There are epistemic barriers to knowing the full range of behaviors a single disposition is capable of. At first blush, a disposition's range of manifestations could be known only by a being with a synoptic grasp of the whole of modal space, for its actual manifestations never exhaust its possible ones. Bacon is well aware of the difficulty, and in the passage above proposes a method for making progress. When a body engages in an '*actus puros*,' it acts in a way that is not interfered with or distorted by another body.[12] Bacon is here making a non-trivial assumption: not only do powers have a set of manifestations

---

[8] NO II.ii: 102. (References to the *Novum Organum* ('NO') are in the following format: Book. Aphorism: page number in (Bacon 2000)). Bacon also rejects the final cause, which is 'a long way from being useful' except in cases of human action (NO II.ii: 102).

[9] NO II.ii: 103.

[10] Lisa Jardine and Michael Silverthorne note the analogy with statute law in Bacon (2000, 103 n. 2).

[11] Bacon writes, '*forms* are figments of the human mind, unless one chooses to give the name of *forms* to these laws of act' (NO I.li: 45).

[12] As Jonathan Bennett suggests in his commentary on the passage; see his <https://www.earlymoderntexts.com/assets/pdfs/bacon1620.pdf>. For another use of the phrase, see NO I.li: 45.

essentially, but those manifestations are invariant. They are stable, whether we are looking at the power in relative isolation ('*actus puros*') or as making a contribution to a hugely complex series of events.

Now, most bodies will be composites of the simple natures, such as heat. So the task of natural philosophy is to separate out these simple natures and learn their laws. So far from being a rhetorical flourish, Bacon's laws are at the heart of his program:

> [I]n this *Organon* of ours we are dealing with logic, not philosophy. But our logic instructs the understanding and trains it, not (as the common logic does) to grope and clutch at abstracts with feeble mental tendrils, but to dissect the powers and actions of bodies and their laws limned in matter [*in materia determinatas*].[13]

But what exactly *are* laws? At first sight, Bacon's thoughts are a disappointing mishmash of unanalyzed jargon. He tells us that substances have laws,[14] that laws govern acts,[15] that laws are forms, which in turn are forms of natures,[16] and, to make matters worse, that forms derive natures from essences.[17] Sometimes he speaks as if laws govern simple natures like heat; at other times, as if the laws just were the simple natures; and in one startling sentence, he does both at once: 'when I speak of forms, I mean nothing more than those laws and determinations of absolute actuality which govern [*ordinant*] and constitute any simple nature, as heat, light, weight, in every kind of matter and subject that is susceptible of them.'[18]

To clear this up, we need to work out the relationships among natures, forms, laws, and essences. Throughout Book II, Bacon collapses some of these into others.[19] Aphorism iv tells us that form and nature always go together: when the form of heat is present, so is the nature, and when the form is absent, so is the nature. By the time he reaches xvii, Bacon has decided to identify these.

> When we speak of forms, we mean simply those laws and limitations of pure act which govern ['*ordinant*'] and constitute a simple nature, like heat, light,

---

[13] NO II.lii: 219–20. [14] NO II.iv: 104. [15] NO I.li: 45, II.ii: 103, II.v: 106.
[16] NO II.ii: 102–3. [17] NO II.iv: 104. [18] NO II.xvii: 128.
[19] In fact, Bacon goes further and identifies a thing with its form. See, e.g., NO II.xiii: 119: 'The Form of a thing is the very thing itself. And a thing does not differ from its form other than as apparent and actual differ, or exterior and interior . . . and hence it follows that a nature is accepted as a true form unless it always decreases when the nature itself decreases and increases when the nature increases.'

or weight, in every kind of susceptible material and subject. The form of heat therefore or the form of light is the same thing as the law of heat or the law of light, and we never abstract or withdraw from things themselves and the operative side. And so when we say (for example) in the inquiry into the form of heat, *Reject* rarity, or, rarity *is not of the form of* heat, it is the same as if we said, *Man can* superinduce *heat on a dense body*, or on the other hand, *Man can take away heat or bar it from a* rare *body*.[20]

Forms, then, are nothing more than laws. But how can Bacon at once identify the laws with the simple natures and claim that the laws *govern* those natures? Clearly he can't have it both ways. The problem here is an ambiguity in Bacon's notion of a nature. In its usual use, 'a nature' is a non-dispositional feature or property that we can experience, such as light or heat.[21] Natures stand in need of explanation. This explanation might take two forms: one might give the material cause, which is the underlying stuff and its organiza-tion. In this sense, the nature *heat* is expansive motion. But as we have seen, Bacon regards this as being of relatively little interest to the natural philoso-pher. What counts is the formal cause, that is, the dispositions or powers. From a contemporary perspective, it is tempting to go further and claim that the dispositions described by the forms/laws are grounded in the structure of the matter in question. And that may very well be what Bacon has in mind, though it is not strictly speaking there in the text. Putting it together: a word for a nature like 'heat' might equally well refer to (i) the categorical property we experience, (ii) the forms/laws or dispositions that go along with that property, or (iii) the ultimate micro-structure that might ground those laws.[22]

So when Bacon says that forms/laws govern or organize a nature, he means nature in sense (iii), the micro-structure. And when he says that forms/laws constitute a nature, he means nature in sense (i), the property we experience. And, trivially, when he says that forms/laws are natures, he is using the term in sense (ii). It is annoying but entirely natural for Bacon to slide from one of these senses to another. From here on, I'll use nature in sense (i) only, as the explanandum rather than explanans.

---

[20] NO II.xvii: 128, italics in original. For consistency, I am rendering '*ordinant*' as 'govern' rather than 'organize,' as this translation of NO has it.
[21] Here I ignore the difference, which Bacon registers, between heat as felt and heat as it is in the world. See NO II.xx: 131, where Bacon claims that '[h]eat as felt is a relative thing . . . and it is rightly regarded as merely the effect of heat on the animal spirit.' Throughout, I use 'heat' in Bacon's second sense, that is, as the non-relative nature that exists independently of its being felt.
[22] For (iii), see esp. NO I.li: 45, where Bacon recommends the study of 'matter, and its structure (*schematismus*).'

To sum up: we have a perceived nature, such as heat, whose form just is its laws, that is, the set of dispositions that characterize its behavior. Finding the form amounts to true and active knowledge because knowing those dispositions tells you how to control the thing whose form it is. Although there is nothing in the physical world beyond bodies and their acts, the primary goal of science is the discovery of forms/laws, not of the micro-structures that might underwrite them. These forms are captured in propositions that describe possible states of affairs and could equally be cast as conditionals. What makes these conditionals true will be the powers of the objects that figure in them. Bacon does not take a position on whether the forms/laws are grounded in the micro-structures of the bodies that obey them, though he might well believe that. The important, and perhaps startling, point is that it is the laws *qua* forms, not micro-structures, that are the true *explanans* and the proper focus of empirical inquiry.

Note that Bacon can accommodate the governing intuition about laws. There is a clear sense in which an object is governed by a law. It is not that it obeys it, as a subject obeys a king; instead, the object's behavior is fixed by the laws that define its powers. The laws are 'inscribed' in things, not dictated from above.

3. Spinoza

If Bacon presents a tantalizing first glimpse of what an Aristotelian story about laws might be, it remains merely tantalizing. What's missing is any notion of how to accommodate the rules of calculation, such as the laws of planetary motion or of fall, developed by his near-contemporaries, Kepler and Galileo.[23] Precisely because he remains wedded to the basic idea of investigating nature by investigating nature*s*, he offers no real way of incorporating laws that are

---

[23] Bacon sometimes speaks of fundamental and universal laws. For example, at NO II.v, Bacon writes that the inquiries into such things as the voluntary motion of animals 'are concerned with compound natures, or natures which are joint members of a structure; and they have regard to special and particular habits of nature, not the fundamental and common laws which constitute Forms' (NO II.v: 106). Gaukroger reads this as an endorsement of 'general laws of nature' (2001, 141); Bacon's point, in part, is that it is not enough to know the general laws of nature in order to explain the motions of animals. But there is no suggestion in the text that there is a single law, or set of laws, that governs all of nature. When Bacon talks about the 'fundamental and common laws which constitute Forms,' he is referring to the powers that constitute simple, as opposed to composite, natures. So the contrast Bacon draws is between simple and composite natures, not between specific laws (say, of mechanics) and more general ones.

both capable of mathematical formulation and meant to apply to all bodies whatsoever.[24]

So it makes sense to turn the clock forward a bit and look at Baruch Spinoza. We know that Spinoza read Bacon with some care.[25] Bacon's influence is evident in Letter 32, from 1665, which Curley dubs 'The Worm in the Blood.' Spinoza responds to a question from Oldenburg: 'how [do] we know how each part of Nature agrees with the whole to which it belongs and how it coheres with the others'?[26] Whatever Oldenburg means by 'coherence,' Spinoza stipulates that '[b]y the coherence of parts, then, I understand nothing but that the laws *or* {*sive*} nature of the one part so adapt themselves to the laws *or* nature of the other part that they are opposed to each other as little as possible.'[27]

Spinoza begins by insisting that we do not know *how* this happens, though we do know *that* it does. Even if there is only one substance, the natural world still seems, from the human point of view, to be divided into parts:

Concerning wholes and parts, I consider things as parts of some whole insofar as the nature of the one so adapts itself to the nature of the other that so far as possible they are all in harmony with one another. But insofar as they are out of harmony with one another, to that extent each forms an idea distinct from the others in our mind, and therefore it is considered as a whole and not a part.[28]

Whether a thing is a whole or a part is a function of harmony and coherence. When we consider two things as two, that is, as two wholes, that is because their disharmony produces two distinct ideas in the mind. Given monism, such disharmony must ultimately be an illusion, as is the two-ness of the two, or the *n*-ness of the *n*. Nevertheless, at this initial stage, we attribute distinct

---

[24] A further barrier to Bacon's progress might have been his inductive method. Later on in the *New Organon*, Bacon entertains a hypothesis about the rate of fall, namely, that 'the proportions of quantity equal the proportions of power,' so that a lead ball weighing two ounces would fall twice as fast as a one-ounce ball (NO II.xlvii: 190). Since that is false, Bacon concludes that we 'must look for these measures in the things themselves, not on the basis of likelihood or conjecture' (II.xlvii: 190).

[25] See, e.g., Letter 37 (10 June 1666), in (Spinoza 2002, 861). References to Spinoza's *Ethics* and Letter 32 are to the Curley translation in (Spinoza 1994). References to other works are to Shirley's translation (Spinoza 2002) unless otherwise noted. When referring to the *Ethics*, I first give the part, proposition, and then (S) scholium, (L) lemma, (D) demonstration, (A) axiom, or (C) corollary, if applicable. Thus '2p13L1' refers to Part II, proposition 13, Lemma 1. The Latin text is from volume 2 of the Gebhardt *Opera* of 1926.

[26] Spinoza (1994, 82).

[27] Spinoza (1994, 82). Note that Curley's italicized 'or' translates *sive*, which indicates an equivalence rather than an alternative.

[28] Spinoza (1994, 82–3).

and competing natures and laws, that is, distinct powers, to distinct parts of the natural world.

We can now confront the central puzzle of Letter 32: what could Spinoza mean by speaking of the laws of a thing's nature? Once we see the Baconian context, and construe laws as powers, Spinoza's text comes into focus. The laws of a thing's nature just are the powers it has. This is why Spinoza can speak indifferently of laws *or* natures and laws *of* natures. Either way, what makes heat *heat* is what it does in such-and-such circumstances.

Spinoza tells us that the laws of different objects 'restrain' each other.[29] The blood might be disposed to move in a certain direction, but this disposition can be interrupted by the presence of competing causes. In just the same way, a match can be disposed to light when struck, and yet not light in an oxygen-free environment. Any given event, then, will be the result of the interplay among the laws or dispositions of the objects concerned. As we'll see in the contemporary setting, such a picture affords a greater prospect of accounting for *ceteris paribus* clauses. On this sort of view, such a clause marks the possibility of an interfering disposition.

This is hardly the place to embark on a re-construction of Spinoza's metaphysics in the *Ethica*. I propose to draw attention to just two salient features: the role of laws in God's actions, and their role in scientific explanation. It is useful to think of the Parts of the *Ethica* as moving from a metaphysical image to a manifest or human image. We begin with the deepest metaphysical facts and only later recover the world we inhabit. In Part I, after proving God's existence, Spinoza tells us that

1p16: 'From the necessity of the divine nature there must follow infinitely many things in infinitely many modes . . .'

1p17: 'God acts from the laws of his nature alone, and is compelled by no one.'

In these propositions, we begin to see just how far Spinoza is from the Cartesian concept of extension. Spinoza's extension is not an inert lump awaiting a divine shove; it is active of itself.[30] Note how Spinoza puts it

---

[29] 'There are a great many other causes which restrain the laws of the nature of the blood in a certain way, and which in turn are restrained by the blood, it happens that other motions and other variations arise in the particles of the blood . . .' (1994, 83).

[30] Tschirnhaus (Letter 80) presses Spinoza on just how extension by itself could entail all the infinity of finite modes that are supposed to follow from its nature (E1p16). In answer, Spinoza distances his concept of extension from that of Descartes: '[F]rom Extension as conceived by Descartes, to wit, an inert mass, it is not only difficult, as you say, but quite impossible to demonstrate the existence of bodies. For matter at rest, as far as in it lies, will continue to be at rest, and will not be set in motion

when it comes time to say how God acts: he acts '*solus suae naturae legibus*,' from the laws of his nature alone. Spinoza's point is that there is no competing substance with a different nature, whose laws (that is, powers or dispositions) could interfere with God's.

Later on in the *Ethica*, Spinoza uses the notion of law to knit together the human and metaphysical images:

> . . . Nature is always the same, and its virtue and power of acting ['*virtus et agendi*'] are everywhere one and the same, that is, the laws of nature and the rules according to which all things happen,[31] and change from one form to another, are always and everywhere the same. So the way of understanding the nature of anything, of whatever kind, must also be the same, namely, through the universal laws and rules of Nature.[32]

Change in form is to be attributed to the laws of the one nature that is instantiated in the physical world.

I suggested above that Bacon gives us no real way to carry through the identification of Newtonian, Cartesian, or even Galilean laws with powers. Let's take just one example of how Spinoza might do this. For all their differences, Descartes and Newton each present something like a law of inertia: bodies change their states of motion or rest only when interfered with. Spinoza embarks on a 'physical digression' in 2p13, and constructs his own law of inertia. Lemma 3 states, in part, that a body that moves or is at rest must be determined to do so by another body, and so on to infinity. Bodies, Spinoza argues, are individuated by motion or rest (2p13L1).[33] Since each finite mode has another finite mode as its cause (1p28), and since we have ruled out cross-attribute causation, a particular body's motion-or-rest must have as its cause another body. Given the way bodies are individuated, we can

except by a more powerful external cause. For this reason I have not hesitated on a previous occasion to say that Descartes's principles of natural things are of no service, not to say quite wrong' (Letter 81). In Letter 83, Spinoza adds that 'matter is badly defined by Descartes by means of Extension' and that 'it must necessarily be explicated by means of an attribute that expresses eternal and infinite essence' (quoted in Schmaltz (1997, 220)). That doesn't take us very far toward understanding Spinoza's notion of extension, but it does mark his view off from Descartes's.

[31] Here I depart slightly from Curley's translation. As he has it, Spinoza says that 'the laws and rules of nature, according to which all things happen, are everywhere the same.' I can understand the desire to avoid repetition, but it might be significant that the Latin phrase is '*naturae leges et regulae secundum quas omnia fiunt*,' that is, 'the laws of nature and the rules according to which all things happen.'

[32] Cf. *Treatise on the Emendation of the Intellect*, section 101, in Spinoza (2002, 27).

[33] Note that we are dealing with simple bodies (i.e., those individuated solely by motion-or-rest), not composite bodies—which are only introduced at the end of A2″. It is also important to note that the natures of body that Spinoza speaks of in the discussion of A3″ and following are not Aristotelian natures but merely hard, soft, or fluid characteristics, all of which are explained by mechanical means.

infer that this body itself must be in a state of motion-or-rest. And we can then run the same argument: there must be yet another body to account for its motion-or-rest, and so on *ad infinitum*.

This is a bottom-up picture if ever there was one. What accounts for inertia is the nature of body and its definition. Like Hobbes, Spinoza rejects the whole picture of a God outside of nature; whatever laws there are must spring from the natures of bodies themselves. And like Hobbes, Spinoza gives us no way of understanding conservation laws. To test the powers account on such points, we would do better to leave the modern period and advance to the view in its current state, which we will in due course. Still, it's important to note that Spinoza never calls the principles he defends in the Physical Digression *leges*. It may well be that he sees the laws of motion not as laws-cum-powers but as facts that are made true by powers. That is, although he clearly identifies laws of nature(s) with powers, he may well think that the laws of physics are not laws in that sense at all.

4. Euler and Shepherd

In Bacon, the powers view starts off the seventeenth century with a simple one-to-one pairing of laws, built on Bacon's own way of exploiting the legal metaphor. From the eighteenth century on, it's much more common to find a more relaxed picture, suggested at least by Spinoza, which does not demand of any given law statement that it report on a single power. In terms of our three axes, these are 'one-to-many' accounts.

In 1750, roughly a generation after s'Gravesende's popularization of Newton's top-down view, Leonhard Euler takes a bottom-up approach in his *Reflexions sur l'espace et sur le tems*. After stating the inertia law in two forms (one for rest, one for motion), Euler declares:

> These two truths, so indubitably established, absolutely must be founded on the nature of bodies. As it is Metaphysics, which occupies itself with the study of the nature and properties of bodies, the knowledge of these truths can serve as a guide in its thorny investigations.[34]

Just how these truths about inertia, let alone the inverse square law, are supposed to be founded in the nature of bodies is not easy to discern from

---

[34] Euler (1750, 324), my trans.

the text. If we look at Euler's 1768 letters to a German princess, we find a considerably more detailed account.[35]

There, Euler rejects the picture—also privately scorned by Newton—that makes gravitational attraction an 'essential' or 'intrinsic' property of bodies.[36] Nor does he think, with Christian Wolff and some of the Leibnizians, that bodies are endowed with a power of changing their own states. Force, then, is never in the patient, only in the agent: the force has to come from outside the body moved. But, as the text from the *Reflexions* shows, neither is Euler sympathetic to Newtonian occasionalism. Instead, he locates the forces of bodies in their inertial tendencies:

> I say, then, that however strange it may appear, this faculty of bodies, by which they are disposed to preserve themselves in the same state, is capable of supplying powers which may change that of others. I do not say, that a body ever changes its own state, but that it may become capable of changing that of others.[37]

One way in which bodies persevere in the same state is their resistance to being split up by other bodies. When body A collides with body B, body B is changed by the force of body A. But it always in turn changes body A, in the sense that it deflects it from its initial course. And it does *that* solely by virtue of its ability to resist being penetrated by another body. Euler concludes, 'the impenetrability of bodies, therefore, contains the real origin of the forces, which are continually changing their [that is, the bodies'] state in this world: and this is the true solution of the great mystery, which has perplexed philosophers so grievously.'[38]

Euler does not explicitly connect this position with his account of gravity, but we can infer that it explains his preference for the hypothesis of subtle matter connecting all bodies in the universe.[39] His account of force in general remains somewhat mysterious, since he does not venture an account of impenetrability itself. The prior century had witnessed a debate over the cohesion of the parts of a single body which might account for its impenetrability: Glanvill and Locke are both skeptical that any real explanation for cohesion can be found.[40]

---

[35] Alongside, it must be said, the preposterous folk tale of Newton being hit with an apple, which Euler's British translator, Henry Hunter, takes the liberty of correcting in a footnote.
[36] See Euler (1802, 290–1).   [37] Euler (1802, 297).   [38] Euler (1802, 299).
[39] Euler (1802, 214–29).   [40] For more on this debate, see my (2009, 135–7).

Moving into the early nineteenth century, we find Mary Shepherd defending a powers view against the assault of Hume. Her remarks on laws of nature tend to come in the course of stating the positions of Berkeley and Hume, but those she makes *in propria persona* are instructive.

For Shepherd, it is chemistry, not physics, that serves as the exemplar of natural philosophy.[41] It is natural, then, that she should neglect the Newtonian laws and come at the project of natural science in a quite different way:

> The science of chemistry has now discovered that the whole of the universe that is within the reach of experiment is composed of a few elementary substances; and there is reason to suspect that these substances which we term elementary, may perhaps be compounds, and reduced to fewer still, if the methods of analyzing them were discovered. In considering causation, I have therefore been led to consider the world as one whole, composed of a few elementary parts . . . the whole of the qualities and properties, or powers of matter, are derived from the qualities, properties, and powers of these elementary substances.[42]

Just how laws are meant to fit in, if at all, is unclear. What the passage shows, however, is at least that Shepherd takes the powers of elementary substances to be responsible for the course of nature.

No less than the scholastics, Shepherd thinks powers are necessarily connected to their effects: if present in the proper circumstances, they cannot but have their effects. There is no sense to be made, then, of a world that shares our laws of nature, in whatever sense one likes, but contains objects with different powers. This comes out in Shepherd's brief discussion of Newton's idea, floated in the Queries to the *Opticks*, that God might vary the laws of nature:

> God no doubt may vary the laws of nature, &c., that is, create, arrange, alter the capacities of objects, by means adapted to those ends. But to understand God aright, he cannot work a contradiction; he cannot occasion the same objects without any alteration amidst them supposed to produce dissimilar effects.[43]

---

[41] See esp. LoLordo (2019).

[42] This passage is from an anonymous work, commonly attributed to Shepherd (Anon. (1819, 45–6)), quoted in LoLordo (2019, 2).

[43] 'That mathematical demonstration, and physical induction, are founded upon similar principles of evidence' (1827), in Shepherd (2020, 149).

What plays the role of a 'law' in her system, if anything does, is the fundamental metaphysical claim that similar causes are attended by similar effects: 'there is *but ONE law* which can experience no change whatever, namely, that similar qualities in union necessarily include similar results.'[44] I'm unsure whether Shepherd takes these qualities to somehow ground powers, or whether she thinks the powers are reducible to the qualities at issue.

Despite these reservations, I think two things are clear in the work of Euler and Shepherd, for all their differences: both offer resolutely bottom-up accounts and neither offers a straightforward identification of laws with powers. Instead, they take whatever scientific laws there may be to supervene on bodies and their powers.

5. Helmholtz

Within the bottom-up realist camp, Bacon's original one-to-one laws-to-truthmakers picture is gradually eclipsed by the less stringent one-to-many approach. But it does recur near the end of the nineteenth century, in the work of Hermann von Helmholtz.

Delivered in 1869, 'On the Aim and Progress of Physical Science' is among the popular lectures Helmholtz gave throughout the latter half of the nineteenth century, and one of the relatively rare discussions of the nature of laws in the time period. Helmholtz begins in a Millian vein, claiming that a law 'is nothing more than the general conception in which a series of similarly recurring natural processes may be embraced.'[45] This is the user-facing side of laws—propositions or sentences that play a role in science. But Helmholtz goes on to insist that a law of nature is not a mere '*memoria technica*' that allows us to systematize and recall facts. Looked on as a feature of the world, a law is 'an objective power':

> For instance, we regard the law of refraction objectively as a refractive force in transparent substances; the law of chemical affinity as the elective force exhibited by different bodies toward one another. In the same way, we speak of electrical force of contact of metals, of a force of adhesion, capillary force, and so on. Under these names are stated objectively laws which for the most

---

[44] 'That human testimony is of sufficient force to establish the credibility of miracles' (1827), in Shepherd (2020, 167). Note that Shepherd's candidate law is very similar to those proposed by Walter Charleton two centuries earlier; see above, Chapter 2, section 4.
[45] Helmholtz (1995, 208).

part comprise small series of natural processes, the conditions of which are somewhat involved. In science our conceptions begin in this way, proceeding to generalizations from a number of well-established special laws.[46]

Helmholtz illustrates his identification of laws with power with Newton's second law.

> [F]orce is only the law of action objectively expressed . . . .The actual meaning of [$F=ma$] is that it expresses the following law: if such and such masses are present and no other, such and such acceleration of their individual points occurs. Its actual signification may be compared with the facts and tested by them.[47]

Helmholtz's lecture is clearly aimed at a popular audience; nor is he especially concerned with the issues that trouble us. His identification of laws with forces quickly gets awkward, when he has to introduce the law of the conservation of force as energy.[48] It is hard to judge from his text how that law could itself be a force: a force that acts on other forces, to make sure they are conserved?

Nor is there a clear answer to the problem of *ceteris paribus* clauses. Helmholtz insists that each law be exceptionless and confirmable through its instances, and then goes on to illustrate his idea with a law that has no instances at all, as he casts it ('such and such masses are present *and no other*' is not a state of affairs we can meet in the real world).[49] Fortunately, we can now turn to a more recent and developed version of the one-to-one view.

6. A Contemporary View

In *Laws in Nature*, Stephen Mumford offers a 'Central Dilemma' against the existence of laws: laws either play the governing role or not. If one agrees that they do not govern, then one has endorsed Mumford's own lawless metaphysics: a law that doesn't govern isn't worth having around, a gear that turns nothing else in the mechanism. If they do govern, we face another dilemma: the governing must be either external to the objects governed or internal to

---

[46] Helmholtz (1995, 209).   [47] Helmholtz (1995, 209).   [48] Helmholtz (1995, 212–14).
[49] 'Before we can say that our knowledge of any one law is complete, we must see that holds good without exception, and make this the test of its correctness' (1995, 209).

them.[50] Mumford rejects external stories about governing, such as the universals view. Nor, at this stage, does he think any internal account of governing can be made to work.

In response to the Central Dilemma, Alexander Bird argues that '[w]e may regard laws as identical with or supervenient on potencies.'[51] Mumford counters that laws that supervene on powers are not 'metaphysically substantive': 'they don't provide anything that the potencies haven't already delivered.'[52] If laws supervene on powers, they cannot be said to govern and so hardly merit the name 'laws' at all.

More recently, however, Mumford has found a way to preserve the governing of laws within the context of a powers view. The key move is to identify laws with powers.[53] Unlike Bird's supervenient laws, these meet Mumford's criterion for genuine lawhood: they govern their subjects. They are not merely 'along for the ride.' As Mumford notes, Mill sometimes talks as if laws report dispositions; Bacon and Spinoza, as we've just seen, are more distant antecedents.[54]

Mumford promises to fill in the lacuna we found in Helmholtz: a clear answer to the problem of *ceteris paribus* clauses. Above, I suggested that Bacon and Spinoza have an incipient answer to that problem. When a law-cum-power apparently fails, it is only because it is being interfered with: it is being prevented from manifesting its characteristic effect. Laws compete with each other.

From one point of view, such thwartings of the law/power might count as exceptions. And exceptions, of course, are what *ceteris paribus* clauses are meant to allow for. The problem, as we've seen, is that there is no obvious way to cash out the *ceteris paribus* clause without falling into vacuity. The one-to-one powers view has a ready response: such events are not exceptions at all. The law statement reports a disposition or tendency, and that tendency is there even when it is not manifested.

Consider Mumford's treatment of gravitation. The inverse square law, whether understood in terms of brute forces or the curvature of space-time,

---

[50] For this statement of the Central Dilemma, see esp. Mumford (2004, 144–5).

[51] In Bird et al. (2006, 448). Bird offers reasons we'll explore below for rejecting any identification of laws with powers; here we'll be concerned with the notion that laws supervene on powers. For Bird's picture of laws, see esp. Bird (2007a, 200–3).

[52] In Bird et al. (2006, 464).

[53] For a similar move, see also the recent work of Brian Ellis. Ellis writes, '[t]he displays of elementary causal powers are . . . always law-governed. They accord with what I call the "the laws of action of the causal powers"' (2021, 279).

[54] See Mumford (2018, 215–17). Since Mill is a Humean with respect to powers, I discuss his view in the context of the Best System Analysis below.

reports a disposition of bodies to attract each other. It is not falsified by cases in which two bodies fail to attract each other, nor does it require repair or supplementation by a *ceteris paribus* clause. As Mumford puts it, the law 'is to be understood dispositionally rather than occurrently. It is about the tendency between the two objects rather than about the manifestation of that tendency. The law tells us nothing about the actual movements of any two objects, except that such movement will be in part determined by attraction.'[55] The law neither states nor entails any universal generalizations at all, so, trivially, it does not require those generalizations to be qualified in any way. When we do insert a *ceteris paribus* clause into the statement of a law, it only 'indicates this kind of *sui generis*, dispositional relationship between being F and being G, which does not entail a strict regularity in actual events.'[56]

The basic problem with this maneuver is that it seems just to re-position the cp-clause inside the law itself. Dispositions have their own manifestation conditions, together with all the ways in which they can be masked and finked. And listing all of those is no easier or harder than cashing out the cp-clause. Lipton calls this 'Hume's revenge.'[57]

Above, I argued that Descartes appeals to dispositionality in his system of laws. But the differences could not be more stark. For Descartes, individual laws report on dispositions in they sense that they state what God intends to do in certain circumstances. Descartes's appeal to God's mental states insulates him from Hume's revenge. For Descartes can actually specify the conditions

---

[55] Mumford (2018, 215).

[56] Mumford (2018, 217). As Mumford notes, Lipton makes a similar point: 'we don't know when all things are equal, but the whole point of the dispositional view is in a sense that we do not need to know, since the disposition is present regardless' (Lipton 1999, 166).

[57] Lipton (1999). See above, Chapter 2, section 5. Markus Schrenk (2007a) does a fine job of presenting Hume's revenge. Schrenk mounts a separate argument against the appeal to powers: 'the difference between an object whose disposition's manifestation is masked or counteracted against and one whose disposition is lost (because the basis is lost) has not been properly distinguished. The latter case leads to a sort of ceteris paribus clause which cannot be accommodated by dispositionalism easily' (2007a, 226). As I understand it, the basic problem is that, at least outside of fundamental physics, there are plenty of properties that only confer dispositions given certain other conditions. To use Schrenk's example: hemoglobin binds to oxygen, but a hemoglobin cell can be damaged and lose this capacity without thereby ceasing to be hemoglobin. The powers theorist really needs something like: 'It is a law that for any object x if it has feature F it has, ceteris paribus, the capacity C' (2007a, 244). *Ceteris paribus* clauses like this one cannot be hidden inside the disposition. Schrenk imagines a 'mad strategy' for the powers theorist to employ: go for something like 'It is a strict law that Fs have the capacity C+ to have the capacity C' (2007a, 247). In fact, I think Mumford—though not engaging explicitly with Schrenk's point—follows E. J. Lowe (2005), who takes just this kind of tack. Applied to the hemoglobin example, the idea would be that being hemoglobin does *not* confer the disposition to bind to oxygen, but only the disposition to be disposed to bind with oxygen. Being a raven, to use Mumford's (2018, 218) example, confers the disposition to have the disposition to be black. This move stops the proliferation of cp-clauses at the price of proliferating dispositions. That works only if the dispositional approach to cp-clauses works. So I think the 'mad strategy' is ingenious but doesn't help with the problem of Hume's revenge.

under which any one of his three laws fails to issue in a regularity: those conditions are given by the other two. Of course, Descartes is massively mistaken on the details, as well as the rest of his ontology; but his view at least has the merit of being such that if it were true, it would solve the problem of *ceteris paribus* clauses.

Does Mumford's appeal to dispositions have the same virtue? For him, of course, laws-as-powers are instead inscribed in the things themselves. Nor does he think we need to do what Descartes does, and specify the conditions under which things are not in fact equal. For Mumford, that's the chief advantage of going dispositional: '[w]e do not have to engage in the task, which might not be completable in any case, of specifying the actual conditions in which the cp-laws hold and the conditions in which they don't.'[58]

I fully agree with the epistemic point: you can know that a disposition is present without being able to fully specify its causal profile. We know our own intentions, which are dispositions, without being able to do so. I don't yet see, though, that going dispositional helps with the metaphysical issue. In effect, we seem to have embraced one horn of the Lange/Cartwright trilemma, and ended up with laws that are immune to falsification by virtue of having a blanket *ceteris paribus* clause buried in them.

Quite apart from the issue of *ceteris paribus* clauses, there are non-trivial reasons for worrying about the tenability of the one-to-one match between laws and powers. The first is a question of ontological category: are powers the right *kinds* of things to be laws? Bird argues no: 'it seems wrong to say that laws *are* properties. Coulomb's law is not identical to the property of charge. Rather it concerns that property and its relations to other quantities.'[59] I can understand how the inverse square law might be made true by the powers of bodies. I have a harder time understanding how that law can just *be* that power. The inverse square law at least seems to state a relationship among variables, not a power.

A second worry is, to my mind, more pressing. I don't want to assume that *all* laws are laws of fundamental physics, but surely some are. Whatever the laws of the ultimate Theory of Everything turn out to be, do we have any a priori guarantee they will neatly match up with powers? If we accept the thin concept's requirement that laws play the axiom role, I cannot see any good reason to think so. For all I know, the 'TOE' will need laws that cannot be mapped one-to-one with powers. This problem is especially worrisome if we think the TOE will inevitably have at least one conservation law. We've already

---

[58] Mumford (2018, 217).   [59] Bird (2007a, 200).

seen Hobbes and Spinoza struggle with—or, really, simply ignore—conservation laws. So it seems that the most easily defended account of the relation between laws and powers makes it a one-to-many picture. For any given law, we should allow that there might be multiple powers that it reports on.

We're now in a position to turn to the details of the contemporary powers view. What does endorsing it buy us that we can't get from its competitors, the top-down positions? Before we can evaluate the powers view, of course, we need to see just what ontology it is committed to. I'll argue that the powers view incurs some debts that will be hard to re-pay.

8

The Powers Ontology

1. Whose Powers?

What, exactly, is the powers view? In outline, it's easy enough to state: natural events follow the course they do because of the powers of the objects involved. That metal expands when heated is to be explained by the powers of heat and metal acting in concert. I'll continue to use macro-level examples for illustration, but the best prospects for real-world powers will be at the micro-level. The arguments for powers are most persuasive when the fundamental entities of physics are concerned.[1]

Still, the powers view gains its initial appeal from experience at the macro-level. In everyday life, we're often acting in light of our beliefs about the dispositions of the things around us. Every child who has stuck a fork in an outlet knows that we build our picture of the world by mapping what will happen to what.[2] Although working at a different level, science, on this view, is ultimately trying to do the same kind of thing: what capacities do different molecules or particles have, and how can we manipulate them? Arguing in this vein, Richard Corry has produced a 'transcendental argument' for the existence of powers.[3] Powers, on his view, are necessary presuppositions of the most successful means of explanation we have available: drilling down to the component parts of a system, seeing how they behave, and using this knowledge to predict and explain the behavior of the system as a whole. Such a Baconian line of thought sits nicely with the 'new mechanism' of Stuart Glennan, Peter Machamer, et al.[4] More ecumenical are arguments that infer powers as the best explanation of the goings on around us.[5] Of course,

---

[1] Alexander Bird (2016, 348) makes this point well.
[2] Most powers views hold that an activated power necessitates its effects. Mumford and Anjum (2011) and Mumford and Anjum (2018) reject this view, arguing for a kind of modality in between necessitation and mere chance. For criticism, see Lowe (2012). I aim to be neutral on this issue.
[3] Corry (2019, 43).   [4] Glennan (2017); Machamer, Darden, and Craver (2000).
[5] Nancy Cartwright seems to me to run this kind of argument, albeit as a *modus tollens*, when she writes, '[a] world of Humean features alone would be a strange and under-populated place, devoid of so much that makes up the world that we experience. There would be no pushings; no pullings; no teachings or learnings . . .' (2019, 35).

The Metaphysics of Laws of Nature: The Rules of the Game. Walter Ott, Oxford University Press. © Walter Ott 2022.
DOI: 10.1093/oso/9780192859235.003.0008

arguments like these are only as good—or as intelligible—as the explanation they purport to offer. And as we'll see, understanding what a power is supposed to be turns out to be surprisingly difficult.

In later chapters, I'll argue that there have been versions of the powers view that do not include all of the features we'll look at below. These six features are a consequence of philosophers' choice in the last century to go 'back to Aristotle.' In 1975, when Rom Harré and E. F. Madden tried to resuscitate the powers view, it was the Aristotelian variety they picked, and it's that variety I sketch below.[6] After canvassing the six features of Aristotelian powers, we'll turn to the question of ontology: what, if anything, could exhibit all of these features? What would the world have to look like, with such powers in it?

2. Essential and Invariant

To start, we can say that each disposition or power has some distinctive and typical manifestation. There is no such thing as a disposition full stop; there are only dispositions *to* do or undergo something. And there are conditions that bring about or trigger the disposition: the Corning glassware was fragile, but without the right conditions (namely, my dropping it), its disposition to shatter would never have been activated.

What counts, though, as a manifestation of a disposition? Where forces are concerned, this quickly gets complicated. The Earth's gravity might count as a power that is manifested when my lawn dart is drawn back to the ground. So we might (however crudely) describe that power as the power to pull things to the ground. But there are cases when the power is activated and doing something even when such a state of affairs fails to result. Consider the Shanghai maglev train, which is designed to allow the magnetic force to counterbalance the gravitational force. The train 'levitates' between 0.39 and 3.93 inches above the track. We might say that the power of gravity is exercised, even though it is not manifested, since it is not pulling the train's wheels into contact with the track. We may sometimes want to distinguish between the exercise of the power and the result that its exercise typically produces.[7]

---

[6] One notable outlier is Andreas Hüttemann (2021), who argues that powers need to be understood as characterizing entire systems, rather than any individual element within the system. As Hüttemann argues, such a view would hardly represent a return to Aristotle.

[7] See Cartwright (2019, 33).

One advantage of this scheme (power, manifestation, and exercise) is that it allows us to give a clear account of situations relevantly similar to the maglev example. When two forces are counterbalanced, it can be tempting to think that both forces *are* producing their characteristic manifestations, with magnetism and gravity both producing the forces they would were they unopposed. We can calculate the result using vector addition. But I'm sympathetic to Cartwright's worry about the composition of forces: in what sense are these putative forces there to *be* composed in the first place? It's not as if gravity moves the train down with a force x and magnetism moves it up with a force y. The composition of forces is of course a useful and necessary tool for calculation; but we should be wary of assuming that it describes the metaphysical situation accurately.[8] The powers view can account for the situation equally well by saying that gravity and magnetism are both being exercised in this case—they are not lying dormant, as is the fragility of the glass when undisturbed—but neither are they producing their manifestations. (In what follows, unless the distinction is directly relevant, I'll use 'manifestation' to cover both manifestation and exercise.)

The first feature of powers is implicit in the fragility and maglev examples: a disposition needs to have its manifestations essentially.[9] What makes a disposition the disposition it is must at least include its characteristic or typical manifestations. This seems close to an analytic truth: how could fragility be *fragility*, unless its manifestations included breaking? There's no need to require each disposition to be 'single-track,'[10] with just one particular manifestation whenever it is triggered. Some, maybe all, powers can be multi-track, with different manifestations produced under different stimulus conditions. But we cannot even begin to individuate and identify powers without mentioning their manifestations.

We've already looked at the objections powers theorists lodge against quiddities.[11] They claim that a quiddity is unknowable, by dint of having a causal profile that varies with the laws. I don't find those arguments persuasive, but they do help us to see why the powers theorist would want to insist on

---

[8] As Cartwright (1983a, 59) puts it, 'We add forces (or the numbers that represent forces) when we do calculations. Nature does not "add" forces. For the "component" forces are not there, in any but a metaphorical sense, to be added . . .' Creary (1981) takes issue with Cartwright on this point; see above, Chapter 2 section 5.2.

[9] As with much else regarding powers, George Molnar (2003, 82) has a nice discussion of this point.

[10] McKitrick (2018, 19) traces the multi-/single-track distinction (though not the terminology) to Gilbert Ryle (Ryle 1949, 43–4).

[11] See above, Chapter 5 sections 3–5.

essentiality. Without it, we would have to no way of identifying and re-identifying individual powers.

As we saw at the start, the powers view might be motivated in part by explanatory and predictive considerations. If I know that I'm in the presence of a flammable gas, I'll know to run if someone strikes a match. Conversely, if I know an explosion occurred and flammable gas was present in large quantities when someone lit a match, I'll be able explain the explosion. We've seen some reason to think that case explanation relies on counterfactual reasoning. To explain an explosion by reference to flammable gas and yet insist that had the gas not been present, the explosion would have happened anyway, is close to self-contradiction.

But to get these benefits, we have to build considerably more into powers than just the essentiality of their manifestations. In parallel with the notion of a wide nomic profile, we can develop the notion of a wide dispositional profile. Suppose we have an apparent counterexample to essentiality: some power P whose possible manifestations differ from one world to the next. It has one set of dispositions, call it Gamma, in this world, and another, Mu, in another world, even given the same stimulus.[12] But now it's open to us to take Gamma and Mu as subsets of P's total set of possible manifestations. P now meets the requirement of essentiality: its possible manifestations do not in fact vary from one world to the next. Such a perverse power would not get us the explanatory benefits the powers theorists want to claim. We would no longer be able to say that, had the stimulus not been present, the disposition would have had the result it in fact had. There is also a metaphysical worry about perverse powers. What *is* power P? It can be defined by its range of manifestations across Gamma, Mu, and any others we want to mention. But as for P itself, there's nothing to be said, any more than there is for a quiddity whose causal contributions depend on the laws.

If we want to advertise the powers view as a source of explanation and as metaphysically superior to any view that commits itself to quiddities, we need what I call strong essentiality: a power must have exactly the *same* set of stimulus-manifestation tuples across and within all possible worlds. That's compatible with the disposition's being multi-track, but not with its being perverse, as power P is.

To strong essentiality, we would need to add yet a further condition: invariance. Richard Corry asks, 'if the behavior that a given capacity produces

---

[12] Note that our perverse power P is not a multi-track disposition. Such dispositions have different manifestations under different conditions. P has different manifestations under the *same* conditions.

can change from one circumstance to another, then how is it that learning about the exercise of a subsystem's capacities in isolation can tell us anything at all about how these capacities are exercised in the differing circumstances of the compound system?'[13] Suppose we want to explain the behavior of a system in terms of its component parts and their dispositions. It then makes sense to isolate the parts and map their dispositions. This is exactly Bacon's proposed method: isolate a disposition and see how it behaves in 'actus puros.' But without a guarantee that those dispositions will have invariant manifestations when their owners are arrayed in the system as a whole, this procedure is pointless. Put differently: there is no explanation of macro-level behavior in terms of the dispositions of constituents without the assumption of an invariant manifestation. Strong essentiality plus invariance is a substantial commitment, but I cannot see how to avoid it without sacrificing the putative advantages of the powers view.

Another way to put the same point is in terms of the metric of whole explanation. From the space of all possible worlds, how likely is it that ours would turn out to be as uniform and predictable as it is? Not very, if there are only powers that have their manifestations essentially, for they might all be perverse powers. It becomes much more probable if we build strong essentiality and invariance into our powers from the start.

3. Independent

Although powers have their manifestations essentially, powers do not depend on the presence of these manifestations. That a power is instantiated in a given world is no guarantee that its manifestations are.[14] This feature is enshrined in our everyday talk of dispositions: when I say that the glass is fragile, I'm not saying that it has, or indeed ever will, actually break.

Familiarity hardly entails perspicuity, though, and this is one feature worth puzzling over. Combined with essentiality and invariance, independence makes the identity of powers depend on the doings of other possible worlds. Powers have their manifestations necessarily, and yet can exist without manifesting. So powers rely for their very identity on non-actual states of affairs.

---

[13] Corry (2019, 31).

[14] It is logically possible, though I think unappealing, to deny that powers exist unmanifested. Such a view is called 'Megaran' or 'Megaric' actualism, after a group of Greek philosophers of roughly Aristotle's era. For discussion, see esp. Molnar (2003, 94–8). Stathis Psillos (2006) uses the existence of unmanifested powers to mount a regress argument.

This brings out the tight connection between powers and other possible worlds that will keep popping up throughout the next few chapters. A power has to somehow 'point to' its characteristic manifestations. Unlike a compass needle, however, a power points not just to something real but to things unreal as well. Independence is the source of the 'modal inversion' charge: the actual world depends on the whole space of possible worlds, rather than the other way around.

Powers theorists can bite the bullet: modal inversion is a feature, not a bug. And as we've seen, they can mount a *tu quoque* argument against the universals view. Its N-relation, after all, has to be defined in terms of *possibilia* as well. But there's one interesting way of avoiding the modal inversion charge I want to entertain.

Take the whole history of the universe. Trivially, every power the world ever features will exist at some stage of that history. Now suppose that each power's existence raises the probability of its manifestation to a level above zero. Since we're defining manifestations as types, not as time-indexed tokens, we seem entitled to suppose that, given an infinite amount of time, every power will be manifested at some point. That's a lot of supposing and seeming. But if this strategy worked, we would get both independence (since a power can exist at a given time without manifesting at that time) and essentiality/invariance (since each power is defined in terms of its manifestations) without requiring that either powers *or* their manifestations exist in other possible worlds. We might call this view 'non-Megaric actualism.' Unlike its namesake, it doesn't deny that a power can exist unmanifested. But it doesn't require that these manifestations exist in other possible worlds.

4. Intrinsic

Independence must be distinguished from a closely related feature: intrinsicality. That a power is 'ready to go,' awaiting its stimulus to action even if that stimulus never arrives, does not by itself show that the power is had by a single object. A power that depends on multiple substances for its presence might be independent of its manifestations. The powers theory, then, undertakes a novel commitment in saying that powers are intrinsic properties.[15] As a

---

[15] Helpful recent treatments of the intrinsicality of powers include Yates (2016), McKitrick (2018), and Psillos (2021).

rough-and-ready characterization, we can say that an object has a property intrinsically just in case it would have that property even in a possible world of which that object was the sole constituent.[16]

I've tried to motivate the powers view by appeal to our everyday methods of dealing with our environment. Some go even further, and suggest that, if powers were not real, there would be no way of pursuing scientific explanations, at least those that require us to examine the parts and structures of bodies to see how they interact. But of course even those views that allow only thin dispositions, such as the universals view, can offer their own reconstructions of such practices. Making powers intrinsic to their bearers is the main way proponents of the powers view can distinguish it from its competitors. The N-relation, for example, gets us plenty of thin dispositions, but none that are intrinsic to the things that have them.

What exactly does it mean to say that a *power* is intrinsic to its bearer? A commonly cited elucidation comes from David Lewis: 'if two things (actual or merely possible) are exact intrinsic duplicates (and if they are subject to the same laws of nature) then they are disposed alike.'[17] In fact, that formulation is far too loose, and suggests that intrinsicality is much less strange and powerful than it is. By building in his parenthetical clause, Lewis has made powers *extrinsic,* by any reasonable definition.[18] If the laws of nature vary, then salt that is water-soluble in the actual world is not so disposed in worlds that differ in the relevant laws. But that picture of laws is precisely what the powers view aims to undermine. For the powers theorist, laws are not top-down features of the world that govern events. So there is no sense to be made of intrinsic duplicates that are not disposed alike in virtue of being governed otherwise. The powers theorist, then, should prefer a simpler formulation: any intrinsic duplicates are dispositional duplicates.

Although widely accepted, the thesis that powers are intrinsic properties does not command universal assent. Jennifer McKitrick makes a good case that a key's disposition to open a particular door should count an extrinsic

---

[16] There is a debate over how to define 'intrinsic'; see, e.g., Langton and Lewis (1998) and Molnar (2003, 39–43). How that debate plays out will not, I think, affect the arguments in this context.

[17] Lewis (1997, 148).

[18] This is not to find fault with Lewis himself, of course; it's to be expected that he would generate a definition of intrinsicality that would fit with the rest of his view, on which dispositions depend on a great deal that lies outside the object said to have the disposition. Lewis's laws depend on the whole mosaic, throughout all time, so any disposition, given his definition, depends on everything that has ever and will ever happen (and indeed what happens in other possible worlds).

property: the same key might exist in a different world but lack the power to open that particular door, if the door in question does not exist. As McKitrick herself notes, this disposition is not a good candidate for a ground-floor element of our ontology, for the simple reason that its ability to open this particular door is parasitic on its ability to open doors with locks of the relevant type.[19] I see no reason for the powers theorist to resist McKitrick's argument; as she notes, the powers theory can preserve intrinsicality by restricting it to the fundamental level.

5. Reciprocal

Although we haven't landed in self-contradiction, we seem to be building a notion of powers from incompatible parts. Powers have their manifestations essentially; yet they are independent of them. Now we can add another pair: powers are intrinsic properties, and yet some of them necessarily mirror others.

Powers come in pairs or bundles, not on their own. Most philosophers recognize the superficiality of any distinction between 'active' and 'passive' powers; the salt has to be able to cooperate in the dissolving every bit as much as the water has to be there to do the dissolving.[20] Although our language encourages us to speak of an individual disposition, we rarely—if ever—encounter a single disposition manifesting itself in isolation. In the vast majority of cases, dispositions have to collaborate.[21]

Here again we have a feature that does not sit well with the others. How can powers be reciprocal in this sense, and yet both independent and intrinsic? Powers have to be at once solely possessed by their bearers and yet knitted together somehow with their mates. How to accomplish this without falling into genuine self-contradiction is far from obvious. This combination of features also allows some bizarre scenarios: could there be a world in which fire had the power to burn paper, but paper had the power to sing in Thai when it met up with fire? If powers are intrinsic properties, then nothing stops such a clash of powers from being possible. As we'll see in Chapter 10 this sort of worry is not as silly as it sounds.

---

[19] See McKitrick (2003a, 168).   [20] See, e.g., Cartwright (2019, 33).
[21] C. B. Martin (2007, 50) puts it well: 'A more accurate view [than one that has a single power in act and a set of background conditions] is one of a huge group of dispositional entities or properties which, when they come together, mutually manifests the property in question.'

6. Irreducible

Strong essentiality-cum-invariance, independence, intrinsicality, and reciprocity are substantive commitments. If powers have all four features, they are already beyond the reach of the original universals theory. On that view, powers are not intrinsic to their bearers. There are plenty of intrinsic duplicates that are not dispositional duplicates, for the simple reason that dispositions accrue to properties by virtue of the N-relations they stand in. Vary the N-relations, and you vary the dispositions.

Still, these four features leave open the possibility that the intrinsic property on which a thing's dispositions depend is not itself a disposition. There might be a 'categorical base' on which the disposition supervenes. However appealing such a line of thought is, there can be no doubt that the powers view developed from roughly the mid-1970s on instead embraces dispositional essentialism: the view that at least some properties have dispositional essences, and that these essences are not themselves the result of any more fundamental, non-dispositional states.[22] Until I explicitly challenge the claim, I'll be assuming that the powers theory is committed to irreducibility.

There are at least two different things the powers theorist might have in mind when insisting on 'irreducibility.'[23] These different senses turn on whether we are talking about the possible reduction of an object or of a property. Call an object reducible when it is made up of parts that can be isolated and exist in their own right. In this sense, table salt or sodium chloride (NaCl) is clearly reducible.

By contrast, we can call a *property* reducible when that property supervenes on some other property. Sometimes we have both in play at once. Salt is disposed to dissolve in water. This macro-level dispositional property is reducible: it supervenes on the properties of sodium and chloride, namely on the electrostatic attraction that results from the charges of the sodium, chloride, oxygen, and hydrogen. So salt is object-reducible, and its disposition to dissolve in water is property-reducible.

[22] As Neil Williams reports: 'One central thesis of dispositional essentialism . . . is that our world contains (or could contain) *baseless* dispositions' (2011, 71). I distinguish between dispositional essentialism and pan-dispositionalism, which holds that the *only* properties that exist are dispositional.

[23] In drawing this distinction, I am indebted to the debate between Mumford (2006) and Williams (2009).

Given this, a macro-level property like solubility is not going to be a good candidate for irreducibility. Instead, the powers theory needs to focus on fundamental entities and claim that they are dispositions. Moreover, it needs to claim that these dispositions are property-irreducible: they do not supervene on non-dispositional properties, nor are they the result of lower-order dispositions.

The recipe would go something like this:

(a) find the object-irreducible particles, whatever they are;
(b) if the object-irreducible entity exhibits a structure of properties, such that some supervene on others, find the subvenient base;
(c) show that the ultimate subvenient base is itself a power.

Step (a) is important because, as the salt example shows, if an entity is object-reducible, its powers derive from the properties of its constituents. (I do not mean to rule out emergent dispositions, but I take it even emergentists would allow that a reducible object's powers 'derive from,' or are a result of, the properties of the constituents; not just any properties emerge from any others.)

Step (b) allows that even a simple object can have properties that depend on one another. Not knowing what the ultimate particles in fact are—at least at this writing, the search for more fundamental particles continues apace— I don't have examples ready to hand. But the powers theory need not rule out a priori such a structure of super- and sub-venience among the properties of a particle. A general idea of such a structure is given by a key's power to open one particular door, which depends on its power to open doors of a given type.

Step (c) ensures property-irreducibility. If the power could be shown to be depend entirely on some non-dispositional property, the power would not be a ground-level element in our ontology. Equally threatening is a scenario in which the power depends both on a categorical property and an N-relation. Irreducibility has to claim that the power is not underwritten by anything at all. Coupled with intrinsicality, we get the thesis that the fundamental power of any ultimate particle is a brute feature of that particle, immune to further analysis.

It's important to note that the powers theorist need not insist that *all* properties, even all irreducible properties, are dispositions. Brian Ellis, for example, mounts a powerful argument for thinking that spatio-temporal location is a categorical property that exists alongside purely dispositional properties.[24] John Heil's dual-aspect theory claims that every property has a dispositional and a qualitative side.[25] So irreducibility does not entail pan-dispositionalism.

7. Intentional

It may well be that powers theorists want to treat powers as primitives, immune from further analysis. That move in itself seems unobjectionable. After all, everyone will be stuck with *some* primitive notion or other, on pain of infinite regress. But we need a way to remove the prima facie incompatibility of the features we've built up over the past sections. Here, I want to focus on just one: the tension between making powers irreducible and intrinsic properties on one hand and insisting that they have their manifestations essentially on the other.

The tension is there at the surface. If a power is an intrinsic property, then it's capable of a lonely existence. If we reject my non-Megaric actualist proposal, then we have to purchase strong essentiality and invariance by tying the power to its manifestations across possible worlds. We haven't yet asked what that tie is. How can an intrinsic property be directed to anything? I plan to develop what I take to be the most promising route for the powers theorist to take before moving on to competitors in the coming sections.

The most popular proposal is that powers are directed toward their manifestations in the same kind of way in which mental states are directed at their objects. Just as a thought about the Eiffel Tower might be an intrinsic state or property of a thinker that 'points' outside itself, so fire's power to burn paper might be an intrinsic property of fire that is directed toward the combustion of paper. I'll argue that this proposal doesn't require any objectionable animism. Moreover, I'll suggest that this sixth feature—intentionality—is an inevitable outcome of combining the five we've already covered. To see that connection, we'll need to dip into the ontology of relations generally.

Let's begin with the proposal itself. That powers might be intentional states has been the subject of much debate, since on its face it seems to project a mental attribute onto inanimate objects. In fact, that accusation goes right back to the modern period. In his Sixth Replies, Descartes describes his own youthful opinion of heaviness: 'I thought that heaviness bore bodies toward the center of the Earth as if it contained in itself some knowledge of it.'[26] This

---

[24] Ellis (2010) and (2021).   [25] Heil (2012).   [26] AT VII 441–2/CSM II 297.

notion of heaviness as a kind of knowledge-directed striving, aiming at a goal outside of itself, is derived from the idea of the mind. It's minds, and minds alone, that act, strive, and know.

Picking up the same theme, Nicolas Malebranche offers this argument against the powers view:

> Well, then, let us suppose that this chair can move itself: which way will it go? With what velocity? At what time will it take it into its head to move? You would have to give the chair an intellect and a will capable of determining itself. You would have, in short, to make a man out of your armchair.[27]

With an update in style, Malebranche might well have been the author of these words, written by Stephen Mumford in 1999 and directed against U. T. Place's version of physical intentionality:

> Does a soluble substance in any way strive to be dissolved? Does a fragile object aim to be broken? There is little reason to think that a material object without a mind is capable of having *aims* and *strivings* for events of a certain kind, because to do so would be for it to act, and attributions of action we reserve for things with minds.[28]

In their defense, proponents of physical intentionality can distinguish it to some degree from its mental counterpart. Although George Molnar takes physical intentionality as 'an undefined primitive' of his theory, he does point out some ways in which it differs from mental intentionality.[29] Physical intentionality is not subject to semantic evaluation: unlike the objects of propositional attitudes, it does not admit of truth or falsity. Nor is it ever accompanied by consciousness. For his part, Place explicitly denies that intentionality is the mark of the mental,[30] so he cannot be accused of imbuing the physical world with mental attributes.[31] Finally, Alexander Bird makes a good case that physical intentionality does not admit of opacity, and so doesn't

---

[27] *Dialogue* VII, in Malebranche (1992, 227).

[28] Mumford (1999, 221). See also D. M. Armstrong's exchange with Brian Ellis in Sankey (1999, 35).

[29] See Place (1999) and Molnar (2003, 81).

[30] Molnar defends himself against the panpsychism charge in just the same way (2003, 70–1).

[31] As it turns out, the parties to this particular debate end up reconciling in an unexpected way: Place demands to know just how Mumford's own theory avoids physical intentionality, and Mumford 'leaves open' whether his then-functionalist theory of dispositions is 'really at odds' with Place's; see Place (1999, 231) and Mumford's preface to the 2008 paperback edition of his (1998).

exhibit little-s intensionality.[32] The only feature the two kinds of intentionality need to share is that of being directed at something outside themselves.

Now, if we were in possession of a crystal-clear theory of mental intentionality on which we could all agree, it would make sense to test whether it had a physical counterpart that shared its directedness; we would then have made progress in understanding physical intentionality by means of the mental. Unfortunately, I am not aware of any such theory. Most naturalistic theories of intentionality exploit causation: from teleosemantics to asymmetric dependence, it is the causal role of a mental state, one way or another, that is supposed to give it its content. No such story, trivially, is available in the case of powers, which are supposed to be part of an explanation of causation, not the other way around. Nor are the anti-naturalistic theories such as phenomenal intentionality going to be much help; even if the 'what-it's-like' of experience somehow ties mental states to their objects in the world, not even the bravest panpsychist would be likely to exploit phenomenal character in explaining physical intentionality.

So the invocation of mental intentionality only takes us so far. What it does get us, I think, is a clear specification of the desideratum for powers: they have to be intrinsic properties that nonetheless point to their manifestations. I'll argue now that this result is forced on the powers theorist in any case. My reasoning here depends on the ontology of relations.

We've seen that relations come in at least two flavors: internal and external. Which kind of relation holds between a power and its manifestations? When we look at the relation between the power of fire and the burning of paper, it becomes clear there is no candidate for an external truthmaker. So the relation cannot be external. External relations are always contingent on something beyond the relata. Even when objects a and b exist, they need not stand in that relation. But given invariance and strong essentiality, this cannot be the kind of relation in which powers stand: if a power is linked necessarily with its manifestations, that relation cannot come and go.[33]

Now, if the relation between a power and its manifestation is to be internal, there must be some feature of both in virtue of which it obtains. The power's nature must somehow include a description of its possible manifestations. If

---

[32] For the debate over little-s intensionality etc., see esp. Bird (2007a, 114–31). I find persuasive Bird's case for the claim that the intentionality of powers shares with its mental counterpart only the features of directedness and possible inexistence of its objects. I take the last of these simply to be the result of combining directedness with the independence condition discussed above.

[33] Note that I am taking powers and manifestations here as types, rather than tokens.

so, then the burning of the paper is 'included' in the power itself. That's the sense in which we have intentionality or directedness.

It's worth noting that the physical intentionality proposal is built into the Aristotelian approach to powers. From the very beginning, powers have been treated as a subset of internal relations, but relations understood in what today must seem a very peculiar way. On an Aristotelian ontology, relations have always been odd man out. A prominent anti-realism about relations, running through Ockham, Peter Olivi, and Leibniz, takes as its chief argument the oddity of relations in a substance–accident world. If a relation such as '. . . is to the left of . . .' were real, it would be awkwardly poised between the substances that saturate it. As Leibniz puts it, 'we should have an accident in two subjects, with one leg in one and one leg in the other, which is impossible.' A relation, 'being neither a subject nor an accident . . . must be a mere ideal thing.'[34]

The response, implicit in Aristotle and explicit in philosophers such as Aquinas and John Sergeant, is to split the relation in two. The proper form of a relation is not aRb but Ra and $R'b$, where R and R' are monadic properties. On the surface, this seems like a plausible move with regard to internal relations. If aRb obtains solely in virtue of the monadic properties of the relata—as when a yellow truck is lighter than a maroon one—there is no need for a 'bridge' to span them. Such a story sounds right to me, although it does nothing to explain external relations. But that is not at all what is going on here.

When medieval and early modern philosophers insist on what we might as well call 'relational properties' such as R and R', they have in mind something much more mysterious. They are claiming that, although R and R' are monadic properties, they nevertheless 'point to' something outside themselves.[35] They have a special kind of being, '*esse-ad*' or 'being toward.' This might be what Aristotle has in mind when he says 'those things are relatives for which being is the same as being somehow related to something.'[36] By the time we reach the medievals, it is the orthodox version of realism about relations.

---

[34] From the letters to Clarke in Leibniz (1989, 339). A fuller discussion of the Aristotelian and medieval background can be found in Henninger (1989).

[35] Stephen Barker helpfully distinguishes two kinds of internal relation, which he calls 'Leibniz' and 'Bradley' relations. A relation R is 'Leibniz-internal' just in case R's instantiation by relata holds in virtue of monadic features of these relata. A relation R is Bradley-Internal just in case 'the relata linked by R are, partly or wholly, constituted by their entering into the relation R' (2009, 247). On these definitions, it would seem that the medieval view takes all relations to be Bradley-Internal, since the relational properties that make them up are defined by their pointing to the other relatum. The wrinkle is that, as intrinsic properties, the relational properties R and R' are capable of lonely existence. So we would have to allow the pointing to cross the barriers between worlds. And that, I take it, is part of the point of the analogy with mental intentionality: what is pointed *to* need not exist in the actual world.

[36] *Categories* 8^a33 in (Aristotle 1984, 1: 13).

Unlike the deflationary treatment of internal relations, this view insists there is an extra property, over and above the yellowness of the truck, that makes it lighter than the red truck. Consider Aquinas's treatment of heaviness or weight. '[I]n a heavy body is found an inclination and order to the center of the universe; and hence there exists in the heavy body a certain relation in regard to the center, and the same applies to other things.'[37] The other intrinsic properties of the heavy body are not enough to make it heavy: that requires a further property that orients the body to the center of the universe, like a compass needle.

The key thing to see is that, for those medieval and early modern realists about relations, there is nothing special about powers. They are monadic properties that exhibit just the same kind of directedness to be found in any other relation. Fire's relation to burning paper is no different in kind from a father's relation to his son. By positing their relational properties, they can have their ontological free lunch: if all relations are internal, they require nothing over and above the relata to obtain. In so doing, of course, they make the natures of those relata mysterious: they have to point to one another.

Nevertheless, if there is such a thing as physical intentionality, it must be a relational property—a *pointing* property—if it is to stand in the internal relations the powers theorist requires to fund essentiality. As Heil puts it, if a given key '"points beyond" itself to locks of a particular sort, it does so in virtue of its intrinsic features.'[38] Now, unlike their historical counterparts, contemporary powers theorists are not committed to giving this treatment of *all* relations. But I do think they're best served by treating the relation between a power and its manifestations in this way. All of this is bound to seem unsatisfying. I've argued that the powers theorist is forced into this position. But other proposals for cashing out the relation between a power and its manifestations have been offered, so we should turn to those next.

8. Functions

An alternative to the mysteries of directedness might be found by casting dispositions as functions.[39] Functional states are causal roles that are realized by first-order properties. We might define fragility as whatever state mediates between a given set of inputs (impacts) and outputs (breaking). If we ask, why

---

[37] *Summa Theologicae* Part I q.28 a.1, in Aquinas (1997, 1: 283).
[38] Heil (2003, 124), quoted in Tugby (2013, 460).
[39] For the functionalist account of dispositions, see, e.g., Prior, Pargetter, and Jackson (1982) and Prior (1985).

is fragility directed toward breaking, we get an answer that is necessary and a priori: fragility is simply defined in terms of its role in taking the states of an object from inputs to outputs.[40] The appearance of any *esse-ad* written into a first-order property vanishes. We are left with a world of familiar first-order properties that 'point to' others only in virtue of realizing second-order properties, that is, causal roles.

There seem to be independent reasons to think that at least some dispositions are second-order properties. Elizabeth Prior et al. mount a multiple realizability argument, familiar from Putnam's attack on identity theory in philosophy of mind, to show that a disposition cannot be identified with its causal basis. The simple and attractive idea is that the same disposition can be realized by multiple causal bases. If that is in principle possible, then the disposition cannot be identified with its causal base but must instead find its home as a second-order property.[41]

The functionalist position threatens to make dispositions causally irrelevant. What's doing the causal work is the causal base, not the disposition itself.[42] When Hamlet picks up Yorick's skull, it's the actor playing the role that does the picking up, not the role itself. I'm not sure this result would be as devastating in this context as it is in the philosophy of mind: it would be bad news if your pain sensation were causally irrelevant to your avoidance behavior, but is it so bad if being poisonous is causally inert? To call bleach poisonous is to report on its deleterious effects, but what have those effects are the chemical properties of the bleach.

A more direct reply might appeal to Stephen Yablo's answer to the exclusion argument.[43] Suppose that a disposition stands to its causal base as determinable to determinate: just as red is a determinate of color, so, it might be said, the chemical structure of bleach is a way of being poisonous. Being a color is not a higher-order property of red. If one property is a determinate of another, then in a sense we don't have two distinct properties, like being colored and being

---

[40] I'm indebted here to Mumford (1998, 198–200).

[41] This is precisely the point Mumford takes issue with; his own view from the late 1990s insists that 'it would be a mistake to describe dispositions as second-order properties' Mumford (1998, 205). Mumford defends a different position in more recent work; for an account of his development, see his (2013).

[42] This is a highly compressed version of Jaegwon Kim's famous 'exclusion argument' against functionalism; see Kim (1998). Jennifer McKitrick (2005, 366) argues that the exclusion principle (roughly, if some property F is causally relevant to an event e, then there is no set of the cause's properties sufficient for e that does not include F) is too strong, as it precludes cases of overdetermination. I suspect that McKitrick's solution to the problem of causal irrelevance is similar to the Yablo-style response I develop below. For further discussion, see Choi and Fara (2018) and the references therein.

[43] Yablo (1992).

spherical. Such a defense could simply take over Yablo's point about determinates/determinables in general. No plausible exclusion principle, Yablo writes,

> can apply to determinates and their determinables—for we know that they are not causal rivals... [A]ny credible reconstruction of the exclusion principle must respect the truism that determinates do not contend with their determinables for causal influence.[44]

Is it the poisonousness of the bleach, or its chemical structure, that makes it harmful to humans? This is a false choice: the chemical structure is the way the bleach has of being poisonous, just as being cube-ish is the way a die has of being shaped. There is no competition between a determinate and its determinable. This strikes me as a promising way of recovering causal efficacy for the functionalist theory. And if we could get a functionalist story to work, we could de-mystify the directedness of powers.

Unfortunately, any variant of functionalism is a poor fit for the powers view. The powers view doesn't just block one-to-one reducibility; it also blocks any view on which a power requires a non-dispositional base. Although functionalism is not a one-to-one reduction, it nevertheless distinguishes between the power and the property that realizes it. Even the determinable/determinate solution sketched above is compatible with the realizer being a non-dispositional property. So functionalist treatments of dispositions inhabit enemy territory: they do not accord powers the fundamental, ground-level place in our ontology the powers theorist wants them to have.

9. Relational Views

The functionalist tries to remove the puzzle of directedness by locating it at a higher level, whether a second-order relation or a determinable. This same basic strategy is exploited by relational accounts of powers. And yet, as we've seen, there's a tension between any such strategy and the thesis of irreducibility. Powers are meant to be monadic properties, independent of both their manifestations and other powers. How, then, can a relational account help?[45]

The default picture, as we've seen, takes the relation between a power and its manifestations to be an internal one, founded on the directedness of the power

---

[44] Yablo (1992, 259).
[45] For a critique of the powers theory generally, and of relational accounts in particular, see esp. Barker (2013).

itself. A power 'points to' its actual and possible manifestations, and to that extent is an intentional state. That leaves us with the mystery of how the pointing happens. How could a power, which by definition is irreducible, point beyond itself?

One solution is offered by the work of Alexander Bird. Rather than making the first-order property intrinsically directed to its manifestations, he locates the relational nature of powers at a higher level: 'potencies may have relational essences but nonetheless be intrinsic to their possessors.'[46] So we have a first-order monadic property whose essence is located at the second order, by consisting in relations to other properties. The pointing or directedness appears only at the second-order level.

If that first-order property were categorical, I *think* I could make sense of this picture. Take a property like being square. Squareness itself might be related to other properties, and those relations might be internal ones and so necessary to squareness. But that's not the picture at all. As Bird puts it elsewhere,

> Dispositional monism is the view that all there is to (the identity) of any property is a matter of its second-order relations to other properties... [T]he second-order relation in question is the relation that holds, in virtue of a property's essence, between the property and its manifestation property...[47]

We appear to have three elements in play:

(1) a first-order property;
(2) the essence of that property, in virtue of which it stands in
(3) second-order relations to other properties.

But then we are told that there is nothing more to (2) the essence, identity, or nature of (1) a first-order property besides (3) its relations to other properties. At that point, I simply have no idea what (1) is anymore. Call this the 'structure without stuffing' objection: it's incoherent to suppose that a power could be made up of nothing but relations. Some first-order, monadic property has to stand in those relations.[48] And to try to give it a nature by appeal to the higher-order relations it stands in is a non-starter.

[46] Bird (2007a, 141).   [47] Bird (2007b, 527) quoted in R.D. Ingthorsson (2015, 537).
[48] I realize some 'structural realists,' such as Ladyman et al. (2007) and Berenstain and Ladyman (2012), will not be bothered by the 'structure without stuffing' objection. Like Cian Dorr (2010), I have a hard time working out the precise ontology Ladyman et al. mean to endorse. It's unclear to me whether structural realists really want to deny that there are individuals at all, or what the metaphysics of such a position would be. As Sider puts it, '[a]lthough some informal remarks suggest such a position

The 'structure without stuffing' objection is distinct from the familiar regress family of arguments.[49] The regress argument applies only to pan-dispositionalism. On that view, there are no categorical properties: every property is a power. Unlike the regress arguments, 'structure without stuffing' applies to *any* position that takes powers to be exhausted by their relations, whatever they might be relations to. Even if we allowed categorical properties to feature in the manifestations of powers, it would be obscure how a power could be nothing but a relation, or a bundle of them.

10. The Inevitability of Physical Intentionality

Our path through these alternatives to the Aristotelian view of physical intentionality has led us back where we started. A philosopher who wants to preserve the irreducibility and intentionality of powers, alongside the other features we looked at, is best off simply embracing the model of relational properties, that is, monadic properties that nevertheless point outside themselves.[50] If we try to turn powers into functions, that is, relations between inputs and outputs, we give up on the intrinsicality of powers. On the other hand, if we try to characterize powers *solely* in terms of relations, we run into the structure-without-stuffing objection. We need to posit a kind of property that is at once intrinsic and directed.

I am far from claiming that physical intentionality is unobjectionable. In fact, it seems to make vivid the mysterious nature of irreducible powers. The real issue is not whether realism about powers requires animism; as we've seen, the only feature the powers view needs to take from mental intentionality is its directedness. All the rest, which make it distinctively mental, can be left aside. Instead, the question is how to understand the directedness that all powers seem to have. The irreducibility of powers makes the question more pressing. How can a property be an intrinsic feature whose nature is exhausted by its pointing to something beyond itself?

[viz., ejecting individuals from the ontology], this is just whistling Dixie. No one has even begun to articulate a serious account of fundamental reality along those lines' (2020, 65). For a statement of the 'structure without stuffing' objection against structural realists, see esp. Greaves (2011).

[49] See below, Chapter 9 section 3.1.

[50] A final alternative worth mentioning is that of Matthew Tugby (2013), which presents a Platonic version of the powers view. Tugby suggests that making powers relations among transcendent universals accounts for directedness. For my part, I have a hard time seeing how promoting the directedness relation to the status of a relation among universals, as opposed to immanent properties, helps. If our original question was how to understand how a particular power points to some manifestation, I don't yet see that we gain much insight by reproducing the same relation at the level of transcendent universals.

The problem can be made vivid if we approach physical intentionality from a different angle: the project of naturalizing the mental. Fred Dretske has argued that there's a kind of non-mental intentionality all around us: the width of the metal legs of a chair indicate ambient temperature; tree rings indicate the age of a tree; a compass needle points to magnetic north. None of this is mental representation, since none of these things is capable of misrepresentation. Still, each of them, Dretske thinks, exhibits intentionality.[51] Armstrong himself, no friend of powers, gives the example of poison as a 'miniature model for the intentionality of mental states.'[52] Poisons can be said to 'point to' their characteristic effects in a cruder and less complex way than that in which full-blown mental states represent their objects.

Notice what all these examples have in common: in each case, there is some *non*-dispositional property or state that carries the informational load. Compasses point to magnetic north by virtue of their needles; trees track their ages by means of their rings. To imagine a power that is exhausted by its pointing is like imagining a compass that points north without a needle. The problem is not Meinongianism, or the modal inversion of properties: I'm not objecting to a property that is necessarily related to inexistents. Modal inversion is concerned with what the powers point to, their manifestations in other possible worlds. The 'missing compass needle' mystery is a problem about what is supposed to do the pointing in the first place.

As we've seen, Molnar insists that physical intentionality is primitive, immune from further analysis.[53] Nor can I see that his companions in this sort of view have done much to remove the mystery of *esse-ad*.[54] Now, my worries about powers hardly amount to a refutation. Every realist view might end up positing something more or less mysterious, one might argue, whether primitive laws, or the N-relation, or powers with *esse-ad*. And the arguments for powers might force them on us, whether we can remove my worries or not. So we should turn to those arguments next.

---

[51] Dretske (2002).   [52] Armstrong (1981, 21).   [53] Molnar (2003, 81).
[54] Here I agree with Tugby (2013, 460), who notes that C. B. Martin's elucidations (1993b) offer little in the way of 'metaphysical elaboration.'

9

The Arguments for Powers

Given the features of powers outlined above, we should ask what reasons there are to believe in such things. Of course everyone has to accept that tightrope walking is dangerous, that some gases are flammable, and all the rest. The powers view is offering one among many possible ways of accounting for these facts: it posits an irreducible, intrinsic property that points to its possible manifestations. But the difficulties in understanding such a property might just be problems we have to live with, if we find the arguments in favor of powers persuasive.

At the risk of artificiality, I propose to group the arguments under three headings. First, we find a conceptual argument. Some argue that the failure of the many analyses of dispositions shows that we should give up and embrace powers. Absent any way of cashing out powers in terms of counterfactuals, and hence analyzing them away, we must be prepared to welcome them into the ontological fold.

A second kind of argument is empirical, and appeals to the putative deliverances of current physics. As Simon Blackburn puts it, 'science finds only dispositional properties, all the way down.'[1] Philosophers legislate against scientific results at their peril. So any view that rules out powers can be accused of hubris: if science posits something, who are philosophers to say such things can't be?

Finally, we have metaphysical and epistemic arguments. How could a world without powers at the fundamental level exhibit higher-level powers? A putative explanation of an event that bottoms out in non-powers is no explanation at all, for the 'oomph' has gone missing. On the epistemic side, we've already seen Shoemaker and others argue that only powers are knowable.

Working through these arguments has an independent benefit: it will help us decide among the various ways to incorporate powers into our ontology. If we find them persuasive, should we try to analyze *all* properties as powers, or

---

[1] Blackburn (1990, 255).

The Metaphysics of Laws of Nature: The Rules of the Game. Walter Ott, Oxford University Press. © Walter Ott 2022.
DOI: 10.1093/oso/9780192859235.003.0009

only some? Or should we say that every property has both a dispositional side and a qualitative side?

1. The Conceptual Argument

The history of thinking about dispositions in the twentieth century is shaped by logical positivism.[2] Anyone animated by a desire to purge scientific and metaphysical discourse of unobservable entities is bound to be bothered by dispositions. How can a piece of glass be fragile, if it is never seen to break? The disposition seems to be a hidden force lurking within the glass.

The natural move is to give some kind of conditional analysis. Suppose we could cash out being disposed to φ by means of a set of conditionals of the form if x encounters stimulus S, x will φ. Rudolf Carnap proposed and undermined such a simple account.[3] For one thing, it lands us with too many dispositions: as long as x is never subjected to S, it meets the criterion for being disposed to φ.

A more sophisticated account appeals to counterfactuals. To say that a piece of glass is fragile is partly to say that, had it been struck, it would have shattered. Against this 'simple counterfactual analysis,' C. B. Martin introduces the notion of a fink.[4] Suppose that anytime the glass is struck, a wizard removes the disposition to break. The circumstances in which the disposition would manifest itself are precisely those that remove the disposition. So it's not true that had the glass been struck, it would have broken; and yet it might be fragile all the same. More straightforward are masks, cases in which a disposition endures but fails to manifest itself. Cyanide is poisonous, but fails to harm anyone who has taken an antidote. Suppose Bobo has in fact ingested the antidote. Even though the cyanide remains poisonous, it's not true that had Bobo been exposed to cyanide, he would have been harmed. The counterfactual analysis fails.[5]

The history of conditional and counterfactual analyses is disheartening. E. J. Lowe compares the situation to attempts to solve the Gettier problem,

---

[2] Jennifer McKitrick provides an exceptionally clear history of the attempts to tame dispositions; see her (2018, 15–41) and (2021), both of which I rely on here.

[3] Carnap (1936).   [4] See Martin (1994).

[5] Bird's way of dealing with finks and masks is worth considering. He explicitly builds in the absence of finks and masks as a *ceteris paribus* clause in a conditional analysis. Where 'D' is a disposition, 'S' a stimulus, and 'M' a manifestation, the analysis runs like this: 'If x has the disposition $D_{(S,M)}$ then, if x were subjected to S and finks and antidotes to $D_{(S,M)}$ are absent, x would manifest M' (2007a, 60). Corry also makes use of this analysis; see Corry (2019, 123-4). And with that move, we are back to Hume's revenge.

which multiply endlessly without succeeding.[6] Not everyone is pessimistic, however, and I think there are some reasons to be hopeful.[7] But let's suppose no such analysis of dispositions will ever win through. What does that mean for the ontology of dispositions?

For Lowe, the lesson is that dispositions serve as truthmakers for conditionals and not the other way around.[8] We should give up on the analytical project and learn to live with dispositions as fundamental ingredients in our ontology. Although not often explicitly articulated, part of the motivation for endorsing powers seems to come from their imperviousness to analysis.[9]

For my part, I don't understand why this failure would vindicate a metaphysics of powers. If we found out that talk about astrology couldn't be reductively analyzed into talk about human behavior and the motions of the planets, would we then be justified in assuming that star-signs and the rest had to be ground-level features of our ontology? In the case of dispositions, we might just as easily conclude that they escape the net of counterfactual analysis only because they are conceptually incoherent. Of course, unlike astrology, disposition-talk is part of our ordinary, everyday scheme. Still, I don't see why that automatically insulates a region of discourse from criticism, or certifies it as a guide to metaphysics. Nor do I see why these failures uniquely select *Aristotelian* powers as the alternative.

But here is where the history matters. In the early days of the analytical project, the goal was to purge our ontology of dispositions by reducing disposition-talk without remainder into talk of conditionals, just as the point of analyzing physical-object talk into talk about sense data was to jettison physical objects from our ontology. So we can say, at a minimum, that powers have survived this sustained attempt to analyze them away. That's a victory for the powers view, but not yet a positive argument for their existence.[10]

---

[6] Lowe (2011).

[7] See esp. Lars Gundersen (2002). I also find Troy Cross's (2005) reply to finkish cases intriguing. Cross's point is that a finkish case 'simply reveals the contextual nature of familiar dispositional predicates. Nor is it troubling that we cannot specify (in a non-circular way) which background conditions are presupposed in a given context. That is the very nature of contextual presupposition and what makes it so useful. If we had always to be able to specify our presupposed background conditions exhaustively and in non-circular terms, ordinary communication would be impossible' (Cross 2005, 325).

[8] See Lowe (2011), as well as Barbara Vetter's insightful (2014).

[9] I think Bird is quite right to suspect that 'the resistance of dispositional locutions to analysis provides a fair proportion of the motivation to think of dispositions as genuine entities' (2016, 349).

[10] I think McKitrick gets it just right when she concludes that 'a major line of argument against real dispositions has failed, and disposition-talk cannot be easily explained away' (2018, 41).

2. Arguments from Science

2.1 The Missing Categorical Properties

Here is Brian Ellis's statement of our first argument:

> The most fundamental things that we know about all have causal powers or
> other dispositional properties, and, as far as we know, they only have such
> properties. Of course, it could be that they have structures we do not know
> about, which are somehow responsible for their dispositional properties, but
> there is nothing to suggest that this might be so, and there is even less reason to
> believe that the causal powers or propensities of the most basic things in
> nature are ontologically dependent on these supposed underlying structures.[11]

Ellis isn't claiming that science has reached the end of its search for fundamental properties.[12] He argues more cautiously: as things stand, the properties
science takes as fundamental are dispositional. Although further investigation
may, and probably will, reveal that these properties are not fundamental after
all, there is no reason to suppose that these new properties will be anything
other than dispositional.

Ellis offers a second, very similar argument: one from the nature of the laws
of nature. 'There are no known laws of nature that are concerned with the
shapes or sizes of things,' and those that do mention such relational states such
as molecular structures 'are dependent on the dispositional properties of [the
structures'] parts.' Those who reject the reality of powers have to face a grim
reality: the laws such pan-categoricalists posit 'simply do not exist.'[13]

Both arguments have the same structure, and, I'll argue, the same flaw. Both
proceed from the absence of categorical properties, whether in the variables
fundamental science quantifies over, or in the laws of that science. If categorical properties were there at the fundamental level, the argument goes, science
would already have found them. It hasn't, so we have no reason to posit non-
dispositional monadic properties.

But are we justified in assuming that physics will find whatever categorical
properties lurk at the fundamental level? Ellis may be right that current

---

[11] Ellis (2002, 74–5); for a similar argument, see McKitrick (2003b, 356).
[12] As Williams points out, 'If current physics is any indication, then we should hardly be convinced
[that science has found the fundamental particles]: millions of dollars are spent each year looking for
smaller particles. These are not the actions of rational people who believe that we have already found
the simples' (2009, 11–12).
[13] Ellis (2002, 75).

physics bottoms out in dispositional properties that have no categorical base, and even in the bolder claim that we have no good reason to suppose this state of affairs will change. But if we can come up with some competing explanation for their absence from physical laws and the vocabulary of physics textbooks, we need not feel any pressure to admit irreducible powers.

And it turns out there is just such an explanation. At the fundamental level, physics is guaranteed *not* to find categorical properties, even if they are there. Frank Jackson puts the point well:

> When physicists tell us about the properties they take to be fundamental, they tell us about what these properties *do*. This is no accident. We know about what things are like essentially through the way they impinge on us and on our measuring instruments. It does not follow from this that the fundamental properties of current physics, or of 'completed' physics, are causal cum relational ones.[14]

Once we move from macro-level dispositions such as fragility and solubility to microphysical properties, we should expect only to get an operational or dispositional characterization of those properties. To infer their irreducible dispositionality would be like inferring that Goya's Colossus has no toes just because they are not in the picture.[15]

More broadly, it seems to me that a given science's operating with dispositional terms does force us to accommodate those dispositions in our metaphysics. But it doesn't uniquely specify Aristotelian powers as the metaphysical account of those dispositions. Quantifying over dispositional predicates doesn't commit one to thick as opposed to thin dispositions; even if it did, it wouldn't tell us whether or not those thick dispositions depend on some non-dispositional base.

2.2 The Ungrounded Argument

Let's turn now to Mumford's version of the argument from science, revised with the help of Neil Williams's insightful criticism.[16]

---

[14] Jackson (1998, 23), quoted in Williams (2011, 78).

[15] As Williams puts it: '[b]ecause the dispositional characterization applied to the fundamental entities is an inescapable consequence of the methodology, that the characterization is exclusively dispositional provides no evidence that the fundamental properties are exclusively dispositional—even if that happens to be the case' (2011, 79).

[16] Mumford (2006), Williams (2009). Williams points out that Mumford uses 'simple' in two senses: not decomposable into further distinct objects, and not dependent on a more fundamental property. My [1] and [2] are designed to capture these distinct senses.

[1] There are sub-atomic particles that are not de-composable into more fundamental particles;

[2] These sub-atomic particles have a privileged set of properties that are not grounded in other properties or objects;

[3] All of these privileged properties are dispositional;

Therefore,

[4] The privileged properties of sub-atomic particles are ungrounded.[17]

Although very much in the spirit of Ellis's arguments, this one has other weaknesses that I think are instructive. Williams points out that no self-respecting Humean would allow [2] to pass unchallenged.[18] For Lewis, a thing's dispositions depend on its intrinsic properties plus the laws of nature. So the idea that a particle might have a disposition that doesn't depend on *anything* else is automatically incoherent. More broadly, anyone who denies the intrinsicality of dispositions would reject [2], and that denial is not the minority view it seems. Like the Humean, the universals theorist also denies [2]. For her, the dispositions of a thing depend on what properties it instantiates and, crucially, the N-relations that hold among those properties.

A less ambitious version of the argument might be immune from these objections. Perhaps one has to first grant the existence of powers before stepping further down the path. The argument would then aspire, not to prove that there are powers, but to prove that such powers as exist must include those of the fundamental particles of physics. Does it even do that much? We've already seen reason to question step [3]: the fact that the fundamental properties as revealed by science are dispositional is, on its own, no reason to think that there are not categorical bases for them. We might, for instance, insist on replacing [3] with

[3]′   All of the privileged properties *so far discovered* are dispositional.

---

[17] I use the terminology of 'ground' because it features in the Williams/Mumford debate. I'm sympathetic to Jessica Wilson's (2014) critique of grounding as a *sui generis* category. 'Grounding' seems to be a grab-bag of older relations (reduction, supervenience, superdupervenience, etc.). Here, we can take an 'ungrounded' property to be a power, that is, an irreducible, intrinsic feature of whatever possesses it.

[18] Williams (2009, 12).

But then the conclusion would have to be similarly qualified: at best, the argument shows that all of the privileged properties that feature in physics as it is today are ungrounded. A stronger attack on [3] would maintain, with Jackson, that in fact physics is guaranteed *not* to find any intrinsic properties for whatever it currently treats as a fundamental particle.

2.3 The Vices of Humility

At this stage, a quite different, meta-level argument suggests itself. We can be sure that science at least leaves open the possibility that the fundamental particles are irreducibly dispositional. One might then argue that it is no business of philosophy's to foreclose that hypothesis.[19] It as if biologists entertained the idea that kangaroos are capable of parthenogenesis and a philosopher came along to tell them to stop it.[20] Philosophers, in short, ought to mind their own business.

The disanalogy with the marsupial case is, however, pretty obvious. In that case, our imaginary biologists are entertaining—we may suppose with good reason—a hypothesis their discipline is uniquely suited to investigating. We can expect that further research will either confirm or disconfirm it. That is not the case with our present question. If Jackson's point is right, the method of inquiry used by physics guarantees that it will never discover the intrinsic nature of the fundamental particles it posits. So philosophers are not over-stepping their bounds when they decline to take this particular lesson in ontology. Remaining epistemically humble in the face of an interlocutor who is unequipped to weigh in either way is hardly a virtue. In sum, I agree with Anjan Chakravartty that appeals to scientific practice are, by themselves, incapable of settling the question of powers.[21] I don't find this state of affairs

---

[19] In the context of qualia inversion, Martine Nida-Rümelin makes this same kind of point: 'No hypotheses accepted or seriously considered in colour vision science should be regarded according to a philosophical theory to be either incoherent or unstatable or false' (1996, 146).

[20] A similar argument is, I once thought, suggested by McKitrick (2003b, 356). At the end of her paper, McKitrick argues that the possibility of bare powers ought to be admitted, because it is an empirical question whether there are any such things (2003b, 368). Williams (2011, 77) replies that McKitrick is equivocating on epistemic and alethic senses of 'possible.' The mere fact that S might obtain, for all we know, does not by itself show that S is possible. McKitrick's recent book makes explicit the connections among the different kinds of modality (2018, 140–3).

[21] Other arguments worth mentioning are Cartwright's (2009) 'dispositional exercise' argument (as Chakravartty calls it): there seem to be cases where dispositions are exercised but not manifested, as when two forces are balanced and cancel each other out. Cartwright argues that only the dispositional realist can make sense of such a situation. But as Chakravartty (2017, 32) points out, no opponent of dispositional realism would agree to that description of things: for the Humean, for example, there are only property instantiations at points in space-time. Another argument appeals to the need for

particularly surprising; the same is true for nearly any conceptual or philo-sophical question.

3. The Metaphysical and Epistemic Arguments

Now we can turn to the negative arguments: rather than claiming support from empirical facts or scientific practice, they are directed against categorical properties. In much of this section, we'll be on familiar ground, having already canvassed the arguments against categorical properties in an earlier chapter. Without repeating myself, I hope to bring out what is distinctive about such arguments when deployed in service of powers.

It's worth making a distinction among them. Some—such as the argument from ignorance—purport to show that there are no categorical properties at all. They support only pan-dispositionalism, the view that all properties are bare powers. Call these 'type-I' arguments. Others—such as the regress argument—have the more limited aim of showing that there must be at least some powers. Call these 'type-II' arguments. Any powers theorist who rejects pan-dispositionalism is of course barred from running type-I arguments.

3.1 A Regress?

This argument is an attempt at a *reductio*: omitting pure powers at the bottom level results in a vicious regress. It's a type-II argument, since it doesn't try to eliminate categorical properties altogether. Although Brian Ellis calls it 'the ontological regress' argument, he conducts it in terms of explanation: roughly, the argument claims that no non-power property is suited to play the role of a fundamental property, because the burden of explanation is never discharged, only deferred to some deeper level property.[22] The buck has to stop with a bare power.

abstraction: scientific laws typically abstract from the world in the sense that they are discovered and tested in 'closed' conditions, where confounding variables can be eliminated. Roy Bhaskar (1978) argues that only by treating the laws as describing dispositions can one explain how one can 'export' causal laws from laboratory conditions to the real world; see also Hüttemann (1998). But as Chakravartty argues, whether the properties involved are dispositional is irrelevant: the challenge of making the inductive leap from laboratory to world remains.

[22] For the argument, see Ellis (2001, 115–16) and (2002, 76). Chakravartty (2017, 29) calls it the 'argument from scientific explanation.'

Here is Ellis's own statement of the argument:

Whenever a causal power is seen to depend on other properties, these other properties must always include causal powers, for the causal powers of things cannot be explained, except with reference to things that themselves have causal powers. Structures are not causal powers, so no causal powers can be explained *just* by reference to structures. For example, the existence of planes in a crystal structure does not by itself explain the crystal's brittleness, unless these planes are cleavage planes: regions of structural weakness along which the crystal is disposed to crack.[23]

One might argue, with Chakravartty, that at most this shows the ineliminability of disposition-talk.[24] It is a further step to get from the necessity of dispositional predicates to that of dispositions themselves. As long as competing views can cash out dispositional predicates without a commitment to powers, their withers are unwrung.

I can imagine Ellis responding: as long as the dispositional predicate is understood to be truthfully applicable in virtue of some non-dispositional property or complex of properties, the regress still threatens. Suppose we say that cleavage planes are not inherently dispositional but are there because of some further categorical property. But then, whatever that categorical property might be, it will not be capable of explaining the disposition of the planes to separate.

And yet I cannot help thinking that Ellis's key premise begs the question. Why believe that only powers can explain dispositions, which might be thin rather than thick? Why shouldn't a thin disposition be present because of N-relations among universals, or patterns in the mosaic? To pass from rhetorical questions to argument, we might pose a dilemma. If the premise already assumes that there are powers—intrinsic, irreducible dispositions— then it begs the question. If it assumes only that there are at least thin

---

[23] Ellis (2002, 76), my emphasis. Since Ellis accepts some categorical properties, his arguments are all type-II, that is, arguments that don't purport to rule out all categorical properties. An important caveat here: for Ellis, such categorical properties as there are turn out to be relational, and so parasitic on dispositions: '[w]e do not say that all properties are dispositional. We think that the intrinsic properties of the most fundamental things in nature are all dispositional. But there are many structural properties of, and relations between, things that depend on spatial, temporal, and numerical relationships that are not dispositional: properties such as shape, size, aggregation, and so on' (2001, 135); see Ellis (2002, 70). Even though all monadic properties are dispositional for Ellis, the relations among them need not be. His view counts as 'dualism' then in a peculiar and limited sense. A more robust dualism would countenance both categorical and dispositional *monadic* properties. So although all his arguments are type-II, they are also meant, I think, to rule out this more robust dualism.
[24] Chakravartty (2017, 29).

dispositions, then it fails to prove that those dispositions are powers. The thin dispositions might hold in virtue of any of the other truthmakers proposed by competing views.

3.2 Ignorance

If there are categorical properties, how would we know about them? By definition, they are not themselves dispositions: but then how do they do anything? These are the sorts of question that animate the argument from ignorance. This kind of argument cannot help being type-I: if it works at all, it rules out all categorical properties.

Proponents of bare powers can define the fundamental properties by their causal profile: what they can do in various situations. They take this to be a virtue; without it, they claim, the fundamental properties would be entirely unknowable. This is a version of Shoemaker's argument, covered above: if we assume that we know only what can act on us, then any property that is not itself a power is unknowable.[25]

I have already produced my own responses in Chapter 5. Let me make two quick points here. First, it is not obvious that ignorance of the intrinsic nature of the fundamental properties is a sin: Lewis, for example, is happy to bite the bullet. 'Who ever promised me that I was capable in principle of knowing everything?'[26] That gives up the game too quickly: as I've argued, there are other ways of knowing besides being causally acted upon. Taking a page from Ellis himself, we might nominate spatio-temporal location properties as candidates for quidditism: they are properties that are not defined by their causal powers.[27] It would be absurd to charge such properties with unknowability. I can know that my bike fell over in a certain location at a certain time, even if locations and times, and the properties that refer to them, don't do anything. So there seem to be properties that are not themselves causal powers and yet are knowable. Beyond such properties, there are many others, such as shape and structure, that are causally relevant without themselves being powers. I think we should be suspicious of the move from 'x is not a disposition' to 'x is causally irrelevant.' Ellis puts the point well: '[t]he shape of an object has no

[25] Versions of this kind of argument can be found in Bird (2005), in addition to those covered in Chapter 5 above. Chakravartty (2017, 34) takes this line of thought to be persuasive, though even he thinks it works only 'from a scientist perspective'; otherwise, we are stuck with a stalemate of intuitions.
[26] Lewis (2009, 211), quoted in Williams (2011, 86).   [27] Ellis (2010).

effects essentially . . . but it does partly determine the shape of the pattern of effects produced by the reflective powers of its surface material.'[28]

Finally, we can make a *tu quoque* reply. [29] How are powers any more within our epistemic grasp than quiddities? There are two points here. First, a power has to be defined by its possible manifestations across all possible worlds (or at least those in which it has any instances). So to know a power—or at least to know it fully—is to know an indefinitely large, perhaps infinite, quantity of information. You have to know what it does and what it would do given any possible stimulus condition. And if powers are reciprocal, this is not as simple as knowing the power's own manifestations; one has to know how it can collaborate with a potentially infinite set of other powers.

Second, some of the original scenarios we encountered with quiddities return. Recall the ghost scenario, in which a categorical property exists in a given world but never enters into any N-relations. Can we be sure that there aren't ghost powers, that are fully real but never meet their stimulus conditions? We also have a problem with a counterpart of the sharing scenario. That scenario envisioned two properties with causal profiles that overlap. If two powers share some of the same stimulus-manifestation conditions, they will be very hard to tell apart. And if the only conditions these powers encounter are those that trigger precisely the same manifestations, it will be impossible. Now, the powers theorist can rule out two powers sharing *all* the same manifestations; they would then be one, not two. But that doesn't help if there's even one stimulus-manifestation pair to individuate them. So I see no advantage for the powers view on the battlefield of epistemology.[30]

4. Varieties of Dispositionalism

None of the arguments for powers we've looked at is persuasive. That doesn't mean, of course, that the powers view is doomed. It might be motivated by other considerations. One might argue that it fits with one's metaphysics as a whole better than its competitors can. Indeed, some philosophers aspire to

---

[28] Ellis (2010, 140).
[29] Williams (2011) explores this kind of reply in different ways than I do here.
[30] The powers theory is welcome to adopt my remedy for these scenarios (Chapter 5, section 4), and might be best served by doing so. But it then surrenders any claim of an advantage over the universals view.

unite all of epistemology, ethics, and metaphysics in a grand Aristotelian synthesis.[31]

Still, some of the failures of the arguments are in fact fruitful. At the moment, we have an array of positions on the table, from the view that all properties are powers to mixed views of various kinds. We can run one of the arguments in reverse to help winnow the range of plausible powers views. I'll argue that the inversion of the argument from ignorance shows that pan-dispositionalism faces some serious problems.

4.1 Pan-dispositionalism

Running type-I arguments, pan-dispositionalists claim that categorical properties are an ontological excrescence, by definition causally irrelevant to events and so permanently unknowable. Instead, I suggest we put this argument from ignorance in reverse. We can call it the 'argument from knowledge.'[32] The rough idea is that knowing that a power is manifested require knowing the spatio-temporal properties of both power and manifestation.

(0) If a power is manifested, there must be some place(s) and time(s) at which it is manifested;

(1) If S knows that power P is exercised, it must be in principle possible for S to know where and when P is exercised;

(2) Spatial and temporal locations are not themselves powers;

(3) Being located in a certain place, or exercised at a certain time, are properties but not powers; so

(4) At least some properties are not powers.[33]

Premise (0) does not prohibit powers that are exercised over vast regions of time and space. It's also consistent with the possibility that a single power is exercised at two distant points in space-time (as might be the case on some interpretations of quantum entanglement). Premise (2) should be acceptable to the pan-dispositionalist. The pan-dispositionalist claims only to treat all

---

[31] Perhaps the high-water mark is Greco and Groff (2012). For a chastening of this ambition, see esp. Bird (2016).

[32] This argument expands on that of Ellis (2021, 180–1).

[33] The apparent slide from an epistemic premise to a metaphysical conclusion is licensed by the 'in principle' clause: the only way for the condition expressed in (1) to fail is for there to *be* no spatio-temporal locations.

properties—that is, all repeatables—as powers. And spatio-temporal locations are not repeatables.[34] This is why we need step (3). Locations are not properties, but being at a given location seems to be. At a minimum, it's a repeatable: the point one foot in front of me right now has been occupied by many different objects and events over the course of time. So being located at that point is a property many objects will gain and lose over the course of history.

The heavy lifting is done by premise (1). At the everyday level, it seems obvious: what would a state of affairs look like, in which S knows that fire's power to burn paper is exercised, but is in principle incapable of figuring out where and when that power is exercised? Anything that could be done to make S's ignorance of the spatio-temporal location plausible would at the same time undermine S's claim to know that the power has been exercised at all. Of course there's plenty of room for vagueness and uncertainty: it might be that the most diligent observation only reveals a region of space-time in which the power was manifested. So we might need to ratchet (1) back accordingly. I don't see that doing so would undermine the knowledge argument. The only way I can see of resisting this argument would be to insist that (3) is false. But if even being located in a certain region of space-time counts as a power, I lose my grip on what a power is supposed to be.

I think the argument from knowledge is the best case against pan-dispositionalism. But the idea that all properties are powers may well labor under further difficulties. Some kind of regress seems to threaten. The regress argument can take many forms, but the most powerful, to my mind, is the identity regress.[35] Every property is borrowing its essence from every other one, and so no property's identity can ever be fixed. Not *every* property can live by taking in another one's washing.[36] A closely related charge is Martin and Armstrong's 'always packing, never traveling' objection.[37] Given Independence and Intrinsicality, powers may lurk in their possessors without being

---

[34] As Mumford and Anjum (2018, 8) point out.

[35] R. D. Ingthorsson (2015) documents no fewer than four distinct regress arguments. What I have in mind is what he calls the 'identity regress' from Robinson (1982, 114) and Lowe (2005, 138). As Lowe puts it: 'no property can get its identity fixed, because each property owes its identity to another, which in turn owes its identity to yet another—and so on and on, in a way that, very plausibly, generates either a vicious infinite regress or a vicious circle.' Bird's reply (2007a, 139–41) appeals to Randall Dipert's (1997) graph theoretical treatment, which individuates properties solely in terms of their position in a structure. For an insightful critique of Bird's way of dealing with Lowe's regress, see esp. Barker (2009).

[36] Ingthorsson (2015, 537) makes substantially the same point: 'But what, I ask, is the first-order essence in virtue of which the second-order relation holds, if the second-order relation is all there is to the essence of any property?' Heil claims that 'the real difficulty [with pan-dispositionalism] lies not in the threat of a regress, but in the fact that qualities play a central role in the identity and individuation of powers. Strip away the qualities, and it is no longer clear what you are talking about' (2012, 71).

[37] See esp. Martin (1993b, 178) and Armstrong (2005, 314).

manifested. What is it, then, for a power P to manifest itself? Only, it seems, for P to bring about some effect E. So far, so good: the power of a magnet to attract iron filings is there all the time, just waiting for some iron filings to go to work on. When they come within its sphere of influence, the magnet produces its effect: it changes the location of the iron filings.[38] But according to pan-dispositionalism, the effect E is just another power, P′. So what actually has happened, on the pan-dispositionalist view? As Armstrong puts it, '[c]ausality becomes the mere passing around of powers from particulars to further particulars.'[39]

The pan-dispositionalist should object that the argument begs the question. Why not think that the presence of a power is indeed a genuine and fully real effect? After all, on other theories that are not obviously silly, causation is the 'mere' passing around of energy or force or momentum or whatever quality you want. And as we've seen, the pan-dispositionalist doesn't claim that *everything* is a power, only that all properties are. So the effect of a power might be reflected in these other elements.

This reply helps us move closer to the core of the 'always-packing-never-travelling' objection. McKitrick argues that its real gravamen is that a power as such is not directly observable. When I see water quench fire, I don't (or don't *just*) see fire's powers being extinguished; I see *fire* being extinguished. When the fire catches on to the curtains, I don't just see the curtains acquiring a new power which they can then transfer to the ceiling: I see them catching fire.[40] Even a manifested power is observable only indirectly, through its effects; but if its effects are themselves only powers, no power is ever observed, even indirectly.

In other contexts, proponents have insisted that dispositions are, after all, observable. We have to be careful here. The claim cannot merely be that we observe the categorical properties brought about by a disposition and in *that* sense 'observe' the disposition. The claim has to be that the power is itself observable. And I have a hard time understanding that. Take a body's ability to reflect light. When light strikes it, the body's power is activated. That's what allows you to see it. Now, you can infer that the body has that ability even if you bump into it in the dark. But, again, that's not the same thing as observing the disposition.

---

[38] For the sake of clarity, I'm working here at the macro-level. The same points in principle apply to micro-physical properties.
[39] Armstrong (2005, 314).   [40] McKitrick (2009, 200–1).

Perhaps the most famous defense of the observability of non-categorical properties is Shoemaker's. He argues that experience is not a good guide to the ontology of properties. We might not experience something as relational or dispositional when it really is. Here is how he puts it:

> [C]onsider being heavy. What feels heavy to a child does not feel heavy to me. Reflection shows that instead of there being a single property of being heavy there are a number of relational properties, and that one and the same thing may be heavy for a person of such and such build and strength, and not heavy for a person with a different build and strength. But when something feels heavy to me, no explicit reference to myself, or to my build and strength, enters into the content of my experience. Indeed, just because one is not oneself among the objects of one's perception, it is not surprising that where one is perceiving what is in fact the instantiation of a relational property involving a relation to oneself, one does not, pre-reflectively, represent the property as involving such a relation.[41]

To my mind, there's a confusion here. 'How heavy something feels *to me*' looks like a relation. There's no doubt that my own strength helps to determine how heavy something feels, for example. But that's a causal factor, not something that figures in the object of the experience. Just being due to multiple causal factors doesn't make a property relational, else all properties would, trivially, be relational. The shape of something is due in part the strength of the gravitational field it's in, how and of what material it was made, whether someone is compressing it, and on and on. That doesn't make shape relational.

To put it more bluntly: I think I can just see categorical properties. I'm looking at a rectangular book, and the property of being rectangular is part of

---

[41] Shoemaker (1994, 28). Related strategies (though chiefly employed in reply to the epistemic regress argument from (Swinburne 1980)) include Bird's appeal to mental properties (2007b, 518) and Anjan Chakravartty's appeal to perception (2007, 136). Briefly, Bird argues that a mental state K can stop the regress by being brought about by some dispositional property J, and the subject can know she is in K. But how can either K, or the knowledge of K, stop the regress, unless one or both are *non-dispositional* properties? I read (Ingthorsson 2015, 531) as making the same point. Dealing with a variation of the regress argument, Chakravartty claims that 'the epistemic buck stops with perception' (2007, 137). To use his example: '[c]onsider the everyday use of simple measurement devices. One attributes properties such as ambient temperatures and pressures by appealing to effects registered on such instruments as thermometers and barometers. The properties one associates with these effects (specific states or settings of the measurement devices) constitute what one might call perceptually direct properties . . .' (2007, 136–7). I agree, but only on the assumption that those properties (the mercury's being at a certain level) are categorical properties, not irreducible dispositions. I can't work out how to understand a perception that is a perception *of* nothing but an irreducible disposition.

what I observe. The pan-dispositionalist has to say that that property is merely a power. But a power to do what? Perhaps to cause experiences of rectangularity in me (among a great many other things). What is meant by 'an experience of rectangularity,' if the property I experience is merely a power? The alleged object of my experience has vanished into the realm of possibility. And if we say that the experience of rectangularity itself is a power, then we really do embark on a regress. What would count as a manifestation of that property? Perhaps the disposition to say 'rectangular' when prompted to describe the features of my experience. But then that utterance itself is only a power, or more carefully, it is a particular (an event, say), whose only properties are powers. But now we can ask the same question again: a power to do what? There seems to be no way to invent a description of my initial experience such that it is itself nothing but a power to bring about other powers or alter particulars. We have to bring in some non-dispositional property.[42]

4.2 Dual-sided and Neutral Monist Views

If pan-dispositionalism can safely be ruled out—and I think it can—we should look at views that try, one way or another, to draw a distinction between powers and quiddities while keeping them together somehow.

In earlier work, C. B. Martin defends a two-sided view: each property has a power side and a categorical side.[43] This is an attractive way to make sense of things. A property like the squareness of a table has both a qualitative side—it's square—and a dispositional side—it's disposed to reflect light at various angles, for example. Critics have wondered how the two sides are related.[44] A contingent relation would undermine all the alleged advantages of pure powers: squareness might have been joined with the ability to reflect light as a sphere does. I take Martin to be saying that the relation is necessary. But if each side of the quality necessitates the other, we need some account of why that would be the case. Even those who welcome brute necessities in other contexts might pause before populating an entire ontology with them.

---

[42] I think this argument also tells against Ellis's view in his (2001) and (2002), since he there rejects monadic non-dispositional properties. I would claim that I observe non-relational, intrinsic properties that are not dispositions.

[43] See Martin (1993b) and (1993a).

[44] See esp. Molnar (2003, 150) and Armstrong (2005, 314).

The simplest response is to absorb one into the other, such that there is only a conceptual distinction between them. A single quality is both a power *and* a quiddity.[45] Armstrong replies with an incredulous stare: to identify a power and a quiddity is a category mistake, like identifying a dog and the square root of seven: '[t]hey are just different, that's all.'[46]

I think we need to go a bit further. If Armstrong is entitled to his N-relation by appeal to our sense of 'natural piety,' why might the neutral monist not make the same appeal? Yes, powers and quiddities are different; but that's because they're two ways of looking at one and the same thing. That way of putting things invites Molnar's rebuke: neutral monism introduces a bizarre kind of mind-dependence. The view sacrifices the objectivity and intrinsicality of powers.[47] But the neutral monist can reply that there is no loss of objectivity: every quality really is, as objectively as you like, *both* categorical and dispositional.

Although I don't think that there's a knock-down argument against neutral monism, I do share others' puzzlement. If we can do better, we should. Let's turn, then, to the last region of logical space: a dualism that embraces both categorical and dispositional properties without identifying them with each other.

4.3 Dualism

Perhaps the most natural view—in the sense of accommodating our pre-theoretical ways of speaking—is one that distinguishes between categorical and dispositional properties and allows both.[48] It has a bad name among powers theorists chiefly because their view is so often motivated by a wholesale *rejection* of categorical properties. Here, let me deal directly with a type-II argument that is directed explicitly at the dualist's view, and not at categorical properties *tout court*.

Armstrong argues that combining quiddities with powers produces a very unattractive picture. Here is the core of his argument:

---

[45] For this move, see Martin (1997), Mellor (2000), Heil (2003), and Heil (2012). Mumford also once held the neutral monist view; see his (1998) and, for why he moved beyond it, his (2013, 13).

[46] Armstrong (2005, 315). Armstrong's example of a category mistake is identifying a raven with a writing desk. While certainly a mistake, it seems to me to fall short of Ryle's (1949) definition: ravens and writing desks are still physical objects, and so belong to the same logical category.

[47] Molnar (2003, 155).   [48] For proponents of this view, see esp. Molnar (2003).

Given a power theory, the masses of bodies must surely be powers. But now suppose that the *distances* of the bodies are non-powers. The gravitational forces exerted are inversely proportional distance. So the distances must surely get into the act. But how do they do this? It would seem that the gravitating bodies involved must somehow be sensitive to their distances. Sensitivity is a causal notion. And since the distances are not powers, given the theory we are looking at, it would seem there is no argument for the contribution of distance being necessary. But if there is a contingent factor involved in gravitational causation, then gravitation is a contingent matter.[49]

It isn't obvious just what Armstrong's complaint is. He ends by emphasizing the contingency of gravitation. But he can hardly complain about that, since on his own view, all laws of nature are contingent. I suspect he has in mind something like the following. If the power of gravitational attraction is sensitive to distance, and sensitivity is a causal notion, then the distances themselves cannot be categorical properties: they have to be causal powers, too.

Why should the dualist accept the claim that sensitivity, in this context, is a causal notion? The locations themselves don't have to collaborate, as independent agents, with gravity to produce the attractive effects. Instead, the powers theorist ought to say that the role of locations in gravitation is built in to the power of gravity itself. Attracting bodies to a degree inversely proportional to the square of the distance between them is just what it is to *be* gravity.

Since I take this objection to fail, alongside the others directed at quiddities themselves, I can only conclude that dualism about powers and quiddities is a live option. If the arguments against pan-dispositionalism go through, that view is best ruled out. Dualism can also claim the advantage of perspicuity over neutral monism: there is nothing intrinsically odd or spooky about the relations between powers and quiddities if we take them to be distinct kinds of properties. For now, at any rate, dualism is left in possession of the field of battle.

5. The Best Powers View

We've made some progress in developing the best powers view. First, it will be one that takes powers to be monadic properties of their bearers that nevertheless exhibit primitive *esse-ad*, in spite of the mystery this creates. And when

---

[49] Armstrong (2005, 313).

faced with the question of the extent of powers in the world, we should opt for a dualism that embraces both categorical and dispositional powers without collapsing the distinction between them.

The question remains why we should be willing to endorse the view that results. I've argued that the considerations in favor of the power view lose much of their force under scrutiny. If the powers theorist accepts the knowledge argument, and so abandons pan-dispositionalism, she equally gives up any type-I arguments, such as the argument from ignorance. She can no longer attack quiddities as condemning us to ignorance, or with countenancing properties that are epiphenomenal, or any of the other familiar charges we've looked at. What remains? Only the arguments from science, and these seem to me especially weak.

But if the arguments for the powers view are not especially impressive, bigger problems are just around the corner. It's now time to look at a line of thought that threatens to upend the whole powers ontology. The argument is hardly original to me: in fact, it goes right back to the first widespread rejection of the powers ontology, in the seventeenth century.

10

Facing up to the Moderns

1. Doubts

When twentieth-century philosophers decide to resurrect the powers view, it's the Aristotelian picture they have in mind.[1] To most of the moderns, such a turn of events would be not just surprising but crushing. We've had this debate before, the moderns might say, and powers lost. Among the ironies here is the appeal powers theorists make to contemporary science. Back in the seventeenth century, it was in part because the powers view was the legacy of a pre-scientific era that it had to go. If we hope to resuscitate a moribund view last widely popular in the late middle ages, we had better have an answer for the challenges the moderns pose.[2] Those challenges, it seems, have largely been forgotten.[3]

Here and there in the literature, we can see the modern's doubts creeping back in—a wheel being re-invented spoke by spoke. I shall argue that at least one contemporary problem—Neil Williams's problem of fit—is an instance of the broader argument Descartes and Malebranche mount against powers. Equally important is the alternative picture some of the moderns propose: sensitive to the charges levelled at Aristotelianism, Boyle and Locke offer a way of 'sanitizing' powers for the modern age.

2. Little Souls

Our discussion of physical intentionality brought up one of the moderns' core objections to powers: how can a monadic property 'point to' or be directed at a state of affairs, actual or otherwise? It's as if each object were endowed with a

---

[1] By the 'Aristotelian picture,' I mean the notion of powers fixed by the six features covered in Chapter 8.

[2] I don't propose to discuss every argument the moderns lodge against powers, or even most of them. I set aside, for example, Malebranche's explicitly theological arguments (e.g., that attributing powers to bodies is tantamount to paganism and should result in the worship of leeks and onions), as well as arguments that depend on God's relation to the world (e.g., the divine concursus argument).

[3] A notable exception is Hill (2021).

The Metaphysics of Laws of Nature: The Rules of the Game. Walter Ott, Oxford University Press. © Walter Ott 2022.
DOI: 10.1093/oso/9780192859235.003.0010

'little soul' that allowed it some measure of mental intentionality, an idea mocked by Descartes and Malebranche. But there's much more to the little souls objection than that. Consider this argument from Malebranche:

> [S]uppose this body truly has the power to move itself; in what direction will it go? At what speed? You fall silent again? 'I mean that body possesses enough freedom and knowledge to determine its own movement and its rate of speed: that it is master of itself.' But watch out lest you embarrass yourself. For, supposing that this body were surrounded by an infinity of others, what must it do when it encounters a body whose speed and bulk are unknown to it? It will give to it, you say, a portion of its moving force? . . . But what part?[4]

How will the body 'work out' which way to move, and how much of its force to impart to another body if they collide? The problem isn't just that the power has to be directed at its possible manifestations; it's that those manifestations themselves will depend on the *other* powers that are in play. Malebranche ups the stakes by supposing that each body is surrounded by an infinity of others. So an infinity of powers must each 'take notice' of each other, and stand ready to work out just what each one will do.

Although Malebranche doesn't mention Descartes's conservation of motion, his questions become more pressing when we ask about such laws. How will two instances of kinetic energy know how to calibrate their activities so their total quantity is conserved? How does a pair of entangled particles know that if one acquires spin-up, the other must get spin-down? How do they each decide which one gets which?

The 'problem of fit' is, if not just the little souls argument over again, its contemporary descendant.[5] Roughly, it asks why, given the indefinitely large range of possible powers, the ones instantiated in our world match up with each other. The argument's recipe requires only a handful of ingredients; each is a feature of powers discussed in Chapter 8. First, we have independence and intrinsicality: an object can have a power even when it is not being manifested, and the power is a monadic property of its possessor. To this we add essentialism: the set of possible manifestations of a power is essential to it. We need only one more element: reciprocity. Powers come at least in pairs or bundles, not on their own. What we treat as background conditions are typically other powers that cooperate to produce the effect. But let's take a simple case: fire

---

[4] Vol. x, 47–8, in Malebranche (1958), my translation.
[5] Williams (2010) and Williams (2019). For further discussion, see Segal (2014) and Bauer (2019).

has the power to burn paper and paper has the power to be burned by fire. Since we have bought intrinsicality, we cannot treat these claims as two ways of saying the same thing. We have to be attributing numerically distinct powers to the fire and the paper: the power to burn and the power to be burned.

But we don't get this for free. Why should it be that paper's power is to be burned by fire, rather than to turn into a chicken when touched by the flames, or to produce the sound of C#, or whatever you like? Once we add in what we used to think of as background conditions, we find that the situation becomes all the more puzzling. For it is not just *two* powers that have to be calibrated to one another; now there's an indefinitely large number of powers in play, all perfectly—and perfectly mysteriously—suited to their roles. The problem is not that powers have to have their manifestations essentially, or that these manifestations must be invariant. Instead, the problem concerns *which* powers get instantiated.

Many, I assume, will react with incredulity: how can *this* be a problem? Isn't it just obvious that powers 'cohere' with each other the way they do? I have some sympathy with this reaction. In fact, Locke and Boyle will try to dissolve, rather than solve, the problem: at least one feature of Aristotelian powers must be jettisoned. But if we want instead to resurrect Aristotelianism, it's no good insisting that powers obviously *do* fit. The question is, in virtue of what?

To that question, the Aristotelian has no obvious answer. That powers fit in the way they do looks very much like a miracle. A theist coming across the problem of fit would have new ammunition for a design argument: how else could it turn out that every power in our world is perfectly in tune with all the others?In fact, Nancy Cartwright's recent work suggests a version of that move. Like Malebranche, she wonders how powers can have enough information to be sensitive to all their neighbors: '[h]ow does the power of gravity, doing what it does because it is gravity, make an outcome happen that also requires Coulomb input?'[6] In response, she introduces a startling addition to her view: nature or God. Powers on their own, of course, do constrain what can happen, just in virtue of their own nature: electromagnetism can move an iron filing, but it can't directly change the filing's mass. That leaves open how far and in what direction the filing moves, especially when other forces such as gravity are at work (as of course they always are). Something—'Nature or God'—must be on the scene to make things happen, and make the powers fit

---

[6] Cartwright (2019, 38).

together as they do.[7] For my part, I think one would have to be very attached to the powers view to take this option. And from the perspective of someone opposed to the powers view, it can only seem a *deus ex machina*.

The explanatory burden is actually higher than it already seems. For the fit of powers should not turn out to be a contingent matter. It's hard to believe that it's an accident that our world is one in which the powers fit together. And indeed, Williams seems to endorse the necessity of fit: 'the powers of the objects involved must cohere'; 'to fail to have the correct fit is to describe an impossible situation.'[8] The *necessity* of fit only makes the puzzle tougher to solve. The comparison with theism can be illuminating here. A design argument that appeals to the Earth's distance from the sun (a little bit further away, and it's too cold for life; a little bit closer, and it's too hot) is unconvincing for the obvious reason: the presence of life guarantees that the Earth is suitably situated to sustain life. But now suppose that the Earth's position were not contingent but necessary: in every possible world in which it exists at all, the Earth is just where it is. That really *would* call for a supernatural explanation (or else full-blown necessitarianism).

So we have two intertwined problems: why powers fit together as they do, and what makes it impossible for them to fail to do so. I'll argue that neither one can be answered by the Aristotelian. Although there are non-supernatural answers available, nearly all of them either fail to solve these problems, or are enormously implausible, or both.

It's important to see that neither fit nor its necessity is the problem of structure without stuffing, which besets purely relational treatments of powers. Even if powers have some stuffing—say, because they are monadic, first-order properties that are not exhausted by their relations to other things—the problem remains. Nor is it the regress argument against pan-dispositionalism: even if one stops the regress by allowing powers to be related to quiddities, for example, the problem of fit remains. That's because, just like the moderns' little souls argument, the problem exploits the key features of Aristotelian powers.

2.1 Holism

Let's leave aside what guarantees that powers fit and just ask how they can fit at all in the first place. If we accept the core features that generate the problem,

---

[7] Note that Cartwright isn't explicitly addressing Malebranche or Williams.
[8] Williams (2010, 87) and (2010, 89).

we seem to be stuck with a pretty implausible position: what Williams calls 'power holism.' Here is how he puts it:

> In order to provide the fit of powers, we must set up the powers so that they always match. How can this be done? One way is to cram all the information about every other property into the power, thereby 'building' powers accord-ing to a plan—a plan that includes what kind of manifestation would result from each and every possible set of reciprocal partners... [E]ach property contains within it a blueprint for the entire universe.[9]

To get powers to fit, it's not enough to demand that they have their manifesta-tions essentially and invariantly. For the question here is about *which* powers get instantiated. We need them to be knitted together in a kind of network. The holist solution is to define each power in terms of the others; each power depends for its identity on all the other instantiated powers.[10]

Far from being unique to Williams, holism seems to be the default view among powers theorists. For example, Mumford embraces holism, writing that '[t]he properties that are real in a world must . . . form an interconnected web.'[11] In the work of other writers, it sometimes seems as if the problem of fit is being used to motivate holism, lurking just beneath the surface. When John Bigelow, Brian Ellis, and Caroline Lierse argue that some laws and powers result from the world's falling into the natural kind that it does, they do so partly because 'it is implausible that [the natures of the fundamental particles] should turn out to be independent of each other.'[12]

We ought to be suspicious of holism from the start. Williams explicitly models it on semantic holism, and, unsurprisingly, the powers version suffers from many of the same flaws.[13] As Mumford and Williams recognize, holism entails that if any one power were different, every other power would be different, too.[14] Add or subtract even a single power, and the identities of every other power are changed! This is the analogue of a well-known problem

---

[9] Williams (2010, 95); see also Williams (2019, 89).

[10] With this move, we might seem to be back at the relational theory of powers, covered in Chapter 8, section 9. But the proponents of holism do not want to abandon the intrinsicality of powers, or follow Bird's maneuvers. Williams, for instance, is clear that, on his view, powers are not relations (2010, 96).

[11] Mumford (2004, 182). Other powers theorists who seem to hold some version of holism include Chakravartty (2007, 140), Martin (1993b), and Heil (2003, 97).

[12] Bigelow, Ellis, and Lierse (2004, 158).

[13] For a critical take on semantic holism, see esp. Fodor and Lepore (1992).

[14] 'A possible world that appeared to have all the same fundamental properties as ours, but had one additional fundamental property X, would thereby have none of the properties from our world' Mumford (2004, 215–16).

for semantic holism, and it seems no more palatable in this context. Even prima facie unrelated powers are linked in a highly improbable fashion. Suppose Earth in world w_1 contains water, with the power to dissolve salt. Earth in w_2 is exactly like Earth in w_1—it has salt, water, and so on—but w_2 also has, in some unimaginably distant region of space, a single X-particle, with its own powers not instantiated anywhere in w_1. Does w_2 instantiate the power to dissolve salt? The holist has to say no. But this seems a very high price to pay. For salt, water, and everything else is just as it is on Earth and throughout the Milky Way.

The powers theorist can of course say that water in w_2 still has *some* power or other, among whose possible manifestations is the dissolving of salt. But whatever that power is, it is not the very same power that is instantiated in our world. There must be, in the space of possibilia, an infinite number of waters, each with its own distinct power-to-dissolve-salt, since all you have to do to generate a new power is vary one other power, no matter how distant in time or space, no matter how causally isolated.[15]

Williams isn't exaggerating when he says that every power has to contain 'a blueprint for the entire universe.'[16] Powers are like Leibnizian monads, in that each one goes its own way under its own steam, and yet each one unfolds in accordance with all the others. Leibniz can tolerate this because he has God in the picture to set up the pre-established harmony. Absent theism, such an explanation for the fit of powers would be absurd.

But the chief weakness in power holism is, at bottom, just Malebranche's objection all over again. In its spirit, we should ask how exactly all the 'information' Williams speaks of is going to get 'crammed' into each power.[17] 'Information' is ambiguous. Sometimes information means symbols

[15] In conversation, James Reed has suggested an interesting maneuver the holist might make. Just as some semantic holists claim that some meanings are more resistant to change than others—e.g., the logical connectives are not going to shift meaning when we re-define the sound 'donkey'—perhaps some powers are more or less independent of the others. I think is a promising way to go, but I see problems ahead. First, the semantic holist can explain this resistance in concrete, practical terms, whether psychological or sociological. No such grounding is available to the powers holist. Second, it's hard to see how the powers theorist could carry this line through without sacrificing the holism: each power has to depend for its nature on the others, and there doesn't seem room for this dependence to vary by degrees.

[16] Williams (2010, 95).

[17] This 'cramming' of information also seems to be a feature of Corry's 'causal influence' view, which, as we've seen (Chapter 8, section 2) supplements traditional powers with the requirement that their contributions be invariant. On Corry's picture, such causal influences can ground 'laws of influence,' true generalizations about how these influences behave. But he also requires 'laws of composition,' which govern the relations among distinct powers, and 'cannot be so grounded' (2019, 132). But when it comes time to explain how these further laws can obtain, Corry tells us that they 'can be grounded in the essential nature of the influences involved' (2019, 132). How these essential natures can do this work is, as far as I can tell, just as mysterious on Corry's account as it is on Williams's.

with semantic content, in the sense that those symbols exist in a context of a convention that imbues them with meaning. In this sense, the phone book does, but grass does not, contain information. Presumably that cannot be what is meant. In other contexts, 'information' can mean anything that *could* be interpreted by someone as evidence of something. In this sense, grass contains information about recent weather patterns, the moisture in the ground, and so on. But this kind of 'information' is everywhere: every state of affairs can be grounds for *some* kind of inference. So that cannot be quite is what meant either. Finally, one might think that the information required is something like a computer program, a set of if-thens that tell the power what to do in every possible state of affairs. Programs in the literal sense require realizers: the circuitry of your computer. Powers, however, are not themselves objects that admit of that sort of structure. So once again, we have failed to cash out the relevant sense of 'information.' One needn't be Malebranche, and still less Hume, to wonder whether power holism, and the 'information'-cramming it requires, is intelligible.

2.2 Monism

Even if holism were intelligible, it would not be a complete solution to the problem of fit. For the second question asks, in virtue of what do powers *necessarily* fit? It's not enough that each power be stocked with the appropriate information. There has to be some further reason why the powers that fit are the only ones that can get instantiated. But holism still has to respect the intrinsicality condition, which means there is no a priori guarantee that the powers fit.[18] If each power is a monadic property of its bearer, then there cannot be a logical contradiction in the instantiation of incompatible powers. Something must prohibit such instantiations. But what?

Williams opts for monism. Powers must fit together 'in virtue of their all being properties of the same particular.'[19] The power to burn and the power to be burned are not properties had by discrete objects. If they were, it would be a mystery why they were instantiated in the same world. Instead, there is only a

---

[18] At times, Williams comes close to denying intrinsicality. He writes, 'No power depends for its existence on any other power, and powers are not relations, but all powers within a system contribute to the nature of all other powers, such that each is set up for the appropriate fit with one another' (2010, 96). In what sense can powers be independent of each other, if each one contributes to every other one's nature?

[19] Williams (2010, 100).

single object—call it the 'Blobject'—that instantiates the whole set of properties, powers among them.

For my part, I cannot see how monism helps with this second question.[20] Why should being possessed by a single object explain the necessity of fit? Well, it might be the case that, if the powers did not fit, it would be impossible for them to be co-instantiated. That sounds right. But the task is to explain *why* it is impossible for non-cohering powers to occur in the same world. The explanation cannot merely be that they then would be conflicting powers had by the same object. If it isn't impossible for such powers to be instantiated in the same world, neither should it be impossible for them to be instantiated in the same object.

Here it's vital to keep in mind that, on the present view, the existence of conflicting powers is not prohibited by logical necessity. Given intrinsicality, there is no logical contradiction in a world that contains fire (with the power to burn paper) and paper (with the power to turn into a chicken on encountering fire). That assumption is precisely what opens up the necessity gap: in virtue of what is such a state of affairs impossible? The gap cannot be closed by appealing to monism. More carefully: monism closes the gap only if one already assumes that conflicting powers cannot coexist in the same object. But there is no reason to think that, apart from the idea that such powers cannot populate the same possible world. And that claim is exactly what stood in need of justification. Monism does nothing to get us the necessity of the fit.

2.3 The Blower

I can see only one way out. In addition to making all powers, powers *of* the one object, the powers theorist ought to postulate only a *single* power had by the 'Blobject.'[21] We could call it the 'Blower' (rhymes with 'power'). We need power monism, not just substance monism. If this view is right, the problem of fit can no longer arise, for there are no powers—plural—to fit or not. There is no network of inter-defined powers, each neatly slotted in to place; there is

---

[20] I owe this point to Will Harris.

[21] I suppose one could endorse the Blower while rejecting the Blobject. On such a view, although there's only one power in the universe, it's instantiated by multiple discrete substances. I don't think that's a particularly winning combination of views, for at that point I wonder what it is that unites all the powers of the distinct substances into a single power.

only the one power. What we call individual powers are just the glimpses we have of the working of the Blower.[22]

Now, I have no in principle problem with otherwise fanciful mereologies. But this one comes at a cost. Back when we started down the powers route, I tried to advertise its virtues: the powers theorist can take everyday processes, which certainly do admit of description, at least, in terms of powers and their manifestations, and leverage that into a story about how the world unfolds as it does. The project of science, then, would be continuous with the everyday project we all have of navigating the world: working out which powers do what when. Science just does that at the fundamental level. But the Blobject-cum-Blower view sacrifices all that. The everyday causal transactions we see are not the outcomes of individual powers, perfectly calibrated to one another, but the workings out of a single power within a single object. The initial motivation has been mortgaged in response to the moderns' objections. Much else has been mortgaged, too. Pan-dispositionalists often complain that quiddities are unknowable. But the Blower itself is unknowable, if anything is. In principle, we could never get a complete description of the Blower, for it fixes—and is defined by, given essentialism—every actual and possible state of affairs.

How should the Blower view account for laws? We've covered one-to-one views, which map individual laws onto individual powers, as well as one-to-many views, which simply say that laws hold in virtue of the distribution of powers. Neither works in this context. The only way to preserve a one-to-one relationship would be to insist that there is only one law of nature, made true by the Blower. On the bright side, we would need no *ceteris paribus* clauses: there would be no exceptions to the one law. Still, the price for this one-to-one move is too high. Such a law would be indefinitely, if not infinitely, complex; no single statement of it would ever be available to play the axiom-role in any branch of science.

Instead, like the revised universals view, the Blower account should be a many-laws-to-one-truthmaker view. In this case, the view would say that there are lots of laws that hold only in virtue of the one power. Such a view should

---

[22] My proposal bears some resemblance to that of Bigelow, Ellis, and Lierse (2004). On their view, a great many laws—such as Maxwell's equations—are 'directly concerned with essences,' in this case, the essence of the electromagnetic field (2004, 155). Others, however, such as the conservation laws, characterize 'not natural kinds *within* the world, but the world *as a whole*' (2004, 155). In my terms, they posit a power had by the entire world plus lots of other powers, which depend on the kinds instantiated in it. On my view, their proposal doesn't go quite far enough: as long as there are independent powers for particular natural kinds, the problem of fit will arise.

exploit the web-of-laws approach, and insist that only when applied in concert do the individual laws issue in accurate predictions and explanations. Each law would be implicitly web-qualified, its promissory note cashed by its relationship with the others.

Despite its lack of intuitive appeal, we should note that the Blower not only solves the problem of fit but the problem of conservation laws as well. Such laws are mysterious if you have to get multiple powers to cooperate so that the total quantity of whatever you're concerned with—angular momentum, spin, energy—is constant. In fact, conservation laws might count as positive evidence for the Blower: the reason we see these quantities conserved is a direct result of events' being the working out of a single power. If I'm right, nothing short of the Blobject and the Blower will save the powers view from the little souls argument and its descendant, the problem of fit. It's time to revisit the features that generate those problems and consider rejecting some.

3. The Moderns' Way out

3.1 Locks and Keys

If the Blower is the cure, one might want the disease back. But just as the little souls objection is the moderns' invention, so is a reply to it. Instead of embracing Aristotelian powers, some moderns reject two of its core features: intrinsicality and irreducibility. What I develop below, then, will be ruled out of court by the contemporary powers theorist: these two features are non-negotiable.[23] Still, rejecting them does not automatically sacrifice the initial appeal of the powers view. Indeed, it might do a better job than the Blower of preserving them. For it will still be the case, or so I'll argue, that these 'sanitized' powers provide a bottom-up picture: laws supervene on the distribution of such powers, with no need to apply metaphysical glue from above in the form of N-relations among universals. Nor is this just Humeanism in powers garb: there is still metaphysical glue. Taxonomy aside, I think the moderns' alternative is well worth developing. For Boyle and Locke, among others, the scholastics' mistake was to treat powers as irreducible, monadic

---

[23] Molnar, for example, claims that 'Dispositions are intrinsic properties of their bearers. This is one of the crucial appearances which has to be saved by an analysis' (1999, 3), quoted in McKitrick (2003a, 155).

properties of their bearers. I'll try to clear this long-neglected path of debris and follow it to its end.[24]

Before beginning, it's worth thinking about what *kind* of solution the problem calls for. To settle that, one needs to decide what kind of necessity mandates the fitness of powers. Williams treats it as a kind of metaphysical necessity and then looks to monism to ground it. Something has gone wrong right at the start. It ought to come out as analytically true that any world contains all and only powers that fit.

It's only the artificial division among powers that generates the problem in the first place. We are invited to think of *the power of flame to burn paper* and *the power of paper to be burned by flame* as distinct properties, intrinsic to the things that have them. And of course once we accept this invitation, we have to explain why paper and fire instantiate just these and not any of the competing, incompatible powers we can think up. But surely at this stage one begins to suspect that there is something amiss with the problem. The distinction among these powers is an artificial one. Although it's perfectly fine to go on talking about *the* power to burn paper, what makes such statements true will not be an isolated property in the fire. Instead, it will be the whole complex of properties relevant to the event. That's the basic line of thought behind the Blower maneuver. But the moderns take it in a very different direction.

Rejecting intrinsicality is the first step in the moderns' attempt to de-mystify powers. Robert Boyle presents his case in terms of locks and keys:

> We may consider, then, that when Tubal Cain, or whoever else were the smith that invented *locks* and *keys*, had made his first lock . . . that was only a piece of iron contrived into such a shape; and when afterwards he made a key to that lock, that also in itself considered was nothing but a piece of iron of such a determinate figure. But in regard that these two pieces of iron might now be applied to one another after a certain manner . . . the lock and the key did each of them obtain a new capacity; and it became a main part of the notion and description of a *lock* that it was capable of being made to lock or unlock by that other piece of iron we call a *key*, and it was looked upon as a peculiar faculty and power in the key that it was fitted to open and shut the lock: *and yet by these attributes there was not added any real or physical entity*

---

[24] In what follows, I take 'extrinsic powers' to be relations. Some predicates (such as 'is tall') might be counted both monadic and extrinsic at once. But I take 'is tall' to abbreviate a more complex predicate, '. . . is taller than the average instance of its kind . . .', which involves a relation, rather than to refer to a monadic property.

either to the lock or to the key, each of them remaining indeed nothing but the same piece of iron, just so shaped as it was before.[25]

Boyle asks us to imagine the very first lock. When Tubal Cain then makes a key to fit it, he is endowing that lock with a new power—the power to be opened by a key. If powers were intrinsic to their owners, then the lock takes on a new, intrinsic property. But that is a mistake: what has happened to the key is a mere Cambridge change. The proposition 'the lock has a new power' is surely true; but what makes it true is not a change in the lock.[26]

In fact, George Molnar unwittingly directs us back to Boyle's view in his very argument for intrinsicality.[27] Responding to Boyle's argument, Molnar points out that what we might call the congruence of the lock and key is 'a comparative. Comparatives are founded relations that supervene on the properties of the relata . . . these properties have to include some that are part of the nature of the key and the lock respectively, and are therefore intrinsic to their bearers.'[28] Molnar is of course right: the congruence of lock and key is a function of their intrinsic properties. Like internal resemblance—that is, resemblance in respect of intrinsic properties—congruence is a relation that one gets for free: fix the relevant properties of lock and key and congruence will ride in their train. That does nothing to show that the key's power to open the lock *is intrinsic to the key*, and that is what's in question. Instead, Boyle thinks, the power is founded on intrinsic properties of both the lock and the key.

---

[25] 'The Origin of Forms and Qualities according to the Corpuscular Philosophy,' in Boyle (1991, 23), last emphasis mine. For an excellent treatment of Boyle on these issues, see esp. Dan Kaufman (2006). In an interesting recent paper, Lisa Downing (2021) takes issue with my (2009) reading of Locke on relations and powers. (I defend and elaborate on my view in my (2017).) On my view, both Locke and Boyle hold a thoroughly traditional, and in their day commonplace, view of relations, which denies that they have any reality over and above the existence of the relata. A true ascription of aRb requires only that a and b each have the appropriate monadic, non-directed property. On this sort of view—common currency in Locke's world—all relations are internal, and there is no need to posit Sergeant-style 'relational' properties. As far as I can tell, no one in the modern period entertains the notion that relations might be extra elements in one's ontology, over and above the existence of the relata, except to mock it (as Leibniz does). It would be extraordinary if Locke's scattered and brief remarks on relations were intended to blaze a new trail.
[26] Note that this view goes beyond McKitrick's limited defense of extrinsic powers. Her example is a key's disposition to open a particular door (2003a, 168). By contrast, the present proposal aims to treat *all* powers—not just those directed at particulars as such—as relations.
[27] Molnar (2003, 103–5). At first, Molnar seems to grant Boyle's point about locks and keys, and insists that Boyle's claim cannot be generalized. But two pages later he attacks Boyle's example on its own merits. I focus on that argument, since I cannot make out why Molnar thinks the claim cannot be generalized.
[28] Molnar (2003, 105).

3.2 Extrinsic and Reducible Powers

The Boylean view claims that powers are not intrinsic to things that have them. 'The power to open locks' does not refer to a single property had by a given key, or even a given kind of key. It is at best a slightly misleading way of stating the facts of the case: this key, and others like it in relevant respects, are such that, when they are applied with sufficient force to this lock, and others like it in relevant respects, the lock is opened. It's no surprise that ordinary language has found ways to abbreviate this claim. But those linguistic short-cuts can't change the facts of the case.

Although the property of a key of kind-K to open locks of kind-L is relational and so extrinsic, it is nevertheless a reflection of an *internal* relation: any world with K-keys and L-locks will be such that individual K-keys will have the power to open L-locks. Since we've denied intrinsicality, we have to deny that any K-key, in any world whatsoever, will have the power to open L-locks. That depends on whether there are any L-locks about. So intrinsic duplicates are *not* dispositional duplicates. Still, if the power internally relates just two objects, then any pairs of those objects will in fact be dispositional duplicates. The real world is much more complex, of course, and there may be no powers that relate just two objects. But if n objects figure in the power, then any duplicates of those n objects will be dispositionally identical. The central point is that this is still a bottom-up view. We haven't surrendered the self-direction of objects to laws imposed from above.

Just as important is the rejection of irreducibility. There's nothing more to the key's power to open the lock than the shapes and sizes of the key and the lock. Gone are the Aristotelian's powers. True, we've given up on reducing a power to the monadic properties of a single object. But there is nothing especially puzzling about multi-lateral reducibility.

The best way to develop this view is by considering objections. First, one might worry that it has Megaric consequences. Relations, after all, depend on the existence of the relata. So when all of the locks a given key fits are destroyed, the key loses its capacity to open these locks, even though it does not change in itself. The result is unappetizing: powers become 'mere Cambridge' properties, as Sydney Shoemaker puts it.[29]

But here it's vital to keep in mind that the power in question was never a power *of the key alone*. That's just what it means to deny intrinsicality. If we insist on construing the power as belonging to the key alone, then the key gains

---

[29] See Shoemaker's 1980 paper 'Causality and Properties,' in his (2003, 206–33).

and loses it depending on what locks there are. But that's not the view. The position claims that the power is a relation among all the relevant relata, however many there turn out to be. To make it the case that a power comes or goes, you have to alter or destroy at least one of the relata that make up the power relation. Consider an analogous worry: I can change the number of pages in a given book by ripping one or more of them out. The number of pages changes just in virtue of changing one of the pages, leaving the rest untouched. But the number of pages is not a mere Cambridge property for all that.

I suspect there's also a type-token confusion lurking here. Shoemaker himself distinguishes between powers as token-distinguished and as type-distinguished. What the key-lock aggregate loses when the lock is destroyed is the power P1 whose manifestation is the opening of lock 1 by key 1. That's the token version. By contrast, the type-identified power P2, defined as the power whose manifestation is the opening of locks of type-L by keys of type K, can still obtain in virtue of the relations in which those types stand. It's just this type/token confusion that allows Locke to say that porphyry in the dark has no color, and Boyle to say that the key's capacity comes and goes.[30]

Second, one might worry that the view invites a regress. If a power is a relation among, say, three objects or properties, in virtue of what does that relation obtain? Must there not be some further relation to guarantee the presence of this one? But as will already be clear, the regress has to assume that the relation in question is external. The Boylean view instead maintains that powers are internal relations: they obtain, if and when they do, in virtue of the intrinsic properties of the relata. What secures the presence of the relation is not some further relation but the properties of the things related.

This response invites a third objection: what exactly are the relata? They cannot themselves be powers, on pain of regress. But if they are not powers, they are, by definition, not capable of making a causal contribution to events. In short, we seem to have arrived at a view of powers that, paradoxically, deprives them of their 'oomph' (to use the technical term).

I think this objection begs the question against the relational view. Note that it has to assume that anything that is not a power is causally irrelevant. We've already seen good reasons to reject that assumption. But we don't need anything as sophisticated as those replies. On the present view, the power just is the relation: it is not reducible to any *one* of its relata. So one cannot

---

[30] See *Essay* II.viii.19: 139. References to Locke are in the following format: Book.chapter.section: page number in Locke (1975).

then complain that each relatum on its own is not itself powerful: that's just what the view claims.

Here a further clarification is in order. I have been talking as if the only intrinsic properties necessary for a power's exercise belong to the objects locally present when the event takes place. This is clearly false. At a minimum, some of those properties themselves will owe their existence to the exercise of other powers. (Gravity is a case in point.) As Locke puts it, 'Things, however absolute and entire they seem in themselves, are but Retainers to other parts of Nature.'[31] Is the true subject of all power attributions, then, a single thing, namely, the world as a whole? Are we back to the Blobject and the Blower? To my ear, this question asks for a decision rather than a discovery. If the dependence of some causally relevant property P on the exercise of some further power P' is enough to make one want to count P' as one of the truthmakers for propositions about P, that's perfectly fine. If one instead wants to count only the locally relevant properties (as defined by whatever scientific theory one is currently deploying), that's fine, too. The decision is a pragmatic one.

How does the Boylean view help with the problem of fit and the little souls problem generally? Recall that the problem of fit presupposes that we are dealing with powers intrinsic to a single actor. The Boylean view has a simple answer: there just is no such thing as *paper*'s having the power to be burned. Nor does a flame have the power to burn anything. What powers there are, are internal relations among intrinsic properties of things. We have given up on the whole project of somehow stocking each individual power with 'information' about its co-instantiates. And we have, as a result, removed the thorn of the little souls argument.

If I'm right, it was a mistake to try to go 'back to Aristotle.' To do so is to leap over the early moderns and their formidable case against powers, along with the innovations they make in response to it. Whatever its faults, the Boylean view reminds us that the realist bottom-up picture is not the exclusive property of the Aristotelians. One might well reject the top-down 'governing' concept of laws without endorsing Aristotelian powers. It would be a shame if the proponents of Aristotelian powers drove this part of logical space from the map altogether.

3.3 Problems

Much detail remains to be filled in. How can there be monadic properties that, just by standing in internal relations with others, help to explain how an event

---

[31] *Essay* (IV.vi.11: 587).

THE MODERNS' WAY OUT 195

happens? It seems to me that Boyle and Locke are working with a geometrical model of causation.[32] The view had already been bruited on the Continent, in the counter-attack on Malebranche's occasionalism. Malebranche argues that there is no necessary connection between any two putative cause–effect pairs, with the sole exception of God's will and its effects. Hume was to take over the argument and make it his own. For both philosophers, it's just supposed to be obvious that one can conceive of any alleged cause without its characteristic effect; and since all parties grant that conceivability entails possibility, there is some possible world with one and not the other. By means of this simple imaginative exercise, anyone can see that, as Hume would put it, 'solidity, extension, motion; these qualities are all complete in themselves, and never point out any other event which may result from them.'[33]

For a direct response to this line of thought, we have to look, not to Locke and Boyle, but to their French fellow-traveler, Pierre-Sylvain Régis, who writes,

> We should not say that we see no necessary connection between the secondary causes and the effects we attribute to them, such as we see between the first cause and its effects. For unless we renounce the senses and reason, we must admit that we see an obvious connection. We see, for example, that the production of flour is necessarily connected with the way in which the mill changes the motion of the water and wind . . . We see, again, that a house one builds is necessarily connected with the way in which the motion of the stones is modified.[34]

From Régis's point of view, only someone in the grip of a theory could deny that the shapes of things determine the way motion is transmitted. Locke makes the same kind of point in the *Essay*, using Boyle's lock and key example. Although Locke of course denies that the science of his day is complete and correct, he does give us a model for the *kind* of thing a completed science would reveal:

> [I]f we could discover the Figure, Size, Texture, and Motion of the minute Constituent parts of any two Bodies, we should know without Trial several of

---

[32] For more evidence and detail, see chapter 20 of my (2009).

[33] *Enquiry* VII.8 in Hume (1999, 136).

[34] Régis (1996, 415–16), first published 1704, my trans. I leave aside Régis's concurrentism: Régis believes that God is the true source of motion at all times, and that 'secondary causes' modify this motion. It is among these second causes that Régis sees a connection no less necessary than that between God's will and its effects.

their Operations one upon another, as we do now the Properties of a Square, or Triangle... The dissolving of Silver in *aqua fortis*, and Gold in *aqua Regia*, and not *vice versa*, would be, then, perhaps, no more difficult to know, than it is to a Smith to understand, why the turning of one Key will open a Lock, and not the turning of another.[35]

The promise is that at the microscopic level, there is something that makes as much sense to us as keys turning the tumblers of a lock. Whatever features are revealed will simply fit together of their own natures—with no God or Nature, and no Blower, needed. More important, there is no mysterious *esse-ad* with its missing compass needle. Adapted to a contemporary context, the view would claim that the fundamental properties of all objects, whatever they turn out to be, will have—or better, just be—natures that stand in internal relations such that the instantiation of one or more of them brings about the instantiation of others. Talk about powers is just disguised talk about these properties.

The obvious problem is that no analog of the lock-and-key picture is forthcoming from contemporary science. Nor does there seem to be any good reason to hope for one. Although I had high hopes for the Boylean view at the start, I can't see any way around the big problem: the geometrical model is a historical curiosity, not something we can reasonably expect to be accurate. I have to conclude, then, that the powers view is stuck with *esse-ad*; there's no way to remove, or analyze away, the directedness of the monadic properties it postulates. The missing compass needle mystery cannot be solved; it's a vice proponents of the powers view must tolerate for the sake of the virtues.

Still, I think we've made progress in this part. We've come some distance toward the best version of the powers view. To start with, it will accept two distinct kinds of properties, powers and categorical properties. That's the best way to avoid the problems that beset pan-dispositionalism. The powers cannot be purged of their intrinsic directedness, so we ought to embrace that feature as a primitive. The most substantial revision was forced on us by one version of the 'little souls' argument: even if we can tolerate *esse-ad*, we need an explanation for how each power can be precisely calibrated to all the others. I argued that the best move here is to embrace power monism, and insist that there is only one power in the world, the Blobject's Blower. We also need the

---

[35] *Essay* IV.iii.25 in (Locke 1975, 556). Locke's position is more complicated than I have the space or inclination to discuss here; for more, see my (2015).

Blower to get our conservation laws. Our view of laws-to-lawmakers then becomes many-to-one. The Blower makes it the case that force, mass, and acceleration are related as they are. Although these revisions are substantial, they seem to me the best way to preserve the powers view in the face of the objections we've looked at.

PART IV

THE REGULARITY THEORY AND BEYOND

11

Origins

Hume and Mill

1. A New Family

So far, we've been toiling the fields of realism about laws of nature, where 'realism' means positing an enforcer for the laws. Whatever their differences, all such views provide a metaphysical system that makes the phenomena come out as they do. It's not a brute fact that energy is conserved, or that temperature, pressure, and volume in gases vary in proportion to each other: these are the result of some more fundamental entity. Such realism will always be purchased at the cost of accepting some primitive or other. In the most extreme case of the top-down family, the laws themselves; for some bottom-up theorists, monadic properties exhibiting *esse-ad*.

Our last family of views rejects realism in this technical sense. We begin with its most prominent member, the regularity theory. Rather than taking N-relations or powers as their primitives, the regularity theorist takes the mosaic itself: there is no explanation to be had for the patterns we find among the instantiations of properties across space-time. What we call 'the laws of nature' are just those regularities the statements of which play a certain role in our scientific theorizing. There's no getting behind the regularities to find some extra entity turning the gears. As Lewis puts it, the core motivation is to 'resist philosophical arguments that there are more things in heaven and Earth than physics has dreamt of.'[1]

A helpful way into the regularity theory is provided by Helen Beebee.[2] She asks us to imagine a god who wants to tell us all the facts that there are, and so provides us with 'God's Big Book of Facts.' In the first draft, the book is just a list of every fact there is. But that is hopelessly complex for minds like ours. So instead, God looks for regularities: suppose that God notices that every time there's an x which is F, it is also a G. So he deletes all those particular fact

---

[1] Lewis (1994, 475).   [2] Beebee (2000).

The Metaphysics of Laws of Nature: The Rules of the Game. Walter Ott, Oxford University Press. © Walter Ott 2022.
DOI: 10.1093/oso/9780192859235.003.0011

statements and enters instead: '$\forall x(Fx \to Gx)$.' The relation of law to particular matter of fact is summary to subject matter, not governor to governed.

The regularity theory has a long history, from Hume's original formulation, through Thomas Brown, Auguste Comte, J. S. Mill, and Ernst Mach, to Lewis's 'Best System Analysis' ('BSA') in the last century and the more recent changes rung on it.[3] Reading the seemingly endless catalog of objections, and the epicycles of reply and revision they create like ripples in a lake, can lead one to think that we have reached a stalemate.[4] Realists (in my sense) insist that the regularity theorist's laws aren't *really* laws: they don't support counterfactuals, or explain their instances, or support induction, or what have you. The regularity theorist says she can do just as well in meeting some of the desiderata the realist sets out, and rejects the others as the relics of an unenlightened age. The realist replies that the best the regularity theorist can do is offer *ersatz* 'laws' that don't *really* do all they are meant to. And so it goes.[5]

It would be dispiriting to embark on an exploration of the regularity theory on this note. So let me say that I don't think that *all* the objections to the regularity theory are like this. I'll argue there is one in particular—what I call the 'central tension'—that is internal to the regularity theory. It is not a matter of holding the regularity theorist to a standard she rejects. Instead, the central tension arises from the commitments of the theory itself. Let me explain what I take the tension to be before tracking it through the development of the regularity theory.

No regularity theory takes *all* regularities to be laws. The sun appears to rise and set; every game of senet is played by a human being. These might be regularities, in some sense of the term, but no one counts them as laws. Opinions differ on what the extra ingredient is, but the core idea, first clearly articulated by J. S. Mill, is that the laws of a science are those regularities that play the axiom role: they are fundamental, in that they are not derivable from still wider regularities, and they permit inferences symmetrically in time and space.[6]

The central tension is simple, and it arises when we combine the regularity requirement with the demand that laws serve as axioms. Newton's laws of motion explicitly play the axiom role in the *Principia*, and there is good reason

---

[3] We'll explore the BSA in detail below. For now, the crucial point is that the BSA, in whatever form, takes laws to be regularities. Lewis (1994, 479) leaves no room for doubt: 'like any regularity theory, the best-system analysis says that laws hold in virtue of patterns spread over all of space and time.'

[4] A brief catalog of the most damaging criticisms of the BSA would have to include Armstrong (1983), Cartwright (1983b), Maudlin (2007), Lange (2009), Woodward (2013), and Woodward (2018).

[5] Perhaps the best account of this back-and-forth is Loewer (1996).

[6] I see no reason for the BSA to bar regularities that are merely fundamental to a given science, and not fundamental full stop, from lawhood.

to think that, however far from Newton physics wanders, science requires its laws to function as axioms in roughly the same way. Is there any guarantee that the most general statements of regularities will be able to play the axiom role?

I don't think so. In fact, I argue below that almost none of the regularity theory's early proponents think so, either. Hume's laws are a subset of causes, where a cause is the first member of a constant conjunction. Had Hume tried to match up his laws of nature with Newton's, he would have found them a Procrustean bed. In the work of J. S. Mill, the tension shows itself in his tendency to speak of laws in incompatible ways. In Chapter 12, I turn to the contemporary BSA, and argue that some of the objections it faces spring from this central tension. Chapter 13 presents a general argument that displays the incompatibility of regularities and axioms in any modestly complicated world such as ours. I then move on to present an alternative anti-realism, one that retains much of the BSA but jettisons the regularity requirement.

For now, I want to turn to Hume and Mill. I begin with a brief treatment of the problem of unobservables: if we require any cause–effect pair, and a fortiori any law of nature, to include all and only observable objects or properties, we have to reckon with the all the forces that seem to operate behind the scenes. While fascinating in itself, the problem vanishes for con-temporary regularity theorists, who simply drop the observability require-ment. Their regularities can range over any properties or events there might be, observable or not. Such a catholic approach is not open to Hume. His regularities are not just any true universal generalizations, but observable sequences of properties, objects, or events. In the rest of the chapter, I turn to the issue that will be with us throughout this part of the book: the central tension between laws as axioms and laws as regularities.

2. Force and Gravity

Textbook accounts typically, and rightly, draw attention to the subtitle of Hume's *Treatise*: 'Being an attempt to introduce the experimental method of reasoning into moral subjects.' This is clearly a reference to Newton, and it would not be going too far to suggest that Hume aims to do for the mind what Newton did for the physical world.[7] It seems likely that he hoped his analysis

---

[7] Note, though, that there is little evidence that Hume ever studied the mathematical parts of the *Principia*. As Michael Barfoot (1990, 161) puts it: 'if Hume's explicit statements about Newton and

of laws could accommodate Newton's three laws of motion. If so, those hopes would have been unfounded.

It's remarkable that, for all his talk of laws of nature, Hume never once defines the phrase. We have to pick up clues where we can find them. Of the twenty-five occurrences of 'law(s) of nature' in the *Treatise*, not one concerns the topic we're interested in. Instead, Hume uses the phrase to denote the artificial rules of justice.[8] Only in the *Enquiry* does Hume use the phrase in our sense. And yet the basic metaphysics is unchanged when we move to the *Enquiry*. That alone should signal that the expression is not introducing a new category.

In Section 4 of the *Enquiry*, Hume argues that 'all the laws of nature, and all the operations of bodies without exception, are known only by experience.'[9] I take these to be equivalent: the laws of nature just are the operations of bodies. Nowhere does Hume suggest that science progresses by discovering some new kind of thing, a law, over and above the network of causes. Just the opposite: Hume uses 'law,' 'principle,' and 'cause' all but interchangeably, the only difference being that 'law' and 'principle' should be reserved for 'general' causes. Long before he has defined 'cause' in his sense, Hume lays out his vision of the progress of natural philosophy:

> [T]he utmost effort of human reason is to reduce the principles, productive of natural phenomena, to a greater simplicity, and to resolve the many particular effects into a few general causes, by means of reasonings from analogy, experience, and observation. But as to the causes of these general causes, we should in vain attempt their discovery; nor shall we ever be able to satisfy ourselves, by any particular explication of them. These ultimate springs and principles are totally shut up from human curiosity and enquiry. Elasticity, gravity, cohesion of parts, communication of motion by impulse; these are probably the ultimate causes and principles which we shall ever discover in nature; and we may esteem ourselves sufficiently happy, if, by

scientific procedure [in E] are compared with the wider community of 18th-century texts which discussed such matters, it is clear that there is nothing unusual about them. In fact, it can be argued that his rather brief and undeveloped views were either commonplace or vicarious, and perhaps even inconsistent.' Barfoot goes on to note that, when Hume was a student (in the 1720s), 'Newton may have been at the height of his power and influence in the Royal Society, but his ideas were by no means fully institutionalized elsewhere in Britain' (1990, 162).

[8] T 3.2.2.19. References to Hume are in the following formats. For the *Treatise of Human Nature* (1739–40): Book.part.section.paragraph number in Hume (2000). For the *Enquiry concerning Human Understanding* (1748): Section.paragraph number in Hume (1999).

[9] E 4.9.

accurate enquiry and reasoning, we can trace up the particular phenomena to, or near to, these general principles.[10]

Note Hume's candidates for the ultimate 'causes and principles' we can discover: gravity, elasticity, cohesion, and communication of motion by impulse. These are not new entities alongside the causes of everyday events; they are just a subset of these causes. What makes them into 'principles' is their generality. To say that all motion is subject to the law of gravity is only to say that gravity is among the causes of all motions of objects. Strictly speaking, Hume should say that the law is the cause–effect pair: if the regularity/law is A–B, then A is the cause. Hume ignores this, and speaks of laws simply as general causes, but his meaning should be clear enough.

Hume is very fluid in what counts as a cause; sometimes he speaks of objects or qualities where we would speak of events. But all causes must stand in a relation of constant conjunction with their effects.[11] And a cause must be such as to trigger the relation of custom or habit: given enough repeated experience, our minds move from the impression of the cause to the idea of the effect. Note that a constant conjunction is thus not merely a true universal generalization. Such generalizations are true so long as they have false antecedents. Hume's constant conjunctions build in a psychological element: not only do they have to have true antecedents, they have to trigger the custom or habit of the mind to form an association between the properties or objects that feature in the conditional.

But Hume also requires that any idea be a copy of one or more impressions.[12] What, then, are we to make of laws that apparently refer to unobservables? If gravity is a force underlying and explaining the appearances, Hume has a problem. He has no way even to represent, let alone endorse, such a force. Given Hume's theory of meaning, it is surprising that commentators (myself included) have not done more hand-wringing over his endorsement of gravity as a cause, indeed, a 'universal' cause.[13] Peter Millican suggests an instrumentalist reading, which Matias Slavov also endorses: in Millican's words, even though there is no active power operating behind the phenomena, Hume thinks 'the ascription of powers to objects has considerable instrumental

---

[10] E 4.12.   [11] T 2.3.2.4.

[12] As Hume states the copy principle in T: 'All our simple ideas in their first appearance are deriv'd from simple impressions, which are correspondent to them, and which they exactly represent' (T 1.1.1.7; see E 2.5).

[13] To the other exceptions I discuss, I would add Rosenberg (1993, 80).

value.'[14] For his part, Slavov argues that Hume 'sees [Newton's] physical concept instrumentally *as if* it provides a cause which refers to an effect.'[15]

Whatever its philosophical merits, I don't think Hume's texts can bear this interpretation. On the instrumentalist reading, there is no truth value to the claim that gravity is a universal cause. I take it as obvious that Hume wants to preserve the truth of the Newtonian laws; at a minimum, he gives no sign of being otherwise inclined. And the truthmakers for laws must themselves be observables, in Hume's system. For him, there's no sense to be made of a regularity among unobservables. Without observing the regularities, there is no way for the mind to have its faculty of custom triggered in such a way as to think of one event or property as causing another. This is the core of the Epicurean case against the design argument in the *Enquiry*, and not something Hume can relinquish.

But we needn't read Hume as an instrumentalist. As we've seen, 'gravity' is not univocal in Newton's work; it's ambiguous between mathematical and physical gravity. In the mathematical sense, gravity is a phenomenon we simply observe, and which presents no special epistemological or metaphysical problems. There is plenty of mystery about the 'physical' cause, which is responsible for the phenomenon; about the phenomenon itself, there is none. My suggestion is that Hume takes himself to be giving an account of what Newton would call 'mathematical' gravity. Hume typically treats gravity, not as a hidden cause of phenomena, but as among the phenomena themselves. It's not a force acting off-stage, pushing the players about: it's there to be seen, and meets his criteria of cause and effect. For example, he tells us that 'the most establish'd and uniform conjunctions of causes and effects' include 'gravity, impulse, [and] solidity.'[16] If the conjunction is established and uniform, it must be a conjunction of cause and effect, the cause (gravity) counting as the first member of an exceptionless A–B sequence.

In his discussion of probability in the roll of a die, Hume mentions '[c]ertain causes, such as gravity, solidity, a cubical figure, &c., which determine it to fall, to preserve its form in its fall, and to turn up one of its sides.'[17] The cause of its falling is gravity, just as its solidity and figure cause the die to retain its shape and to land so that it shows one of its sides up. Gravity, here, is no different from figure: both are observable, surface phenomena.[18]

---

[14] Millican (2002, 145).   [15] Slavov (2013, 288), emphasis in original.   [16] T. 1.3.8.14.
[17] T 1.3.12.10.
[18] Although it's hard to tell, it seems to me that Hume in effect treats gravity as weight. That reading is consistent with his other remarks on the subject. For instance, Hume tells us that 'the gravity of a

Although he does not discuss it as such, Hume seems aware of Newton's distinction between mathematical and physical qualities. In a footnote to the *Enquiry*, Hume notes that Newton 'had recourse to an etherial active fluid to explain his universal attraction; though he was so cautious and modest as to allow, that it was a mere hypothesis.'[19] Hume is careful to relegate this hypothesis to the realm of speculation about the hidden springs and wells of phenomena. Gravity itself is an observable regularity: '[t]he production of motion by impulse and gravity is an universal law, which has hitherto admitted of no exception.'[20] As Eric Schliesser notes, this is a very odd way of putting the inverse square law.[21] But it is Hume's way, precisely because he sees no difference in kind between the experience of a billiard ball striking another (impulse) and releasing one's grip on a ball and watching it fall to the ground (gravity).

How can Hume count gravity as a cause, if he requires temporal priority and spatial contiguity?[22] His own rules for judging cause and effect (T 1.3.15) make simultaneous action at a distance impossible. But his examples show that he is always thinking in terms of simple terrestrial events—rolling a die, dropping something to the ground—where the quality of gravity had by a single object is responsible for the ensuing event.[23]

J. S. Mill has a roughly similar approach. Here he is arguing against Herbert Spencer, who insists that gravity itself is unobservable:

If Mr. Spencer means that the action of gravitation gives us no sensations, the assertion is one than which I have not seen, in the writings of philosophers,

---

body encreases or diminishes by the encrease or diminution of its parts, we conclude that each part contains this quality and contributes to the gravity of the whole' (T 1.3.12.16). D. F. and M. Norton, editors of T, refer us to Roger Cotes's 1713 preface to the second edition of the *Principia*, where Cotes writes that 'the attractive force of entire bodies arises and is compounded from the attractive force of the parts...therefore the action of the earth must result from the combined actions of its parts; hence all terrestrial bodies must attract one another by absolute forces that are proportional to the attracting matter' (in Newton (1999, 387)). Cotes's passage does suggest that gravity is an occult quality, a hypothetical posit justified by experience. But what Hume himself means is fairly clear: we find that when we remove part of an object, we lessen its gravity. So we conclude that gravity is a property of each of an object's parts, not just of the object as a whole. Note that Hume replaces Cotes's talk of proportional *forces* with qualities.

[19] E 7.1.25 fn. [20] E 6.4. [21] See Schliesser (2020).

[22] Note that, in E, Hume drops the requirement of temporal priority.

[23] Eric Schliesser (2020) argues that Hume would reject Newton's application of gravity to celestial mechanics: 'There is textual support to suspect that Hume would deny the universal scope implied by Newton's third rule. Hume states a "maxim" in a footnote to section XI of E which argues against inferring new effects from any cause only "known only by its particular effects." This denies Newton's strategy of making ever more audacious inferences (about planetary motions, the tides, the shape of the Earth, comets, etc.) based on the acceptance of universal gravity.'

many more startling. What other sensation do we need than the sensation of one body moving toward another? "The elements of the representation" are not two bodies and an "agency," but two bodies and an effect; viz., the fact of their approaching one another.[24]

Following Hume's lead, Mill takes Newtonian gravity to be mathematical rather than metaphysical: it is there to be observed in the phenomena themselves. Nor is Mill bothered by the lack of a one-to-one correspondence between terms such as 'force' or 'gravity' and individual objects or phenomena: talk of force is just 'an instrument of abridgement.'[25] Despite the terminology, this is hardly instrumentalism: there is no claim that propositions involving force lack a truth value. They have one, and it is the same as whatever proposition (however complex) the statements involving force are meant to abridge.[26]

The question for the abridgment treatment of force is simple: what is a force doing, when it is not in act? Mill ends up defining force as 'not motion but the Potentiality of Motion,'[27] and it is this potential, not motion itself, that is conserved. Any actual motion must be regarded as 'a draft upon this limited stock' of potential motion.[28] This seems to make matters worse: if force is potential motion, it is never by itself observable, and never figures in any regularities in the phenomena. Mill's answer is to insist that the force said to be stored up, say, as gravitational potential energy, is not a 'really existing thing.'[29] Instead, he writes,

A force suspended in its operation, neither manifesting itself by motion nor by pressure, is not an existing fact, but a name for our conviction that in appropriate circumstances a fact would take place.[30]

This is an extraordinary thing for Mill to say. It foreshadows twentieth-century projectivism, and it hardly sits well with the rest of Mill's view. We began with the hope that any law of nature would get cashed out as an exceptionless generalization, where the domain is actual or at least possible objects of experience. If we want to take as a law the proposition 'in all events, force

[24] SOL III.vii.4: 275. (References to A System of Logic (first published in 1843) are in the following format: Book.chapter.section: page number in Mill (1973).)
[25] SOL III.v.8: 345.   [26] For the instrumentalist reading, see Cobb (2016, 238).
[27] SOL III.v.10: 351.   [28] SOL III.v.10: 351.   [29] SOL III.v.10: 353.
[30] SOL II.v.10: 353.

(=the potentiality of motion) is conserved,' we cannot pretend it is a regularity. Instead, it signals an attitude or conviction on the part of the speaker. Mill's remarks are, in other words, a confession: he cannot cash out laws about forces in observable terms.

3. The Tension in Hume

Nothing in the regularity theory per se ties it to the details of Hume's or Mill's own views. In particular, there's no good reason to accept that the qualities that figure in regularities must be observable. Although it's a non-trivial task for the regularity theorist to cash out seemingly dispositional terms such as 'force' and 'gravity,' it's a task we can set aside for the moment. Instead, let's move to a second issue that arises when we try to come to grips with Hume and Mill: what I'm calling the 'central tension.' If I'm right, it's a problem that infects regularity theories as such; but for now, it will do to show that it's a problem for Hume and Mill. And although the tension arises in different forms, given the details of their views, I think it's the same problem all over again.

For his part, as we've seen, Hume takes laws to be a subset of causes. So Hume has to be thinking in terms of a hierarchy of causes, arranged by their degree of generality. To see the difference between ordinary causes and those Hume calls 'laws,' consider Hume's treatment of miracles. Why is it more probable 'that lead cannot, of itself, remain suspended in the air' or that 'fire consumes wood' than the opposite, except that the first is 'found agreeable to the laws of nature'?[31] Hume's answer doesn't concern us. What matters is that these everyday causal claims are subordinated to the laws of nature.

Such a hierarchy of causes and laws makes sense only if there is overlap among them, such that a single event falls under both particular and more general causes. This hierarchy fits with Hume's definitions of cause: 'the constant union and conjunction of like objects' or 'the inference of the mind from the one to the other.'[32] Given Hume's talk of qualities as causes, it seems to me most plausible to read him as treating similarity as either the instantiation of the same property, or, in terms friendly to Hume's nominalism, the exact resemblance of particular properties that are either had by or constitute

[31] E 10.12.   [32] T 2.3.2.4.

objects. Similarity, even in this restricted sense,[33] can still be relatively pro-miscuous, such that detecting wider and wider similarities need not displace our initial causal claims.

Still, someone might argue that Hume leaves himself open to charges of overdetermination. If we say that dropping the iron bar [A] caused it to fall [B], isn't it superfluous to add gravity to the picture? I take Hume's answer to be that [A] isn't the sole description of the cause: the wider regularity that subsumes it is one that invokes gravitation.

The real problem is just around the corner. The [A]–[B] connection is not exceptionless, and hence not a regularity. In Hume's terms, there are plenty of cases where the lower-level cause–effect sequence is *not* 'agreeable to the laws of nature.' Suppose that the bar is very light and is subject to a powerful magnetic force. We would then observe the breach of the connection between events of type-A and type-B.

Hume faces a dilemma: either acknowledge that [A]–[B] was never a regularity in the first place, and, by parity of reasoning, admit that hardly any of our everyday causal claims are true, or account for the apparent breach in a way that preserves those causal claims. There is good reason for Hume to abjure the first option. On that view, everyday claims such as 'heavy unsup-ported objects fall to the ground' do not come anywhere near stating regula-rities. But then his project of accounting for the concept of causation is in jeopardy. His replacement concept of causation is, after all, meant as a piece of psychology: if he aims to account for our concept of cause, he cannot make nearly all applications of it false.

So we should expect Hume to take the second option and try to hang on to [A] as the cause of [B]. That is just what he tries to do. Whenever we find an apparent exception to a regularity, it is because we have not accounted for a competing cause that is interfering with it. As Hume puts it in the chapter on liberty and necessity,

a contrariety of effects always betrays a contrariety of causes, and proceeds from their mutual opposition. A peasant can give no better reason for the stopping of any clock or watch than to say that it does not commonly go right: But an artist easily perceives that the same force in the spring or pendulum has always the same influence on the wheels; but fails of its

[33] I say 'restricted' because there are notions of similarity that require only resemblance in some respect or other, where that respect can include relational properties. On this permissive version, Procopius's big toe and a beer bottle resemble each other (in that they've both just now been used in an example).

usual effect, perhaps by reason of a grain of dust, which puts a stop to the whole movement. From the observation of several parallel instances, philosophers form a maxim that the connexion between all causes and effects is equally necessary, and that its seeming uncertainty in some instances proceeds from the secret opposition of contrary causes.[34]

In short, 'the irregular events, which outwardly discover themselves, can be no proof that the laws of nature are not observed.'[35]

All of that sounds reasonable. Unfortunately, Hume isn't entitled to say it. Given his account of causation, how can he speak of a competition or collaboration among causes? The cause *just is* the regularity, plus or minus the determination of the mind to move from the idea of one to that of the other. The regularity is either there or it isn't. If it isn't, then we do not have a cause at all.

Hume then is pushed back on to the first option, denying that most lower-level causal claims are true. Still, perhaps all is not lost. Given his two definitions of 'cause,' he can say that a subject whose experience includes an [A]–[B] association will undergo habituation such that her mind moves from [A] to [B], even if [A]–[B] is not in fact a regularity. This sort of consideration is what leads Don Garrett to introduce his ideal observer reading: the objective sense of causation is captured by stipulating that an ideal observer, with access to all the patterns of the world, would come to associate all and only those events that are in fact constantly conjoined.[36] Taken objectively, then, Humean causes cannot collaborate or interfere with one another. To speak of causes intermingling as Hume does makes sense only if we take 'cause' in the subjective sense.

There is an obvious problem with this maneuver: it makes a hash of Hume's talk of discovering wider and more general causes-cum-laws. Instead, we would be abandoning our putative causes/laws as we go, rather than subsuming them under more general ones. Taken objectively, there are very few—maybe only one?—genuine causes, since they have to be exceptionless regularities when seen from the point of view of eternity. The unsurprising point is that there are just too few regularities around. The somewhat surprising point

---

[34] E 8.13.

[35] E 8.14. In his chapter 'Of Probability' in E, Hume says that, although the 'production of motion by impulse and gravity is a universal law... there are other causes, which have been found more irregular and uncertain' (E 6.4). But an irregular cause–effect sequence is not a cause–effect sequence at all, given the definition of causation.

[36] See Garrett (1997).

is that Hume himself seems aware of this and tries to solve the problem in ways he is simply not entitled to. To put the point in terms of the central tension: preserving laws as regularities in this way frustrates the attempt to cast laws in the axiom role. I can think of no plausible way to transform Newton's laws of motion, individually or collectively, into a statement of a single, massively complex regularity. Even if we could, such a regularity would be ill suited to playing the axiom role.

We can bring out the conflict between Hume's laws and Newton's in another way. Recall that Humean regularities are not just true universal generalizations; they have to be observable sequences between types of objects or properties. There are then two difficulties: Hume has to find qualities that are the right sort to stand in succession even in a single case; and he needs to make sure there are no exceptions. In neither case are the prospects anything but bleak.

First, Hume needs to isolate two qualities or objects, such that one always precedes the other. Since Hume is taking Newton's laws in the mathematical, rather than physical, sense, he can bypass all the worries about unobservables. But that doesn't mean he's home free. The mold for Humean causation is given by ordinary, macro-level events: fire burning paper, water quenching thirst. Even on the most catholic conception of 'qualities,' I don't have the slightest idea how to shoehorn gravitational causation into that mold, especially since Hume has no place for 'mass' in his catalog of qualities or objects.

After he has found these qualities or objects, Hume needs to ensure that they are constantly conjoined. Even one exception will scuttle the causal connection, by definition. But Hume does nothing to show us how Newton's laws might be exceptionless. At least on its face—though the issue is complex—the inverse square law does not report a regularity. In the presence of a powerful electro-magnetic field, it will give us the wrong results. We'll look at recent attempts to deal with this sort of problem in the next chapter. My point here is that Hume has not made any such attempt.

Even more obviously, Hume will have a hard time making sense of Newton's first law. If that law claims that a body perseveres in its current state so long as it is not subjected to forces of any kind, then it can hardly state a constant conjunction, for it has no instances.[37] 'Unicorns are always followed by rainbows' might be a vacuously true universal generalization, but there's no such thing as a constant conjunction of unicorns and rainbows (unless I am much mistaken).

---

[37] Psillos makes this point especially clearly in Psillos (2002, 143).

A more promising treatment of Newtonian laws is offered by Mill. In some places, at least, he treats laws as a web, which function only as a group to produce predictions and explanations. In doing so, he recovers one of the key insights of the Cartesian tradition. Our question will be whether he is entitled to it.

4. Mill's Web

Mill seems to have been the first to explicitly set out the two key features of the BSA: laws are regularities, the statements of which function as axioms.[38] His work is unusually complicated, and it is doubtful whether a single interpretation can do justice to it. This very complexity is further evidence for my general thesis: those two key features are by their nature in competition. No surprise, then, to see Mill tie himself in knots trying to forge an alliance between them. To make matters worse, Mill mixes in a third notion of lawhood, familiar from Bacon and Spinoza, according to which laws are laws *of* a given kind of thing's nature.

Mill's first definition is thoroughly Humean: a law is a 'partial regularity.' In combination, laws constitute the general regularity in nature.[39] Suppose that 'A is always accompanied by D, B by E, and C by F, it follows that AB is accompanied by DE, AC by DF, BC by EF, and finally ABC by DEF.'[40] The A–D connection is a law by virtue of being a partial regularity. Note that the individual regularities exist unaltered in the final product: A–D is a constituent of ABC–DEF. Given his regularity theory of causation, Mill must regard ABC as the cause of DEF. These regularities persist unaltered, only supplemented, as the laws of a completed science.

A couple of pages later, Mill offers a different understanding of 'law.'

---

[38] The same combination is found in Mill's successors. Forty years after the 1843 publication of Mill's *A System of Logic*, Ernst Mach's *Die Mechanik in ihrer Entwickelung* treats laws as regularities and as rules: 'In nature there is no *law* of refraction, only different cases of refraction. The law of refraction is a concise compendious rule (*Das Brechungsgesetz ist eine zusammenfassende concentrirte*), devised by us for the mental reconstruction of a fact, and only for its reconstruction in part, that is, on its geometrical side' (1883, in Mach (1893, 485–6)). Elsewhere in the same volume, Mach writes, 'The business of physical science is the reconstruction of facts in thought, or the abstract quantitative expression of facts. The rules which we form for these reconstructions are the laws of nature ... The laws of nature are equations (*Die Naturgesetze sind Gleichungen*) between the measurable elements α β γ δ ... ω of phenomena' (1893, 502). For his part, Frank Plumpton Ramsey formulates the laws as regularities/laws as axioms view in 1928 (1990, 143) only to abandon it within a year or so. We'll look at his replacement below, Chapter 14, section 2.

[39] SOL III.iv.1: 315.   [40] SOL III.iv.1: 315.

According to one mode of expression, the question, What are the laws of nature? may be stated thus: What are the fewest assumptions, which being granted, the whole existing order of nature would result? Another mode of stating it would be thus: What are the fewest general propositions from which all the uniformities which exist in the universe could be deductively inferred?[41]

Mill thus appears to hold that the fewest assumptions needed to deduce the whole order of nature will state regularities. Even he himself, I'll argue, doesn't really believe that.

As at least partial regularities, Mill's laws are a subset of causes, where causes are defined as invariable and unconditional antecedents to events. Mill's causes, whether they be laws or not, fail to meet the regularity requirement. The reason is simple: Mill's partial regularities interfere with each other. As we saw with Hume, that is unintelligible so long as causes are regularities.

Let's start by looking at how Mill evades a counterexample to the crude regularity theory: we would have to say that night is the cause of day, since they follow each other. The condition of invariability alone ought to defeat the counterexample. Prior to the formation of the Earth and the Sun, there was no such sequence of events. But Mill appeals instead to unconditionality. Day depends not on night but on the existence of the Sun, the orientation of the Earth, and the absence of any body blocking its light from reaching wherever we are on Earth, and so on.[42] In fact, the list of conditions, positive and negative, that would have to be invoked would be extensive. As Mill says,

> The cause, then, philosophically speaking, is the sum total of the conditions, positive and negative taken together; the whole of the contingencies of every description, which being realized, the consequent invariably follows. The negative conditions, a special enumeration of which would generally be very prolix, may be all summed up under one head, namely, the absence of preventing or counteracting causes.[43]

---

[41] SOL III.iv.1: 317.   [42] SOL III.v.6: 339.

[43] SOL III.v.3: 332. Someone might suggest that this definition of cause 'philosophically speaking' opens the door to an un-philosophical or everyday sense of cause, *and* that it is only in the everyday sense that Millian causes and laws can compete. I agree with the first conjunct but see no evidence for the second: as far as I can tell, Mill does not take himself to be relaxing the standards of causation when he speaks of the collaboration or interference of causes.

What is going on here? Mill has buried a *ceteris paribus* clause inside the definition of a cause, and hence within the very definition of a law (laws being a subset of causes). To do that, however, is to abandon the notion that a law is a regularity at all. 'Other things being equal, bodies continue to move in a straight line' is consistent with bodies *never* moving in a straight line.

Consider Mill's treatment of the composition of causes. In some cases, as we've seen, it's simple addition: the combined effect is just the sum of the two effects each cause would have produced separately. This is the unproblematic cooperation of partial regularities. But in other cases, there's a 'mutual interference of laws of nature, in which, even when the concurrent causes annihilate each other's effects, each exerts its full efficacy according to its own law, its law as a separate agent.'[44] For example, a stream might flow into a reservoir being emptied by a drain, the result being just as if neither had been in operation.[45] In still other cases, 'the agencies which are brought together cease entirely, and a totally different set of phenomena arise: as in the experiment of two liquids which, when mixed in certain proportions, instantly become, not a larger amount of liquid, but a solid mass.'[46]

Here, we have yet a third sense of law. Mill, knowingly or not, follows Francis Bacon, who claims that each natural kind has a law that dictates how its instances behaves.[47] These laws can cooperate, as in the simple additive case, but they can also block or transform the action of other laws. This third concept of laws is at odds with the other two. That the laws might be the axioms of the simplest system needed to deduce all particular facts sits uneasily next to the claim that there are laws of many different natural kinds, all of which can cooperate or compete against each other. There's no conceptual contradiction here, but it is hard to see how the two notions of law could pick out the same propositions, at least in a world with as many distinct natures as ours.

Nor is it clear how laws that compete for supremacy can be regularities. Cartwright's move against the composition of forces strategy of saving the truthfulness of the laws is to argue that forces are not real; they are not there to *be* composed in the first place. Mill's problem is different: the *causes-cum-regularities* are not there to be composed, or to interfere with each other. Mill is, in short, in the same position as Hume. He can retain the language of competing laws only if he relinquishes the claim that laws are regularities.[48]

---

[44] SOL III.vi.1: 373.   [45] The example is Mill's; see SOL III.vi.1: 372.   [46] SOL III.vi: 373.
[47] See Mumford (2018).
[48] An excellent treatment of Mill's web of laws approach is to be found in Psillos (2002). Psillos argues that the regularity theory can accommodate uninstantiated laws, such as Newton's first law; such

At the risk of belaboring the obvious, let me mount a quick argument to show that Mill's laws cannot be regularities. Suppose cause/law L1 is $[A \rightarrow B]$, where '$\rightarrow$' means that what flanks it on the left is an invariable and unconditional antecedent of its neighbor on the right. Now suppose L2 is $[C \rightarrow D]$. If we get a situation where $A \wedge C \rightarrow B \wedge D$, we are in the clear: L1 and L2 are still regularities, and their causes compose additively. But now take a case like the one Mill describes, where two liquids form a solid. In that case, $A \wedge C \rightarrow F$, where 'F' is a result that is never paired with either A or B alone.

Once again, we face a dilemma. On the first horn, we might tough it out and deny that there ever *was* a regularity captured by L1. On that view, A ceases to be a cause at all. The *true* regularity—and hence the true cause—would need a more complex statement, as, for instance: L3 $[A \rightarrow B$ unless C is co-present with A, in which case $A \wedge C \rightarrow F]$. And of course, once we start playing this game, it will run on for a long time: we will also need to incorporate every background condition to the production of F and B by A and other properties, until we end up with a regularity so complex as to be all but unstatable. Such a move just makes the tension with the axiom-requirement all the more pressing. The simplest assumptions needed to produce the whole course of nature will not include laws of the form of L3.

And in fact Mill agrees, and embraces the second horn. He is not willing to give up on the lawhood of our simple $[A \rightarrow B]$ statement. Instead, Mill insists that the laws function together as a web: 'The regularity which exists in nature is a web composed of distinct threads.'[49] I've argued that this web approach is implicit in the Cartesians. Certainly Descartes himself doesn't think that any one of his laws, on its own, will report on or predict the way things actually go. Only as a group can they be used to predict and explain the course of events. Relaxing the standards for lawhood in this way is an important result, one we should keep on board if we can.

But notice that it *is* relaxing the standard. Keeping the web's threads distinct means giving up on treating them as regularities. At most, the distinct threads—the individual laws—tell you what happens when no other law is operating. But, as Mill insists, nature isn't like that—multiple laws apply to every event, and sometimes even interfere with each other. So the laws are not

---

laws gain entry to the best system if they 'arise in the attempt to account in the strongest and simplest way for actual regularities' (2002, 152). That helps us see how moving away from Humean constant conjunctions toward regularities as universal generalizations can help: we can have vacuously true laws, but only those that fit into the best system. I don't think that goes quite far enough in reconciling the web approach with the regularity theory, since it doesn't account for the conflicts among laws that *do* have instances.

[49] SOL III.iv.2: 318; see SOL III.v.2: 327.

regularities. Taken individually, they don't describe what happens: they describe what *would* happen when no other law is around to interfere.

Mill's web of laws treatment marks a crucial departure from Hume's faith in regularities. The web of laws approach grows out of Mill's second definition of laws as the fewest and simplest premises needed to infer the whole course of nature. But it can't be made to fit with the conception of laws as regularities, or of laws as the laws of individual natures. Mill himself, of course, doesn't see this: he shows no sign of backing off from his initial definition of laws as regularities.

To sum up: some of Hume's and Mill's problems come from the clash between their empiricism or theories of meaning and the need to posit regularities among unobservables. Even if we give up their restrictions, rejecting Hume's copy principle and Mill's 'profound theoretical timidity,'[50] the central tension remains. I'll argue that it persists even in the most sophisticated contemporary refinements of the regularity theory. Before getting there, we need to have the full-dress version of the Best System Analysis before us.

---

[50] As Geoffrey Scarre puts it (1998, 136, quoted in Cobb (2016, 243)).

12

Contemporary Best Systems Analyses (I)

1. Two Masters

I've argued that the ancestors of the Best System Analysis ('BSA') suffer from an internal tension. Despite numerous refinements, or so I'll argue, the tension persists. The two commitments that generate the problem—laws are regularities; laws play the axiom role—are there in the canonical statements of the BSA. Consider David Lewis's characterization in 'New Work for a Theory of Universals':

> Following Mill and Ramsey, I take a suitable system to be one that has the virtues we aspire to in our own theory-building, and that has them to the greatest extent possible given the way the world is. It must be entirely true; it must be closed under strict implication; it must be as simple in axiomatisation as it can be without sacrificing too much information content; and it must have as much information content as it can have without sacrificing too much simplicity. A law is any regularity that earns inclusion in the ideal system .... The ideal system need not consist entirely of regularities; particular facts may gain entry if they contribute enough to collective simplicity and strength. (For instance, certain particular facts about the Big Bang might be strong candidates.) But only the regularities of the system are to count as laws.[1]

Like Mill, Lewis uses a statement's function in science to winnow the regularities: only those regularities that would serve as axioms of the best deductive system, where 'best' is a balance between simplicity and strength, should count as laws. The laws of the contemporary BSA thus serve two masters. On the one hand, they have to be true universal generalizations that state regularities. On the other, they must play the axiom role. Here I mean not just that they have to

---

[1] Lewis (1983, 367). Although one reader (for a journal) has objected that it is 'strange' to try to root the BSA in Hume, I think the lineage could hardly be more clear, from Hume to Mill to Braithwaite and Lewis. This is especially clear on the crucial issue of the nature of regularities; see below, Chapter 13, section 4.

The Metaphysics of Laws of Nature: The Rules of the Game. Walter Ott, Oxford University Press. © Walter Ott 2022.
DOI: 10.1093/oso/9780192859235.003.0012

be axioms of Lewis's best system, but that they have to do what axioms of an actual science would do.

This chapter and the next develop and refine the BSA by subjecting it to a barrage of objections. We begin with those that allow for improvements, or show the view at its best advantage, before moving on to more challenging issues.[2] As we go, we'll gradually see the central tension emerge. At the end of the next chapter, I give a direct argument against the BSA that exploits that tension.

2. Anthropomorphism and Better Systems for Us

The BSA is anti-realist in the stipulative sense: it denies that the laws have enforcers. Laws summarize events; they do not govern them. In a broader sense of the term, it aspires to realism: the tiles of the mosaic are themselves perfectly objective. But if lawhood is relative to a deductive system, we might worry that idealism looms. Consider one of the metrics Lewis proposes for the best system: simplicity. That's a virtue only for creatures with cognitive limitations. A Laplacean demon doesn't care about simplicity, nor would it need to formulate or avail itself of laws.[3] Lewis himself worried about the 'ratbag idealist': 'if we don't like the misfortunes that the laws of nature visit upon us, we can change the laws—in fact, we can make them always have been different—just by changing the way we think!'[4]

Lewis's response is twofold. First, he denies that simplicity and strength are entirely psychological matters: there's some objective sense in which a differential equation is less simple than a linear equation. That seems fair enough, but doesn't touch the main problem: we value simplicity to start with only because of our cognitive limitations. Change our cognitive powers, and you change our laws. His second reply relies on a Kantian story about the project of science: we have to hope that nature will be 'kind,' and that, whatever the

---

[2] I leave aside some other objections, both because I think it's reasonably clear that Humeans have produced good replies to them and because they don't contribute to my overall argument. Notable among them are the objection that Humeanism cannot account for chance: see Lewis (1994), Hall (2004), and Kimpton-Nye (2017), and that it cannot justify induction: see Loewer (1996) and Beebee (2011).

[3] For good treatments of the anthropomorphism objection, see Loewer (2007, 325), to whom my thoughts here are indebted, as well as Dorst (2019a).

[4] Lewis (1994, 479).

standards of simplicity and strength, a single best system will emerge that is so far ahead of any rivals as to be uniquely singled out.[5]

For my part, I don't see the threat of ratbag idealism.[6] Even if lawhood is relative to the creatures for whom the best system is developed, the propositions that count as laws have a totally independent truth value. What is mind-dependent is their status as laws, not as truths. So of course a Laplacean demon doesn't formulate laws to itself. Maybe there's an alien intelligence that finds it easier to think of everything in terms of a duodecimal number system. And on and on. All of these scenarios will feature the same Humean base—the same mosaic—and yet have different laws.

Once we let in the fresh air of pragmatic considerations, another problem— Ned Hall's phony constant—can be dealt with. Note that Lewis excludes non-regularities from lawhood: he allows that facts about, say, the big bang might get into the best system by virtue of the strength they add, but they cannot be laws. It isn't obvious, though, that Lewis is entitled to deny them the mantle of lawhood. Arguing in this vein, Hall introduces the notion of the 'phony constant,' a complete description of the world at a particular time, which would produce massive gains in strength that would outweigh any cost to simplicity.[7] The question then is on what grounds such a constant is barred from lawhood. The broader point is that quite a number of propositions, even those not featuring the phony constant, might earn their place in Lewis's system and yet be all but irrelevant to beings with our epistemic goals and limitations. Making this broader point, Chris Dorst argues that

> an efficient summary of the particular matters of fact is not likely to be particularly useful to creatures like us. For example, an efficient summary may end up giving us a lot of statistical facts, such as facts about the average lifespan of stars in this universe, or facts about the standard deviation of galaxy diameters. Facts like this, interesting though they are, are not partic-

---

[5] Lewis (1994, 479). Note that there are two different issues raised by the ratbag, namely, what do we say about a tie, and what do we say about the importance of our standards of simplicity and strength. Lewis seems to run these together.

[6] In this I agree with Hall (2015), though perhaps for different reasons.

[7] Hall (2015); see also Lange (2009, 57). As Hall puts it: '[c]onsider that the state of the world at any time can be coded up, in a very simple way, by a single real number: just take all the coordinates, masses, and charges of all the particles, expressed in decimal notation, and interleave the digits. Suppose we include this number in a candidate system; then, once again, we get an increase in informativeness that shrinks the set of nomological possibilities down to one. Call this the problem of *the phony fundamental constant*' (2015, 269–70).

ularly useful for predictive purposes, nor do they resemble the dynamical form of actual putative laws in scientific practice.[8]

In response, a recent spate of papers has proposed revisions to the BSA.[9] They suggest changing the criteria of lawhood to create a best system that works for us humans, given our cognitive limitations. Michael Hicks, for example, proposes an 'epistemic role' account, which limits laws to propositions that serve our needs of prediction and explanation.[10] For Dorst, 'the laws are the regularities of the systematization of the totality of the particular matters of fact which is maximally predictively useful to creatures like us.'[11] None of these philosophers relinquishes the requirement that laws state regularities; they only want to use a different selection principle than Lewis.[12]

Such revisions have a further payoff, since they can help us respond to a question first posed by Bas van Fraassen: why think that, even at the ideal end of science when all the facts are in, what the best system counts as laws will match what the science of the day counts as laws?[13] Lewis's laws will be efficient summaries from the God's-eye point of view: they maximize the virtues Lewis lists. But it's unlikely that those summaries will feature in a completed science, simply because scientists are humans who need laws to play the axiom role. On this revision of the BSA, the laws set out by the best system are supposed to be guaranteed to match the laws of a completed physics, simply because it 'reverse engineers' the laws from the standards of physics.[14] A summary from God's point of view might be considerably *less* useful to us than one that is limited in time or space. We have to note that the 'reverse engineering' is not complete: even the most radical revisions to the

---

[8] Dorst (2019b, 2662).

[9] I have in mind here Dorst (2019a) and (2019b), Hicks (2017), and Jaag and Loew (2020). The idea of making lawhood relative to a given science is part of Loewer's (2007) 'package deal account' discussed below.

[10] Hicks (2017).

[11] Dorst (2019b, 2663). Here, Dorst is stating the 'Best Predictive System Account,' which he attributes to Hicks and Jaag and Loew, as well as himself. There are of course some differences among these figures; Jaag and Loew (2020) critique Hicks (2017), for example.

[12] For further evidence of continued fidelity to laws as regularities, see Hicks (2017, 995) and Dorst (2019b, 2663). More radical revisions to the BSA also retain the regularity requirement: examples include the perspectivalist accounts of John Halpin (2003) and Michela Massimi (2018). The exception is Jaag and Loew (2020), who do not want to deny non-regularities lawhood a priori (2020, 2529). But the non-regularities they admit are particular facts, such as the Past Hypothesis (the claim that the initial state of the universe was a very low entropy state) (2020, 2546). As they put it, 'the laws are whatever facts are part of [the] best system' (2020, 2534).

[13] See van Fraassen (1989, 59).   [14] Dorst (2019b, 2668).

BSA retain the requirement that a law state a regularity.[15] And whether physics, ideal or not, cares about regularities is for now an open question.

3. Natural Properties

What the BSA should be looking for, then, is not the best compromise between simplicity and strength, but a system that is best for 'creatures like us.' The revision helps us solve another problem. Just as the BSA has to worry about a mismatch between its laws and those of an idealized completed physics, it also has to worry about a potential mismatch between the properties that are allowed to figure in the regularities and those of physics.

The BSA's core is that laws supervene on the mosaic of property instantiations. So far, we've focused on the laws, but now we need to turn to the other half of the equation. What is the mosaic made up of? Lewis rules out any putative laws that do not refer solely to 'natural' properties, as opposed to abundant or grue-ish properties. 'Naturalness' is a primitive for Lewis; it's meant to be obvious that charge is a candidate and, say, not being funny isn't. Even granting that, why think that Lewis's natural properties will include all and only those quantified over by an idealized science? If the ideal Theory of Everything needs properties that fail to meet Lewis's standards of naturalness, it will fail to be the Best System. And at that point, one wonders why we care about Lewis's Best System in the first place.

In the spirit of the other revisions to the BSA, we should follow Barry Loewer and leave the question of naturalness to the sciences themselves.[16] A property will count as 'natural,' and hence suitable for figuring in laws, just in case the relevant science quantifies over it.[17] In this way, we automatically avoid any mismatch between the properties the BSA counts as natural and

---

[15] There is an exception here: as I've noted, Jaag and Loew (2020) argue for the admission of some particular facts as laws, such as the Past Hypothesis. But the essential form of the BSA is unchanged; Lewis himself might admit such facts as axioms, though he wouldn't call them laws.

[16] Loewer (2007). A different worry is posed by the special sciences: as Fodor (1974) argues, it is highly unlikely that the properties invoked by, say, psychology or economics will be reducible to those of fundamental physics. In response, Jonathan Cohen and Craig Callender propose a 'Better Best System' analysis; see their (2009) and (2010). Following Loewer's 'Package Deal Account,' the Better Best System makes naturalness and lawhood relative to the science at issue. Other isotopes include John Halpin's and Michela Massimi's 'perspectivalist best system' account, which allows for multiple best systems, depending on the perspective one adopts, where a perspective is fixed by the standards of adequacy for laws adopted by scientific communities; see Halpin (2003), Massimi (2017), and Massimi (2018).

[17] Travis Tanner posed an interesting question: what if some science quantifies over a property that is regarded by another science as merely a useful fiction? (Frictional forces might be an example, which

those so counted by the sciences. If the ideal Theory of Everything inelimin-ably quantifies over what Lewis would take to be a non-natural property, then so be it. This move cedes significant metaphysical territory to science, but in a way consistent with the goals of the BSA. We can make this move while still retaining the fundamental notion that the laws are fixed by the mosaic, and not vice versa. This is only the beginning of the proposed revisions to Lewis's ontology, as we're about to see.

4. Humean Supervenience

As Lewis himself recognizes, quantum physics poses a problem for his way of understanding the mosaic.[18] Lewis's 'Humean supervenience' claims that 'all else supervenes upon the spatiotemporal arrangement of local qualities instan-tiated by point-sized things.'[19] There's a lot built into that formula that now seems extraneous. Why, for instance, require that the properties be instan-tiated by point-sized things?[20] That seems to be an unmotivated addition. The core of Humean Supervenience, I think, is the claim that each property instance is metaphysically independent of the others. There are no necessary connections among distinct existences.

I don't see that the BSA as such is committed to even this central part of Humean supervenience. But it's no accident that Lewis holds both. The same impulse animates them, namely, the desire to eliminate mind-independent necessary connections. If the motivation for the BSA is dissatisfaction with N-relations and powers, it abandons the core of Humean supervenience at a significant cost. Once we allow necessary connections among distinct exis-tences, why not choose some kind of nomic realism?

Unfortunately, even the core of Humean supervenience seems to run afoul of the phenomenon of quantum entanglement. Consider a spinless particle

can be treated as merely useful approximations for micro-physical facts.) More broadly, is it possible for different sciences to issue incompatible judgments on which properties are to be quantified over? If so, the package deal might be bundling together incompatible propositions.

[18] In 1986, Lewis had already seen the problem; at that point, he was 'not ready to take lessons in ontology from quantum physics' (1986a, 11).

[19] Lewis (1999, 4).

[20] As Elizabeth Miller notes, '...Lewisians need not say the only facts are facts about the qualities of point-sized individuals and their arrangement. In fact, Lewisians who believe some (perhaps all) collections of space-time points have mereological fusions certainly grant that there are non-pointy individuals. Rather, the key claim is that facts about these other individuals supervene on facts about the qualities and arrangement of point-sized elements of the mosaic: there can be no difference in any facts obtaining at two worlds without some difference in the qualities instantiated at or geometrical relations among space-time points in these worlds' (2015, 1313).

that decays into two. Spin is conserved, so if one of the two particles has spin-up, the other must have spin-down. The two particles are 'entangled' in the sense that knowing the spin of one tells you what the spin of the other will be. There is more to the phenomenon of entanglement, but even this feature is enough to cause problems for Humean supervenience, for the spin states of the resulting particles violate the claim that instantiations of properties are metaphysically independent of all the rest.[21]

One possibility is simply to abandon Humean supervenience. We can still preserve the idea that lawhood is relative to a best system, even while rejecting the independence of property instantiations.[22] But by admitting necessary connections among discrete property instantiations, I worry that we give away too much to the realist. Lewis himself claimed his Humean lineage on the basis of the denial of such connections: the leading idea is to have laws without policemen. There are no special entities that go around enforcing the laws. If we admit that quantum entanglement requires the introduction of such necessities, then it seems to me we've arrived at a hybrid powers view. The result is not a powers view simpliciter, since the laws would determine what powers or necessary connections there might be, not the other way around. Still, it makes sense to consider whether the Humean has other options.

If we really want a world of laws without enforcers, we should reject any attempt to posit new metaphysical machinery behind entanglement. A latter-day Hume should be unimpressed by claims of robust metaphysical dependence among properties. Quantum mechanics has shown us a new and important correlation, one that holds, not between hitting a billiard ball and the ball moving, but between spin states of discrete particles. But it's no different in kind from the whole mass of correlations we can see elsewhere in the world. Unlike the motion of billiard balls, entanglement might have a better claim to being a regularity, but that depends on how the physics comes out. Either way, entanglement shouldn't cause us to abandon the metaphysical independence

---

[21] The origin of the concept of entanglement appears to be Schrödinger (1935), but it is significantly developed in Bell (1964). I make no claims to expertise in quantum mechanics. My sketch of entanglement here relies largely on P. J. Lewis (2016), Rickles (2016), and Healey (2017).

[22] The resulting position would be similar to Loewer's package deal account, which not only lets science decide what the natural properties are but also defers to science on the question of necessary connections, and indeed on whether properties are dispositional or categorical. But note that Loewer doesn't take the necessities permitted by the package deal account to be problematic from the Humean perspective. As I understand his view, the claim is that we can construe mass in Newtonian mechanics, say, as either categorical or dispositional. If we take it to be dispositional, that's because we're defining it in such a way that it necessarily conforms to Newton's laws. But that's a kind of stipulative necessity, not a reflection of the way the world is independently of our scientific theory See Loewer (2007), (forthcoming a), and (forthcoming b).

of the properties of discrete objects. Alternatively, someone might argue that entangled states can no longer be treated as distinct states at all, but only as a whole.[23] If that's the case, then the possibility of necessary connections *between* distinct objects or properties vanishes.

What, then, is the right response? I think the proponent of the BSA should let the properties that have to be quantified over be relative to, and fixed by, the science in question; they should come along with the laws of that science as part of the package. Whether the chosen properties are instantiated in point-sized things or not can be left up to the particular disciplines. But we shouldn't defer to science on the metaphysical nature of the relations among these properties. The experimental evidence can of course show us new correlations, which have to be captured by the laws; but that doesn't mean that that evidence is itself evidence of any metaphysical police force. So I propose a modest version of Humean supervenience: with the stipulations just mentioned, all else supervenes on the spatio-temporal arrangement of properties, none of which is necessarily connected to the others.

Now that we have a better idea of what the Humean mosaic has to look like, we should ask about the other end of the supervenience relation we're concerned with: the laws. Once we make lawhood relative to our epistemic goals and cognitive limitations, what becomes of that supervenience claim? The revised BSA has to deny

Nomic HS: the laws of a given world supervene on the mosaic (the spatio-temporal arrangement of properties).

Two worlds can share a mosaic and yet have different laws depending on who's creating the best system for predicting and unifying them.[24] And now we seem back to the ratbag idealism Lewis himself is so anxious to avoid.

But there's a solution that doesn't require much violence to the supervenience thesis. In explicating Lewis's original view, Ned Hall introduces the 'limited oracular perfect physicist' or 'LOPP.'[25] She's limited but oracular in

[23] As Schrödinger puts it, two entangled systems 'can no longer be described in the same way as before, viz. by endowing each of them with a representative of its own' (1935, 555). (Thanks to Travis Tanner for the pointer.)
[24] Michela Massimi's perspectivalism explicitly rejects supervenience; on her account, lawhood is relative to the perspective of the science at issue. There is of course much more to her view; see esp. Massimi (2018).
[25] Hall (2015, 265). Hall uses the idea of a LOPP to bring out one of the 'guiding ideas' of the BSA, and traces it to remarks like this from Lewis: 'I take a suitable system to be one that has the virtues we aspire to in our own theory-building, and that has them to the greatest extent possible given the way the

knowing all (but only) facts about the mosaic, that is, the spatio-temporal arrangement of property instantiations. She's a physicist, and a perfect one, because she's ideally able and situated to formulate and evaluate hypotheses about the laws of physics relative to that evidence. On the original BSA, the laws are just whatever the LOPP says they are.

Now, to accommodate our revisions, we need to stipulate that the LOPP has the epistemic goals and, *modulo* her omniscience with respect to particular matters of mosaical fact, the same cognitive limitations we do. Although we can consider laws relative to a Martian physicist, one with totally different epistemic interests and cognitive limitations, we'll keep the LOPP as human as possible. So we end up with

Human LOPP-relative Nomic HS: the laws of a given world supervene on the mosaic (the spatio-temporal arrangement of properties), and are relative to the epistemic goals and cognitive limitations of a recognizably *human* LOPP.

Given our prior revisions to the BSA, it makes no sense to look at an uninhabited world and ask what laws hold there. To answer that question, we need to fill in the blank: laws *for whom*? So when we consider an uninhabited world, we can only get an answer to our questions about laws by stipulating the interests and cognitive limitations of the law-formulators. It's fine if we want to project ourselves into that situation, of course, and ask which regularities our LOPP would single out for lawhood. But it will be important that we're aware of doing so.

How much objectivity does Human LOPP-relative Nomic HS give away? Not much, and nothing of value. Once you fix the standards and creatures at issue, which regularities count as laws is a perfectly objective matter. We have to hope, with Lewis, that nature will be kind, and provide enough tiles of the mosaic to single out one system as better than all its competitors; but the original BSA already had to live in hope in that respect. More important, which regularities there are to begin with is perfectly objective, regardless of whether anyone, or who, is around to come up with a system of laws to summarize them.

<hr />

world is' (1983, 41), quoted in Hall (2015, 264). Hall introduces the idea of a LOPP to explain, not revise, Lewis's own view. The revisions come when we stipulate features of the LOPP, especially her epistemic needs and interests.

5. The Underdetermination Argument Redux

Earlier in the book, we brushed up against a prominent motivator for top-down views, what I called the 'underdetermination' argument. It features in the work of primitivists such as Carroll and Maudlin, where it's known as the 'argument from scientific practice.' As far as I can tell, that argument originates in the work of Barry Ward, who argues for projectivism, a view we'll consider below. Since the argument is directed squarely against the supervenience claims distinctive of the BSA—a common enemy of both projectivists and primitivists—we'll need to evaluate it more fully here.

The precise version of supervenience the BSA is committed to turns out to be controversial. I've argued that human-LOPP-relative Nomic HS is the best option, and the one needed if we are to accept the 'Humeanism with a human face' revisions to the BSA. On that view, the laws are relative to the epistemic goals and cognitive limitations of a recognizably human ideal observer. Since Ward's argument antedates these revisions to the BSA, he takes a different version of supervenience as his target: 'no pair of worlds like ours can differ without differing in respect of the arrangement of qualities.'[26] I'll present his argument in its original form before extending it.

Ward presents a stripped-down world w. Some parts of w have M-fields, others don't. It has particles that are negatively or positively charged, or neutral. It also has E-particles, with the same mass and charge as electrons in our world. In w, only E-particles ever enter M-fields, and when they do, their trajectory shifts: the E-particle no longer goes in a straight line but shifts to a path that rotates counter-clockwise.[27] Now we ask: *why* does the E-particle's trajectory shift? Nothing about w tells us that. To answer it, we'd need to know what would have happened in w, had some other kind of particle entered the field. Do *all* particles entering M-fields change their trajectory in this way, or just some, or just E-particles?[28] I can't explain what happens to the E-particle if I don't know what would happen to *other* particles when they enter the M-field.

World w provides no means of choosing among these explanations, because it doesn't by itself fix which laws hold in it. So nomic supervenience is false: a single physical state 'may fail to determine' which of a competing set of

---

[26] Ward (2002, 193).

[27] I'm glossing over some details here; for the full thought experiment, see Ward (2002, 195).

[28] Ward is making an important point here about explanation: explaining an event often involves contrasting it with counterfactual scenarios. We've encountered this point above (Chapter 2, section 3) and will return to it below (Chapter 13, section 2).

explanations, each featuring its own laws, is correct.[29] We can come up with an indefinitely large number of competing laws, each of which is consistent with the non-nomic facts of world w. Even if we build in our human LOPP-style revisions, the supervenience claim fails: our LOPP can do nothing but throw up her hands, concluding that multiple sets of laws are equally consistent with the mosaic of w. In later work, Maudlin and Carroll deploy the same argument; it becomes the 'argument from scientific practice.' To merit the new name, Maudlin adds that scientists routinely treat worlds such as w as models of different sets of laws.[30]

Lewis himself foresees the possibility of mini-worlds such as Ward's w. Recall Lewis's response to the worry that the mosaic might fail to select a single system as the best: '[i]f nature is kind, the best system will be robustly best-so far ahead of its rivals that it will come out first under any standards of simplicity and strength and balance.'[31] Lewis admits that there's no guarantee of that. All we can do is hope that, when all the evidence is in, a single best system will emerge victorious. The worlds Ward and Maudlin describe are worlds in which nature is extraordinarily unkind, and no system is able to defeat its competitors in the simplicity-strength contest. So Lewis already allows that some worlds—including, in the worst case scenario, our own—fail to pick a single best system.

Does that mean that nomic supervenience fails? I don't think so. Above, I suggested that if the underdetermination argument worked, it would prove too much: it would undermine many a supervenience claim that is independently plausible and has nothing to do with laws of nature. To see this in more detail, consider an exchange between two unusually reflective but workaday magazine illustrators of the 1980s, Lou and Al.

AL: Funny how I can make an illustration of a cat just by putting ink on the paper. All you need to do is suffuse different regions of the paper with black, and leave others alone, and blammo—cat appears.

LOU: So you're saying that the distribution of ink fixes what the drawing's a drawing of?

AL: Tell me you didn't read that Putnam stuff about the ants and Winston Churchill.[32]

LOU: No, my point is different. Look. (Draws two dots on a piece of paper.) What's that a drawing of?

---

[29] Ward (2002, 196).   [30] See Maudlin (2007, 67) and Carroll (2018, 124).
[31] Lewis (1994, 479).   [32] Putnam (1981).

AL: I don't know. Could be anything.

LOU: Aha! So the arrangement of ink on the paper doesn't determine what the image represents.[33]

AL: You just started, though. You have to finish it.

LOU: You're the one who's saying that the ink on the page fixes what the drawing's of. This could be a drawing of two eyes, or the edge of a dragon's tail, or...

LA: I get it. What's your point?

LOU: Your supervenience claim is just false. If it were true, my two dots would have to be a drawing of something or other in particular. And it isn't. Q.E.D.

AL: Are you putting me on?

Al is right to be incredulous. How does Lou think they do their jobs as illustrators, if not by putting ink to paper? If the underdetermination argument worked against Ward's version of HS, it would *also* work against the supervenience of the representational objects of pictures on the material they're made up of. Now, there are lots of ways to challenge that supervenience claim (as Al's mention of Putnam shows), but this one isn't any good. If the underdetermination argument worked against nomic supervenience, then, it would generalize. But it proves too much: it shows that images aren't fixed by the arrangement of colors that make them up. So we shouldn't give up on supervenience so quickly.

If we want a more formal reply, we need to sharpen up what nomic supervenience is committed to. I'm accepting the revisions to the BSA, which require us to use

Human LOPP-relative Nomic HS: the laws of a given world supervene on the mosaic (the spatio-temporal arrangement of properties), and are relative to the epistemic goals and cognitive limitations of a recognizably human LOPP.

But what does it mean to say that some facts supervene on others to begin with? Although there are nearly as many kinds of supervenience as there are philosophers who've written about it, the core idea is that B-facts supervene on A-facts just in case there is no way to wiggle the B-facts except by wiggling the A-facts. Duck-rabbits aside, you can't make a picture of a brick wall into a

---

[33] Lou's point here has something in common with the so-called paradoxes of rule following, discussed by 'Kripkenstein' in Kripke (1982).

picture of a dog wearing a hula hoop without changing some of the ink on the page.

That doesn't entail that fixing a single A-fact is enough to fix one or more B-facts, which would mean that even a single pixel fixes an image. What we should care about instead is the core idea: the laws are responsive to, *sensitive* to, the tiles of the mosaic. If we keep the human LOPP constant, there's no way to change the laws without changing the mosaic. And even if we don't—if we consider the perspective of some other kind of LOPP—there's no way to change the regularities from which the laws are selected without changing the mosaic.

6. Counterfactuals

Our next set of objections all revolve, one way or another, around the notion of counterfactuals. Let's start by taking the issue head-on, and then see how solutions to it can help with other problems the BSA encounters.

The *locus classicus* of attacks on the regularity theory must be Armstrong's *What is a Law of Nature?* There, Armstrong points to a central difference between the realist and the regularity theorist. On the universals view, a law of the form N(F,G) sustains or supports a related counterfactual, namely, had this x which is not-F been an F, it would also have been a G. Now, if the laws are contingent, as Armstrong thinks, the counterfactual needs to be qualified. If we go to a world where N(F,G) doesn't obtain, then of course it's no longer true that had this x been an F, it would also have been a G. Still, Armstrong insists, it should be obvious that the N-relation underwrites or 'ensures' the truth of the counterfactual. Equally obvious is that no mere regularity can.[34] Suppose everyone in a certain building has been, and always will be, wearing shoes. That doesn't suggest that had you tried to enter barefoot, you would find yourself resisted by a supernatural force, or that, on crossing the threshold, shoes would magically appear on your feet.

There are enormous complexities in the analysis of counterfactuals.[35] For our purposes, we can look at the two main contenders: one truth-functional (the Lewis/Stalnaker possible worlds view) and one not (the Ramsey/Edgington supposition theory). Here's the core idea of the supposition theory, from Frank Ramsey in 1929:

---

[34] Armstrong (1983, 50).
[35] As a glance at William Starr's excellent *vade mecum* (2019) shows.

If two people are arguing "If p, will q?" and are both in doubt as to p, they are adding p hypothetically to their stock of knowledge, and arguing on that basis about q; ... they are fixing their degrees of belief in q given p.[36]

Imagine that we're wondering whether we could have gotten the novel coronavirus from an unwashed bag of Doritos, although neither of us has handled it. We suppose that we have in fact handled the bag and then try to work out what would've happened. There's no way to do that, of course, without appealing to our other beliefs, say, about the ability of various materials to carry fomites. The supposition theory says that we keep these other beliefs constant and hypothetically add the handling of the bag to them.[37] Now, the supposition theory only tells us about counterfactual beliefs. It doesn't treat assertions of counterfactuals as assertions of propositions at all, whether true or false, but, as Dorothy Edgington puts it, as statements 'of the consequent under the supposition of the antecedent.'[38]

The Lewis/Stalnaker approach aims to go further, and provide a semantics for counterfactuals.[39] The basic idea is to evaluate a counterfactual of the form $P{\rightarrow}Q$ by looking at the nearest P-world and asking whether Q is true in it. If it is, our counterfactual is true.

What does 'nearest' mean? We can pull out two notions:

Nomic proximity: Worlds w_1 and w_2 are nomically close just in case the same laws obtain at each.

All-things-considered proximity: w_1 and w_2 are all-things-considered closer than any other pairs just in case they share more tiles of their mosaics than any other pairs.[40]

---

[36] Ramsey (1990a, 155), quoted in Edgington (2020).

[37] For a defense of the regularity theory's account of counterfactuals using supposition theory, see esp. Psillos (2014, 23).

[38] Edgington (2008, 2). This line comes from her statement of the supposition theory of conditionals, which she goes on to apply to counterfactuals.

[39] As Stalnaker puts it, 'the problem [for supposition theory] is to make the transition from belief conditions to truth conditions...The concept of a *possible world* is just what we need to make the transition, since a possible world is the ontological analogue of a stock of hypothetical beliefs...[Here is] a first approximation to the account I shall propose: Consider a possible world in which A is true and otherwise differs minimally from the actual world. *"If A, then B" is true (false) just in case B is true (false) in that possible world* (1968, 33-4), quoted in Edgington (2020).

[40] Here is Lewis's (1979, 472) 'default' ordering considerations in determining closeness of worlds:

'1. It is of the first importance to avoid big, widespread, diverse violations of law.

2. It is of the second importance to maximize the spatio-temporal region throughout which perfect match of particular fact prevails.

Worlds can be nomically close without being all-things-considered close. There might be a world in which humans never evolve language (say, because they die out too soon), with all the differences that would entail, that is nevertheless predictable by precisely the same laws we use in our own. Of course, it would have to have different initial conditions, under the assumption of determinism. By the same token, worlds can be all-things-considered close but differ in their laws. If they differ nomically in virtue of the regularities they feature, they must be out of step with respect to at least one tile of the mosaic.

Let's start with the Lewis/Stalnaker approach and see what answer the BSA can make to Armstrong. The putative counterexample presents us with a regularity—everyone in a given building wears shoes—that fails to support relevant counterfactuals. Now, it would indeed be bad to insist on mysterious agencies enforcing requirements for donning footwear. But our example doesn't involve anything like a regularity that is suitable for lawhood. So let's work with a more damaging counterexample. Suppose our regularity is something like 'heavy unsupported objects near the Earth's surface fall such that they accelerate at a rate of 9.8 meters per second per second.' (Even that, of course, will be a derivative regularity. And note that I'm asking the reader to make this assumption; I don't claim that it's actually true in our world. This lets us focus on the issue of counterfactuals without getting bogged down in the difficult question whether there are indeed many, or any, regularities to be found.)

Consider the counterfactual: had I dropped my bowling ball on my foot instead of sending it down the lane, it would have accelerated at 9.8 meters per second per second. To evaluate the counterfactual, we need to look at the closest possible world. But not just any similar world will do: we have to find the closest one with the same laws. At any such world, the rate of fall will be just the same. So our counterfactual is true: any world that shares the same laws as ours will be such that a dropped bowling ball near the Earth's surface will behave exactly as described. There's an important caveat here: the nearest nomically identical world will have to differ from ours in at least one way. Namely, there will have to be a 'small miracle' in order for the bowling ball to drop (assuming determinism). If the histories of the two worlds are identical up to that moment, there is no way, absent such a miracle, for the ball to drop in one and not the other. Of course, since there's no metaphysical glue tying

---

3. It is of the third importance to avoid even small, localized, simple violations of law.
4. It is of little or no importance to secure approximate similarity of particular fact, even in matters that concern us greatly.'

one moment to the next, this isn't a real miracle; it's just an exception to an otherwise exceptionless generalization.

Even if a realist declines to pursue the technical difficulties that lurk here—especially the question of how to gauge with any precision the proximity of possible worlds—she won't be satisfied.[41] The whole procedure for squaring the regularity view with counterfactuals has the air of the purely stipulative: we get it to come out right by rigging things up so that the only relevant worlds are those that, by definition, agree with the actual world in their respective laws. And in such worlds, of course it's true that the bowling ball accelerates at 9.8 m/s/s.

In the same vein, Stuart Glennan argues that the regularity theory makes the truth of a counterfactual depend on states of affairs entirely foreign to the particular case at issue. On the regularity theory, Glennan writes, 'the facts upon which a singular counterfactual (like the fact that if I were to flip the switch I would turn on the light) depends are facts not just about the particular instance but about the whole class of particulars.'[42] But even in the actual world, the regularity theorist thinks a single case of causation depends, for its status as a case of *causation*, on what happens in the rest of the mosaic. So she has already learned to live with facts external to a single case determining the nature of that case.

To the regularity theorist, this kind of support for counterfactuals is all that is to be had. And to ask for some further truthmaker for them is not dialectically powerful. If the starting point for the regularity theorist is the denial that there is any glue binding distinct events, it can hardly be a surprise that her story about counterfactuals will seem insufficiently robust to a realist. So I think the BSA emerges unscathed from Armstrong's attack.

If we go with Ramsey/Edgington supposition theory, the reply is even more straightforward.[43] If I've observed a pattern such that all Fs are Gs, and I then

---

[41] As Starr (2019) notes, Alan Hájek (2014, 250) critiques Lewis's similarity measures on the grounds that they have no bearing on, or grounding in, scientific practice: 'science has no truck with a notion of similarity; nor does Lewis's (1979) ordering of what matters to similarity have a basis in science.'

[42] Glennan (2009, 319).

[43] As Psillos puts it: 'The supposition view takes it that counterfactuals are not truths about possible words but ways to express an attitude towards a possible state of affairs made within the scope of a supposition. The notion of the pattern strengthens this view since it can ground the thought of the 'neck-sticking-outness' of regularities. The presence of the pattern in a regularity grounds the supposition that if there were a further X, it would be Y. For the addition of the supposition that there is an extra F does not defeat the grounds for accepting the presence of the regularity insofar as there is a pattern in it' (2014, 23).

ask whether, had some particular *a* been an F, would it also have been a G, then I have to answer 'yes.' Given that I believe that all Fs are Gs, nothing about the supposition that *a* is F gives me reason to abandon my universal generalization. Again, it's true that there's no metaphysical backing, no enforcer of the F–G connection, so the realist will remain unsatisfied. But this doesn't seem to me sufficient grounds to reject the regularity theory.

13

Contemporary Best Systems Analyses (II)

1. Mirrors and Nomic Stability

Working through the Humean approach to counterfactuals pays dividends when we turn to our next objections to the BSA. As we saw in Chapter 4, the mirror argument provides a central motivation for competing, top-down views. Some proponents of the BSA claim that the argument rests on illegitimate intuitions about governing, and simply bite the bullet. But the mirror argument's gravamen is nomic stability: laws ought not to collapse under at least some variations of initial conditions. In this section, I'll argue that the BSA must—and can—do better than dig in its heels. We should leverage our account of counterfactuals and muster a reply to the threat of nomic instability.

A brief recapitulation of the mirror argument: In U_1, two particles, a and b, enter a Y-field and acquire spin-up. Even on the Humean view, it's a law (L1) that all particles entering Y-fields acquire spin-up. U_2 is just like U_1, except that particle a misbehaves: it doesn't acquire spin-up when it enters the Y-field. So (L1) is not a law in U_2, because it's not even true. In both worlds, there's a mirror standing by which, were it swiveled, would deflect the a particles.

Now we consider the starred worlds, in which the mirror is swiveled and prevents the a particles from entering the Y-field. It stops a from taking the Y-field 'test.' In each one, particle b enters a Y-field an acquires spin-up,' while particle a bounces off into space. What should we say about U_1^*? It's natural—or so the argument goes—to think that L1 is still a law in U_1^*. Just swiveling the mirror shouldn't change the laws, and a's failure to take the test doesn't mean a wouldn't pass it. Conversely, it's natural to think that L1 is *not* a law in U_2^*. L1 wasn't a law in U_2, and so it shouldn't be in U_2^*, either. If we accept all this, then we have a failure of supervenience. We have two worlds, U_1^* and U_2^*, that are indistinguishable in their non-nomic respects. They differ from their unstarred counterparts only by virtue of the position of the mirror that does the deflecting. Nevertheless,

The Metaphysics of Laws of Nature: The Rules of the Game. Walter Ott, Oxford University Press. © Walter Ott 2022.
DOI: 10.1093/oso/9780192859235.003.0013

236 CONTEMPORARY BEST SYSTEMS ANALYSES (II)

they are supposed to have different laws: L1 is a law in U_1^*, as in U_1, but not in U_2^*.[1]

We might feel we face a standoff. The Humean just does not agree that the laws are stable under these variations, however minor. The laws summarize what happens, and if you change what happens, you can change the laws. To extend Beebee's 'God's big book' metaphor: the Cliff's Notes version of *Moby Dick* would look very different if Ahab didn't lose a leg. Depending on how detailed the summary—or the propositions it entails—turns out to be, it might even be different if Ishmael spent one of his evenings crocheting a baby blanket in the fo'c'sle. This is the kind of reply we've seen Helen Beebee make: simply deny that U_1^* and U_2^* differ nomically.[2] The position of the mirror changes the mosaic and thereby the laws, so that the two worlds end up with the same non-nomic, and hence nomic, facts. The thought that they must differ is merely a reflection of the outmoded intuition that laws govern.

I don't think we can leave matters there. I've argued that the mirror argument needn't rely on intuitions about governing. The real issue is nomic stability, and that is not an outmoded theological intuition. Nomic stability is required if laws are to play the roles we expect them to play. In everyday life, and in scientific practice, we need to consider how changes to the variables we put into our nomic equations will affect the outcome. If laws aren't stable across minor changes like the swiveling of a mirror, then this practice is threatened: in changing the inputs, we are also changing the laws. But we expect the laws to tell us what *would* happen—and indeed what *will* happen—when those inputs are different. If the laws are so fragile, then we cannot rely on such calculations. Whatever else they are, laws should be invariant over whatever we choose to take as our initial conditions.[3]

---

[1] Carroll gives two versions of the argument: the one I have presented here and in Chapter 2, and a more formal one. I think Susan Schneider (2007, 313) is quite right when she directs our attention to the informal version; the formal one, as Roberts points out, depends on a notion of physical possibility the Humean would reject. The principle is this: 'P is physically possible in W iff there is a world having exactly the same laws as W, in which P is true' (Schneider (2007, 312)). Roberts (1998, 428) argues that no Humean would accept this principle, since P, no matter how insignificant it might otherwise seem, might result in very different laws. Instead, as Schneider puts it, the Humean ought to say that 'P is physically possible in a world W iff P is logically consistent with the truth of all the laws of nature in W' (2007, 312). But Carroll has to rely on the stronger principle to derive his modal principle 'SC' and its relatives. Put in English, the relevant versions of SC say that if P is physically possible, and Q is (is not) a law, then Q would still (not) be a law if P were to be true. Like Roberts, Beebee (2000, 588) argues that no version of SC is true on Lewis's view.

[2] Beebee (2000, 590). For more discussion, see above, Chapter 4, section 3.

[3] As Woodward puts it: '[w]e should formulate individual laws in such a way that they are invariant not just over initial conditions...but under changes in other variables as well—e.g., changes in the colors or shapes of the masses in particular gravitating systems or other sorts of "background" changes' (2018, 163).

So the Humean needs a better reply to the mirror argument, because she needs a way to re-capture nomic stability. We have to recover the nomic differences between U_1^* and U_2^*, despite their qualitative identity. My strategy is this: rather than treating U_1^* and U_2^* as simply given, we have to recognize that whatever nomic difference we think obtains between them is a result of considering them *relative to* their unstarred worlds. That suggests that we insist on treating them as counterfactuals, to be evaluated relative to their 'home worlds.' If the strategy works, we'll be able to say that in fact U_1^* and U_2^* *do* differ nomically, despite being qualitatively identical. The difference comes, not from any intrinsic variation between the two worlds, but from the point of view we adopt when we evaluate those worlds. We're taking that point of view into account when we set up the thought experiment. Consider how Carroll motivates the idea that the two worlds differ nomically in the first place: he has to appeal to their origins as variations of their home worlds. I'm willing to believe that U_1^* has the same laws as U_1 *only* because I'm taking U_1^* as a variation on U_1. My strategy asks us to consider these home worlds as the perspective from which the starred worlds ought to be evaluated.

We've looked at two main competitors: the Lewis/Stalnaker view and the supposition theory. If we're dealing with the possible worlds semantics, we should insist on taking the relevant unstarred world as actual. If I'm in U_1, and L1 is indeed a law, what should I say about a world in which the mirror is swiveled? I need to find a world in which the mirror is swiveled *and* a world that agrees with U_1 in its laws. So, trivially, U_1 and U_1^* don't differ nomically. The same goes for U_2^*. If I'm in U_2, where L1 is not a law because it's not true, I need to find the closest possible world in which the mirror is swiveled *and* the laws are the same. So L1 is *not* a law in U_2^*.

Even on this metaphysically rich way of evaluating counterfactuals, we have no choice but to project our own beliefs on to modal space. On Lewis's view, this process takes the form of selecting which features of similarity to count, and how to weigh them against each other. Stalnaker is more explicit; in place of Lewis's metrics of similarity, he offers the 'projection strategy': 'we should try to understand conditional propositions in terms of a projection of epistemic policy onto the world.'[4] Stalnaker is at pains to insist that such projection is compatible with realism about necessities, as well as causal powers; but projection it remains.[5] As Roberts puts it, the Lewis/Stalnaker approach grants

---

[4] Stalnaker continues: 'Conditional propositions should be understood as propositions about features of the [actual] world which justify certain policies for changing one's belief in response to potential new information' (1984, 119).

[5] See Stalnaker (2019, 223).

that 'counterfactuals are made true (at least in part) by our own counterfactual attitudes.'[6] Only by ignoring the role of those attitudes in evaluating the various scenarios can the mirror argument seem to threaten the nomic stability of the BSA's laws.

An even clearer way out would be to adopt supposition theory. On the supposition view, to say that 'had this particle encountered a Y-field, it would have acquired spin-up' is not to report on some possible world but to make a judgment about a supposed state of affairs. Were we in U_2 and asked about the counterfactual world U_2^*, we would of course say that its laws are those of U_2, and just so for U_1 and U_1^*. So the difference between U_1^* and U_2^* is not to be sought in the space of possible worlds, but in the attitudes toward them provoked by considering U_1 and U_2, respectively, as actual.

If I'm right, the BSA's only chance of re-claiming nomic stability lies in pointing to the inevitable entanglement of our own beliefs and epistemic practices in evaluating counterfactuals. The key move is to refuse to take U_1^* and U_2^* as noumenal waifs, awaiting nomic analysis. Instead, we should treat them as worlds considered counterfactually *relative to* their home worlds. We might say that the mirror argument's sleight of hand consists in first appealing to the connection between the original worlds and their starred counterparts and then ignoring that very connection.

2. Explanation—Case

We've distinguished local from global demands for explanation:

(Case)   What is it for a law of nature to explain its instances?

(Whole)   Under what conditions does a theory of laws promise to explain the whole array of property instantiations?

Let's begin with case explanations. I've argued that the thin concept of laws requires that laws be able to explain their instances in the sense of assimilation, or unification. In the tradition of the regularity theory, this has taken the shape of the deductive-nomological model: you plug in your laws of nature and a given empirical fact (say, that a pot of water has reached 100 degrees at sea level), and you deduce a new empirical fact (the water boils).[7] Our first

---

[6] Roberts (2008, 339).
[7] Lewis himself rejects the deductive-nomological model; see Lewis (1986b, 2: 236–8), discussed in Skow (2016, 49–52). For a recent endorsement of the model, see Loewer (2019, 378).

question is whether the regularity theory can give us even this kind of case explanation.[8] Armstrong argues that the answer is 'no,' since any such explanation will be circular.

Take some particular set of observed Fs and Gs such that all Fs are Gs. Why are the observed Fs also Gs? All the regularity theorist can say is, 'it's because *all* Fs—not just the observed ones—are Gs.' But since the observed Fs that are Gs are themselves constituents of the general fact that all Fs are Gs, to appeal to the unrestricted generalization to explain the observed cases is to go in a circle.[9] Armstrong's argument turns on the assumption that no state of affairs can be explained by a further state of affairs of which it is itself a part. As philosophical principles go, that one doesn't wear its justification on its face. Perhaps nothing explains itself; but that's not the assumption at issue.[10]

Even if we accept the principle, we have to disambiguate 'explanation.' We've seen Berkeley insist on the distinction between metaphysical and scientific explanation. For him, to explain an event metaphysically is to display its efficient cause; in his case, that cause will be either God or a finite mind. But '[i]t is not . . . the business of physics or mechanics to establish efficient causes, but only the rules of impulsions or attractions, and in a word, the laws of motion.'[11] Although Berkeley is no regularity theorist, his distinction resurfaces in the work of Barry Loewer.[12] As Loewer acknowledges, there's no easy way to draw the distinction in the contemporary context, but we can point to some key differences. From the point of view of the BSA, scientific explanation is a matter of unifying events, seeing how they fit into patterns we call 'laws.' Metaphysical explanation needn't involve laws at all, and is often a matter of showing how some property supervenes on, or is reducible to, some more fundamental property.

The distinction is crucial to de-fanging Armstrong's objection. Suppose we want to explain why all observed Fs are Gs. The scientific explanation might be that *all* Fs are Gs, whether they are observed or not. True, the observed Fs and Gs are a subset of all Fs and Gs. But that's no barrier

[8] Not every regularity theorist accepts the claim that laws explain events. In his 1822 *Inquiry into the Relation of Cause and Effect*, Thomas Brown writes, '[W]hen we speak of the Laws of Nature, indeed, we only use a general phrase, expressive of the accustomed order of the sequences of the phenomena of Nature. But though in this application, the word Law is not explanatory of anything, and expresses merely an order of succession which takes place before us, there *is* such a regular order of sequences, and what we call the qualities, powers, or properties of things, are only their relations to this very order' (1835, 83). This view fits with Brown's emphasis on causes (construed in Hume's fashion) rather than laws.
[9] See Armstrong (1983, 40).
[10] In any case, I'm not convinced nothing explains itself. Possible candidates for self-explanation are analytic necessary truths.
[11] *De Motu* §35, in Berkeley (1975, 218).   [12] See esp. Loewer (2012) and (2019).

to scientific explanation—it's a presupposition of it. If the observed Fs and Gs were *not* a subset of all Fs and Gs, we wouldn't have the unification that makes for scientific explanation.

We should look at what I think is a much more common kind of case. Instead of explaining the observed by the unobserved, scientific explanation often involves relations among observed properties or variables. Suppose I see some apparently novel series of events, call them A–B events. Suppose I also know, quite independently, of a much more widespread regularity, D–E. Suppose I then come to realize that the A–B sequence is relevantly similar to, and in fact subsumable under, the D–E sequence. Prior to this realization, I did not include A–B in D–E; that's why it was a realization. True, *after* my realization, I lump A–B in with the D–E sequence. That is what produces the illusion that a state of affairs is helping to explain itself. But it *is* an illusion: the explanation was achieved by detecting the very similarity that allows me to erase any distinction between what I initially called the D–E and A–B sequences. When Newton (to simplify the story) realized that terrestrial gravitational phenomena were of a kind with their celestial counterparts, he was not illegitimately using terrestrial phenomena to explain themselves, even though *after* his realization, he classed them all together in the laws. What else do you want him to do—insist that terrestrial phenomena are *not* of a piece with the celestial? That would just be to fail to explain the terrestrial phenomena at all.

If we ask instead after the metaphysical explanation for a particular instance of a sequence, we're brought up short. To say that it's explained by the whole F–G sequence is to give a scientific or unifying explanation, not a metaphysical one. Of course, the particular sequence is part of the overall sequence, and in that toothless mereological sense, it is 'metaphysically explained' by something larger than itself. I say 'toothless' because it seems perverse to insist that one individual staple, say, is 'explained' by the pack of staples of which it is a part. So we might say that there's no satisfying metaphysical explanation for the particular instance of the F–G regularity. But I don't see why the BSA should be bothered by that. Part of the point of the BSA is that there just *is* no deep metaphysical explanation for why a given regularity holds. There might be a deep *scientific* explanation, if we can subsume that regularity under another, broader one. But if the demand is for some force or power or N-relation that guarantees that the regularity obtains, the proponent of the BSA should be unmoved. Like Berkeley, she doesn't think metaphysics and science are in the same business.

Armstrong's circularity charge has lately re-emerged. Marc Lange challenges the use of the scientific/metaphysical distinction, arguing that the two notions are linked by a transitivity principle that allows circularity to pop up again.[13] In place of talk of 'metaphysical explanation,' Lange uses 'grounding.' Here is the transitivity principle:

> If E scientifically explains (or helps to scientifically explain) F and D grounds (or helps to ground) E, then D scientifically explains (or helps to scientifically explain) F.[14]

To use Lange's example: suppose (F) a given flame is yellow. That is scientifically explained by (E) the law that all sodium salts burn yellow. But (E) in turn is 'grounded' in (D) the Humean mosaic. So (D) grounds (E), and (E) explains (F). By the transitivity principle, (D) the Humean mosaic scientifically explains (F) that a given flame is yellow. The problem is that (F) the flame's being yellow is part of (D) the Humean mosaic. So the flame's being yellow has to explain itself: the flame's being yellow is a constituent of the big fact (D) that explains, via law (E), the flame's being yellow.[15]

In the spirit of my reply to Armstrong, I think the proponent of the BSA should reject the transitivity principle. The BSA doesn't hold that metaphysical explanations or 'groundings,' whatever that might mean, aspire to unify particular facts. The sense in which a given law-cum-regularity is 'grounded' in the mosaic is innocuous: it just means that the regularity is part of the mosaic. No self-respecting scientist would appeal to that part–whole relation to explain why a given flame burns yellow. Instead, she would appeal to the law itself.[16] So I think the BSA is perfectly entitled to its version of case explanation as unification or assimilation.

I've flagged a different element of case explanation, one that appeals to counterfactuals. Although controversial, and for that reason not built in to the thin concept of laws, I think it's very plausible that nomic explanations will need to get us more than the deductive-nomological model provides. If the

---

[13] Lange (2013). Elizabeth Miller (2015) provides a thorough and insightful discussion of the transitivity principle. For further treatments, see Emery (2019), Bhogal (forthcoming), and Loewer (2019).

[14] Lange (2013, 256). I've replaced Lange's brackets with parentheses to indicate that what they contain is in his original text.

[15] Lange (2013, 257–8).

[16] I've recently discovered that Dorst (2019b) makes a similar reply to Lange. Dorst also undermines Lange's motivation for the transitivity principle. Lange (2013, 257) appeals to cases like the following: the internal pressure of a balloon scientifically explains its expansion, and that pressure is in turn metaphysically explained or 'grounded' in the various forces at play. I think Dorst is quite right to insist that, from the point of view of the BSA, this is not metaphysical explanation or grounding but plain old scientific explanation, in the Humean sense. So there is no support for the transitivity principle from cases like these; they involve entirely scientific explanations.

temperature of the water is part of the nomological explanation for its boiling, we'd like to able to say that, had the water not reached that temperature, it wouldn't have boiled. Put more generally, singling out a law as part of an explanation for a given event should let us say what would have happened had the variables the law quantifies over been different. It's precisely this extra element of explanation that is often judged missing from deductive-nomological explanations.[17]

But if my defense of the BSA's treatment of counterfactuals is correct, it is entitled to this extra element. When we ask what would have happened had we dropped the bowling ball from a different height, or had its path been interfered with, we are asking about counterfactual scenarios. If we take the Lewis–Stalnaker route, we have to find the nearest all-things-considered world that has the same laws of nature as ours. And we can use our laws to calculate the results in that world. If we take a supposition theory approach, we hypothetically add our counterfactual belief to our current stock of beliefs. Either way, there is no special problem for the BSA with case explanation, even if it does involve counterfactuals. If I'm right, the BSA can claim both elements of case explanation: assimilation and counterfactual dependence.

3. Explanation—Whole

What about global explanations? It's often taken as a mark against the BSA that it can provide no explanation for the whole arrangement of property instantiations that make up the Humean mosaic. Tim Maudlin gives an especially clear statement of this objection:

> If one is Humean, then the Humean Mosaic itself appears to admit of no further explanation. Since it is the ontological bedrock in terms of which all other existing things are to be explicated, none of these further things can *really account for* the structure of the Mosaic itself.... If the laws are nothing but generic features of the Humean Mosaic, then there is a sense in which one cannot appeal to those very laws to *explain* the particular features of the Mosaic itself: the laws are what they are in virtue of the Mosaic rather than vice versa.[18]

[17] As Woodward points out, the lack of counterfactual dependence is a 'common thread' running through many putative counter-examples to the deductive-nomological account (2003, 191).

[18] Maudlin (2007, 172), italics in original.

I read Maudlin as applying Armstrong's circularity worry to the mosaic as a whole. If we are trying to explain, not just why this pot of water boiled at sea level on Christmas Day, 1900, but the whole mosaic itself, one cannot appeal to the laws, since the laws are fixed by the mosaic.

Once again, it matters what kind of explanation is asked for. True, there can be no scientific explanation of the whole mosaic—but why should that bother us? If science works as the regularity theorist says it does, then that's guaranteed by the nature of the scientific enterprise. So the complaint must be that there is no metaphysical explanation of the whole mosaic.

What is this extra kind of whole explanation? Following Tyler Hildebrand, we might cash it out in terms of probability.[19] Imagine, for the sake of illustration, that you have the whole space of possible worlds before you. Some are Humean, some not, and in the Humean worlds, there are no necessary connections among distinct existences. Hildebrand in effect poses two questions: if that's *all* you know about a world, how probable is it that that world contains *exactly* the property distribution of the actual world? It would have to be vanishingly small. Second, how probable is it that that world in fact exhibits *any* uniformities, regardless of the actual pattern? There would have to be vastly more *non*-regular worlds than regular ones.[20] Other views might seem to do better. if you're told, from your cosmic but limited vantage point, that a given world is structured by powers, you might judge that that world is very likely to contain regularities. And the more you know about what those powers are, the more able you'll be to predict which particular regularities there will be.[21] So the BSA will lose this whole-explanation contest to the powers view.

But the proponent of the BSA ought to reject the set-up of the whole-explanation probabilifying test. That set-up presupposes a space of possible worlds that includes non-Humean ones. But if we're ruling out necessary connections among distinct existences in the actual world, why should we

---

[19] Hildebrand is clear that this kind of explanation is not 'scientific explanation'; see Hildebrand (2013, 3).

[20] One worry here concerns just what counts as a 'uniformity.' Trivially, any world, no matter how chaotic, will in principle be describable by some algorithm or other, even if it's massively complicated, and so exhibit regularities. (Thanks to Travis Tanner for raising this issue). By way of clarifying the notion of a uniformity, Hildebrand writes, 'I believe that our mosaic exhibits *natural uniformity.* This is to say that all particular matters of fact participate in natural regularities, patterns in the distribution of properties in the mosaic. I'll simply assume (a) that natural regularities are objective and (b) that not all possible distributions of natural properties exhibit natural regularities' (2013, 3). For the sake of argument, I'm going to grant Hildebrand that there's some pre-theoretical scale, such that some worlds are more uniform than others.

[21] I should note that Hildebrand does not think that one realist position—namely, primitivism—produces an advantage in this respect over Humeanism.

allow them to pass unnoticed among the merely possible worlds? I recognize that Lewis himself takes Humean supervenience to be a contingent thesis, but that seems to me unjustified. The central claim should be that there are no necessary connections full stop, not just in our little corner of modal space.

Even if we accede to the initial demands of the probabilifying test, I am skeptical that any other view improves on the Humean predicament. Powers theorists, after all, can give us no a priori guarantee that their powers don't produce very messy worlds: imagine a very complex power with a huge array of manifestation conditions and possible masks and finks (that is, imagine nearly any power one could plausibly attribute to an object in the real world). Where's the probabilifying element of the powers view, such that the mere fact that it's a powers-world makes it more probably than not that that world is 'regular,' in whatever sense is meant? The same goes for a universals view—nothing in the view rules out N-relations that produce disorderly worlds, and if it did, it would be purely by ad hoc stipulation.[22]

An independent worry suggests itself here. In general, this kind of objection to the regularity theory seems to share the structure—and flaws—of typical design arguments for theism.[23] You just pick out whatever feature F you want to praise and ask, among all the possible worlds, why should the world with precisely *this* feature be the actual world? The answer is: that's just what somebody living in an F-world would say. I can't see the problem with living with the fact that it's contingent. It just turned out that we have the mosaic we do. So it seems to me that, on the issue of whole explanation, we have a standoff. The realist insists on a kind of explanation the regularity theorist thinks is as unobtainable as it is irrelevant to actual scientific practice.

4. What is a Regularity?

I think the BSA can muster good responses to most of the objections it faces. Its fortunes are about to change. To see where the real problem lies, we need to go back to the beginning and ask: what exactly is a regularity, in the first place?

---

[22] A different point is pressed by Stathis Psillos. He argues that any of the putative explainers of regularities would necessarily be just as mysterious as the *explananda* themselves: 'one alleged mystery (the presence of regularity) is not explained by positing another mystery (a supposed productive relation or the like)' (2014, 18).

[23] Indeed, John Foster (2004) runs just this kind of argument. For a thorough and penetrating analysis of design arguments, see esp. Jantzen (2014).

Reading the literature on the BSA, one sometimes gets the uncomfortable sense that 'regularity' is being used in such an elastic way as to be all but meaningless. At such times, one suspects that a 'regularity' is just whatever a law statement states. I want to develop a clear, substantive meaning for the term, grounded in the literature. If others wish to substitute some different notion of 'regularity' for the one I've been operating with and am about to define, they're welcome to do so; but the resulting view will not be the BSA as I understand it.

The lineage of the concept is as clear as anyone could want. For Hume, a regularity is a constant conjunction of property instances across the mosaic. (Since Hume is a nominalist, we would have to cash that out in terms of exactly resembling sets of tropes. But nothing turns on the ontology of universals, so I'll continue to speak of 'F' and 'G' without bothering with a nominalist analysis.) Hume's constant conjunctions become Mill's regularities, which we've already examined. Both terms are preserved by twentieth-century anti-realists such as R. B. Braithwaite, who calls his view 'the constant conjunction view.'[24] Braithwaite becomes the target of Armstrong, who in turn is the target of Lewis.[25] So when Lewis calls his view a 'selective regularity theory,' he is differentiating his own view from a crude regularity theory, which would count all regularities as laws, *not* by introducing a brand new notion of 'regularity,' but by introducing a principle for *selecting* which regularities count as laws.

A regularity might be hugely complex, and spread over vast regions of space and time. But it is nevertheless a series of property instances within the mosaic. Earlier, we looked at Beebee's helpful notion of 'God's Big Book of Facts.'[26] Whatever else it is, it is a book *of facts*. The canonical statement of a regularity, from Lewis on, is a universal generalization: $\forall x(Fx \rightarrow Gx)$. The transformation rules of logic, of course, mean that lots of superficially different propositions can be cast in that form. Given the course of the debate over the BSA up to Lewis, however, it seems clear that a universal generalization stating a regularity needs a specific kind of truthmaker: $\forall x(Fx \rightarrow Gx)$ is true because there *are* Fs that are also Gs, or are followed by Gs. Now, Hume and Mill assume that F and G have to be instantiated, and some want to relax that standard. Some universal generalizations are vacuously true: 'if x is a centaur, x is good at hopscotch' is true. Some philosophers, such as Stathis Psillos, recommend including some vacuous generalizations in the best system, and I have

[24] Braithwaite (1968, 11), quoted in Armstrong (1983, 107).
[25] Armstrong (1983); Lewis (1983).   [26] Beebee (2000).

no objection to that.[27] Nor do I object to calling vacuous generalizations 'regularities,' although it's worth noting that that's quite a departure from Hume himself. In what follows, it's this expanded sense of 'regularity' I'll have in mind, and to which I'll hold the BSA.[28]

Even this notion needs further elucidation. Just as we worried over the subjectivism imported by making lawhood relative to our own standards and cognitive limitations, we might worry about the objectivity of the regularities themselves. If a regularity is a succession of *similar* events or property-instantiations, haven't we introduced an objectionable mind-dependence? In the late nineteenth century, John Venn argued that the degree and kind of mind-dependence the view requires are not enough to make it a form of idealism. Nature, on his view, 'gives us repetitions...of all the important elements, only leaving it to us to decide what these important elements are.'[29] We might add that, as long as similarity is an internal relation whose ground is the intrinsic properties instantiated at different points in space-time, the foundation of the similarity relation will be perfectly objective: one state of affairs instantiates F, the other G, and when their instances are constantly conjoined over the whole history of the universe, we have our truthmaker for $\forall x(Fx \rightarrow Gx)$. It is up to us to pick F and G as salient, but once we do, the universal generalization involving those properties will be objectively true.

Next, we need to know how a regularity is supposed to be related to the instances that make it up. A regularity cannot just *be* its instances, even if it is in some sense constituted by those instances. Psillos brings this out by comparison with an image constituted by pixels on a screen. There's nothing more to the image than the pixels, but the two are not identical: you can remove one pixel and the image persists.[30] For Psillos, the regularity itself, the truthmaker for the universally quantified proposition, is a perduring entity. Coupled with eternalism—the view that all times are equally real—we get a view that allows the regularity to be fully real at all times.

I am unsure whether the regularity theory has to embrace eternalism.[31] But there is no doubt that all versions of the Best System Analysis I'm familiar with

---

[27] Psillos (2002, 152).

[28] It may well be that other accounts, such as Loewer's 'package deal account,' either reject the requirement that laws state regularities or are operating with a totally different notion of a 'regularity.' If so, then we are owed a new definition of the term. It may well be that such accounts, if fleshed out, would end up converging with the post-Lewisian 'rules' theory I entertain below.

[29] Venn (1889, 98), quoted in Psillos (2009, 133).   [30] Psillos (2014).

[31] Among Humeans, eternalism certainly seems to be the default view. For an argument to the effect that Humeanism is inconsistent with eternalism's main competitor—the Growing Block theory of time—see Forbes and Briggs (2017).

take regularities to cover the entire history of the universe in both temporal directions.[32] They do this to secure the immutability of the laws. If laws were formulated to cover only a certain part of the world's history, there is no reason to think the laws *qua* regularities wouldn't change. Marc Lange argues that Lewis's laws get to be immutable only through an arbitrary stipulation, for nothing about the view itself demands that its laws cover the whole history of the universe.[33] For Lange, this is a shortcoming: the regularity theory can only make its laws immutable through an add-on assumption, whereas immutability ought to be built in from the start.

Here again, one might worry that we simply have a clash of intuitions. If you take laws to be governing, you're going to require that they be immutable by their own nature; a view that purchases immutability through an ad hoc assumption is bound to look like an impostor. The thin concept of laws does require them to range indifferently over an entire world-history. But if you take laws to be summaries, Lewis's immutability-on-the-cheap is no sin.

5. *Ceteris Paribus* Clauses

Even if we understand what a regularity is, we face a problem that's been with us from the start: there aren't enough of them. Nothing as neat as $\forall x(Fx \rightarrow Gx)$ is likely to be true where the laws are concerned. As Galileo was well aware, the law of free fall doesn't summarize a long list of cases where falling bodies have plummeted to Earth.[34] Given real-world conditions of friction, wind resistance, and the operation of other forces, it may almost never state the true rate of fall. This is especially clear as we move to finer and finer grained measurements of time and distance. And yet no one has proposed deleting the equation from the physics textbooks.

In many cases in physics, then, we need to append what Hempel calls 'provisoes,' or *ceteris paribus* clauses. Some may simply deny, with Earman and Roberts, that the laws of fundamental physics will in fact need such clauses.[35] But even if we managed to purge physics of laws that need provisos, we would be left with the special sciences, whose laws clearly will need such clauses. Again with Earman and Roberts, one might deny that the special

---

[32] The possible exception might be Hume himself. As Anne Jaap Jacobson (1986) argues, Hume thinks we can learn about constant conjunctions through experience, which isn't the case if those conjunctions range over all times and places.

[33] Lange (2009, 101).   [34] Lange (1993) presents the free fall case, among others.

[35] Earman and Roberts (1999).

sciences deal in laws at all. But I think it's fairly clear that some special sciences use laws that meet our thin concept: they are contentful, general propositions that function as axioms of the relevant science. Peter Turchin makes a good case for a law of population ecology: 'a population will grow (or decline) exponentially as long as the environment experienced by all individuals in the population remains constant.'[36] For Turchin, that proposition functions in his discipline just as the law of free fall functions in physics. And in fact, both laws wear their provisos on their sleeves: the law of free fall building a great deal into what counts as 'free,' and Turchin's requiring that the environment remains constant. I can only conclude, then, that a view that simply rules out any laws with *ceteris paribus* clauses will rule out too much. Even if it spares the fundamental laws of physics—and you don't have to be Cartwright to doubt that it will—it bars from lawhood a host of propositions that meet our thin concept of laws.

The problem of how to deal with such laws is especially acute for the BSA. As regularities, its laws by definition cannot admit of exceptions: any exceptions spoil the claim to being a regularity at all. That's one lesson I take from working through Hume and Mill. Other views of laws, such as Descartes's, can provide each law with a *ceteris paribus* clause that is redeemed by the operation of other laws. The early modern top-down approach simply doesn't require that each law individually state or summarize a series of events. No such move is available for the BSA, since it builds in that requirement from the very beginning.

Let's review our options, with the help of the Lange/Cartwright trilemma: we can believe in laws that we know to be false; qualify them so that they come out vacuously true; or qualify them substantively so that they come out true but apply only to a handful of instances and are all but unknowable.

Which horn should the BSA embrace? David Braddon-Mitchell proposes what he calls 'lossy laws.'[37] This amounts to the first horn: on his view, the BSA should give up on the requirement that its laws be true. Instead, we should aim for a *mostly* true system, one that incorporates generalizations that admit of some exceptions. Just as data compression algorithms can be 'lossy' in the sense that they omit some information in order to purchase the strength needed for compression, 'lossy laws' are false but enable the prediction and explanation of enough events to earn their place in the best system. The problem is that, whatever view one generates by allowing in lossy laws, it's not the BSA. Lossy laws are not regularities and so statements of them are not

---

[36] Turchin (2001, 18).   [37] See Braddon-Mitchell (2001).

candidates for lawhood to begin with. As Backmann and Reutlinger point out, allowing a false law into the system is simply not consistent with the BSA.[38] What's more, such a move doesn't help with laws that have no instances, or only a few. Lossy laws still need to be *mostly* true.

So I think the first option is a non-starter for the BSA. The second— appending a blanket *ceteris paribus* clause—is equally unappealing, for it simply gives up any hope of cashing out a law as a statement of a regularity. One might, for instance, try to use pragmatic considerations and contextual factors to limit the scope of the clause. But what then has become of the proposition's aspiration to state a regularity? What universal generalization is asserted by the claim that 'bodies in motion stay in motion, unless something interferes'?

All of this suggests that the BSA is best served by making do with the third option: insisting that each exception be entered into the law statement. The BSA's laws are going to have to include specifications of all the ways in which things can fail to be equal. Markus Schrenk has developed a version of this 'brute force' move. For a law of the form 'For all x, if Fx, then Gx,' we substitute, 'for all x, if Fx, then Gx, *except for the individuals a, b, c, etc.*'[39] We then get a true proposition.

One worry comes from the nature of explanation. Against Schrenk, Backmann and Reutlinger argue that hand-coding exceptions is no way to achieve genuine explanation. The brute force move does not 'provide an answer to the pressing question why or in virtue of what there are these exceptions.'[40] What is missing is some account of what it is about individuals a, b, c, etc., that exempts them from the universal generalization.

At the level of the special sciences, I find it hard to see why the defender of the BSA should feel the force of this objection. There is no reason not to hope that, as we proceed from the less to the more fundamental, our system explains the exceptions by appeal to some wider generalization. All grizzlies might be brown except for the Albino Family of northeast Wisconsin; but genetics might be able to explain their exemption from the otherwise true generalization.

---

[38] Backmann and Reutlinger argues that Braddon-Mitchell's view becomes a kind of instrumentalism, although Braddon-Mitchell himself doesn't say so (Backmann and Reutlinger 2014, 386). In fact, Braddon-Mitchell suspects the fundamental laws of physics won't need to be lossy, in his sense. And in any case, he still respects the 'accuracy' condition: the laws are aiming for the highest degree of truth (that is, the fewest exceptions) compatible with playing the role of the axiom of a best system.

[39] See Schrenk (2007b, 162–71), discussed in Backmann and Reutlinger (2014), as well as Schrenk (2014).

[40] Backmann and Reutlinger (2014, 382).

Now, when we reach the fundamental level, we will run out of background generalizations; there would then be no deeper answer for why the exceptions are exceptions. Still, I don't see how this damages Schrenk's position: the Humean is perfectly happy to let the features of the mosaic stand as brute facts. The demand that there be something else behind the mosaic, to make it as it is, is precisely the sort of thing the BSA rejects.

So I think the best move the BSA can make in the face of the *ceteris paribus* problem is to bite the bullet: we have to take such clauses as promissory notes, and not genuine features of laws.[41] A true statement of a law will have to include each and every exception as such. Given the BSA's regularity requirement, there is just no alternative.

That doesn't mean that the BSA is home and dry. I said at the start that we would see the central tension—between laws as regularities and as axioms—popping up as we go. And here we find the tension in a particularly stark form. If we are to avoid a mismatch between the BSA laws and the laws of any actual, let alone completed and idealized, science, we need to build in some extra condition such as Dorst's: the laws have to be 'maximally predictively useful to creatures like us.'[42] Any of these friendly amendments to the BSA makes the *ceteris paribus* problem even worse. In pointing to the epistemic role the BSA's laws need to play, these amendments make plain just how far short regularity statements will fall. An indefinitely long universal generalization with individual exceptions hand-coded is not a plausible candidate for such a law, however useful it might be from God's point of view.

We find ourselves in a puzzling situation. We can solve the problem of the phony constant, and the mismatch of BSA laws to scientific laws, only by acknowledging that laws have to do certain things for us: they have to play the axiom role, and they have to play it *for us* humans, with our cognitive needs and limitations. But acknowledging this leaves us wide open to the Lange/Cartwright trilemma. I've argued that the brute force maneuver is the only response to the trilemma open to the BSA. That lands us with regularities that are even *less* suited to playing the axiom role. The lesson seems to be: we can increase the facticity of a law statement only by decreasing its suitability for the axiom role.

---

[41] Note that neither the 'package deal' account, nor its descendant, the 'better best system,' can help with the *ceteris paribus* problem. Cohen and Callender themselves don't take a stand on the issue: they merely point out that there are various responses to the *ceteris paribus* problem and add that they 'won't pick an option here; suffice it to say that there are some' (2009, 25). For discussion, see Backmann and Reutlinger (2014), who argue that nothing about the 'better best system' view entitles it to attach 'non-lazy' *ceteris paribus* clauses to its laws.

[42] Dorst (2019b, 2663).

This dialectic is, I think, a plain result of the central tension: regularities are not suitable candidates for the laws of the best predictive and explanatory system. The criterion that laws state regularities begins to look as much like an unfortunate historical holdover as the intuition that laws govern. Next, I'll mount a direct argument to that effect.

6. The Central Tension

The objection I'll present in this section exploits the tension within the BSA itself. It doesn't complain that BSA laws fail to meet some independent criterion, which the Humean is free to reject. Instead, I'll argue that regularities are not up to the job the best system assigns them. The rocky marriage between regularities and the axiom role ought to be dissolved. Anything one can do to turn laws into regularities will by its nature make them less fit to play the axiom role. In particular, a system of laws that rejects the regularity requirement will win the simplicity and strength contest. In other words, the virtues enshrined in the 'best system' aspect of the BSA are in fact exemplified by a system whose axiom role is *not* filled by statements of regularities.

The obvious barrier is that the BSA, whether its author is Mill or Lewis, explicitly builds in the regularity requirement: laws have to be *true* universal generalizations that report on property distributions across the mosaic. So my competitor laws will be ruled out of court from the beginning. While that is surely *a* response on behalf of the BSA, it's not a very interesting one. The whole question is whether the regularity requirement is justified in the first place, given the desiderata enshrined in the standards of the 'best' system. So let's simply ask: if it were not part of the BSA by fiat, would there be any grounds for sticking to the regularity requirement?

My strategy is to set up toy examples and present competing sets of laws, one that keeps to the regularity requirement and another that follows out the Cartesian web of laws approach while happily shirking that requirement. Which one will win the simplicity and strength contest?

Before we get going, I have to insist on two important points. First, my claim is not that there is *no* possible world in which statements of regularities play the axiom role. Indeed, in the first thought experiment I set up, that's precisely the case. It's not as if there is some kind of conceptual incompatibility between regularities and laws-*qua*-axioms. Instead, I'm arguing that in *any modestly complicated* world, a set of laws that doesn't respect the regularity requirement beats out one that does. More important, my toy world experiments are not stand-alone counterexamples to the BSA. Many such counterexamples have been batted away by proponents of the BSA, who either refuse to grant that there are such possible worlds, or else deny their relevance. I'm working in a different way. The second toy world is a schema for real-world examples, of which there are an indefinite number. To defeat my argument, it's not enough to show that the toy worlds are impossible. One would have to show that the real-world scenarios that conform to the schema are impossible. And since those scenarios are actual, there are no prospects for doing so.

With those points in place, let's begin. Our first toy world has only two kinds of particle; only the three following kinds of collisions occur.

(1) If an F-particle strikes a G-particle from the west, the G-particle is moved 1 inch to the east

(2) If an F-particle strikes a G-particle from the east, the G-particle is moved 2 inches to the west

(3) If an F-particle strikes a G-particle from the north, the G-particle is moved 1 inch to the south.

Now, it's easy enough to formulate statements of regularities—universally quantified reports on facts (1)–(3) above—that allow for the deductions the BSA has in mind. And indeed, in our toy world, there is no competing class of simpler propositions that would do the same (leaving aside notational legerdemain, e.g., piling all three propositions into one).

Add a complication to our toy world, however, and things shift dramatically. Let's move to our second toy world, which is just like the first, with the following differences. Suppose that the world has two kinds of field, Z and Y, and that every part of the world is one or the other. Z-fields increase the motion of G-particles that are struck by F-particles by half an inch, while Y-fields decrease them the same amount. Now, we could capture this in BSA-fashion. But note already that I haven't done so in setting up the example. Still, I might have done it the hard way:

BSA-1   For all particles and properties: In a Z-field, If an F-particle strikes a G-particle from the west, the G-particle is moved 1.5 inches to the east

BSA-2   For all particles and properties: In a Z-field, If an F-particle strikes a G-particle from the east, the G-particle is moved 2.5 inches to the west

BSA-3   For all particles and properties: In a Z-field, If an F-particle strikes a G-particle from the north, the G-particle is moved 1.5 inches to the south

BSA-4   For all particles and properties: In a Y-field, If an F-particle strikes a G-particle from the west, the G-particle is moved 0.5 inches to the east

BSA-5   For all particles and properties: In a Y-field, if an F-particle strikes a G-particle from the east, the G-particle is moved 1.5 inches to the west

BSA-6   For all particles and properties: In a Y-field, If an F-particle strikes a G-particle from the north, the G-particle is moved 0.5 inches to the south.

But that is not the simplest method available to state the facts. True, BSA-1–BSA-6 each describes a regularity, and each allows us to predict new events and deduce old ones. To explain (in the relevant sense) the motion of a G-particle that has been struck by an F-particle in a Y-field, we might enter the initial position of the G-particle, adduce BSA-5, and then deduce the particle's new position.

The problem for the BSA is that there is obviously a simpler way to do it. Instead of the six BSA laws, consider

(L1)   If an F-particle strikes a G-particle from the west, the G-particle is moved 1 inch to the east

(L2)   If an F-particle strikes a G-particle from the east, the G-particle is moved 2 inches to the west

(L3)   If an F-particle strikes a G-particle from the north, the G-particle is moved 1 inch to the south

(L4)   In a Z-field, the motion of G-particles struck by F-particles is increased by 0.5 inches

(L5)   In a Y-field, the motion of G-particles struck by F-particles is decreased by 0.5 inches.

The defender of the BSA might object that (L1)–(L5) is just a convenient shorthand for BSA-1-6. There's an irony here, since the whole point of the BSA is to provide us with such a shorthand for 'God's Big Book of Facts.' The BSA's laws are not just *any* propositions that can allow us to deduce the particular facts; they have to be the fewest and simplest such propositions. By the BSA's own lights, (L1)–(L5) beats out BSA-1-6.

The point is magnified if, as I've suggested, the BSA accepts the 'Humeanism with a human face' revisions. These revisions impose a further requirement on BSA laws: they have to play the right epistemic role for creatures like us, with all our cognitive limitations. Those revisions are crucial to answering the problem of the phony constant and the mismatch problem. And they obviously favor (L1)–(L5) over the BSA laws.

Nor can one reply that the competitor laws fail to meet the axiom requirement. (L1)–(L5) still fit the deductive-nomological model of explanation. Any particular matter of fact in our toy world can be deduced from these laws in combination. More generally, whatever model of case explanation one wants to give, the competitor laws can function in it just as well as the BSA laws.

But of course the BSA as it stands cannot accept our new set of putative laws, since they do not state regularities. Again, a regularity is (at least) a true universal generalization. (L1), (L2), and (L3) are all false: it *never* happens, as per (L1), that a struck G-particle moves 1 inch to the east, given that the entire world is either a Y-field or a Z-field. And yet (L1)–(L3) are part of a *better* system than BSA-1–6 by the BSA's own standard of 'better,' namely, making the same predictions with fewer and simpler laws.

Note that the point is *not* that (L1)–(L3) are vacuous. Laws about how things behave when no other forces operate on them are vacuous, because they describe situations that never in fact happen. As we've seen, Humeans might argue that a suitably catholic understanding of 'regularities' can admit such laws; after all, a universal generalization with a false antecedent is automatically true. Now, that's quite a departure from Hume's own understanding of a regularity. Hume himself takes regularities to be constant conjunctions among events, and an event that never happens isn't conjoined with anything. But even if we acquiesced in this expansion of the term, it wouldn't help in the present case. For (L1)–(L3) have antecedents that *do* describe instantiated states of affairs. But they have false consequents, so they themselves are false. They're not getting into anybody's Big Book of Facts.

Nor are matters improved when we turn to (L4) and (L5). They are even worse off from the point of view of the BSA: (L1)–(L3), taken as statements of regularities, are at least clearly false, and so must make some kind of sense. The BSA can make nothing much out of (L4) and (L5), for they do not describe any particular states of affairs at all. Instead, the fourth and fifth laws state a tendency or force: we might equally well have said that Y-fields retard motion by half an inch. But the only way the BSA can capture this tendency is by laboriously entering what happens in each circumstance.

Let me reiterate an important point that is easily missed. Many a counterexample to the BSA has been rebuffed on the grounds that the Humean would never grant that it describes a genuine possibility. Such a reply would be irrelevant to my argument here. My scenario is not a stand-alone thought experiment, meant to describe a possible world for which the BSA cannot account. It's better regarded as an instance of a *schema* for generating counterexamples, plenty of which can be found in the real world.

Consider a region of the actual world that includes a solenoid and a bunch of free-floating particles. When electricity passes through the coil, it generates a magnetic field in its interior. Outside the solenoid, there is no such field (or, if you prefer, one with the value of zero). Any particle subject to magnetism behaves differently outside of the field than it does in it.[43] This scenario is relevantly similar to our second toy world: it is better described by laws analogous to (L1)–(L5) than any BSA laws. Our world contains many such systems, even if none of them is isolated; the reader can no doubt come up with many such real-world cases of her own. If the Humean wants to insist that even *this* scenario is impossible, I can only invite him to visit my basement laboratory on any Saturday night.

My claim, once again, is not that there is a conceptual incompatibility between statements of regularities and axioms. In some world, 'all taxis in Athens are yellow' might well function as both.[44] But that world would have to be a far simpler world than the actual one. I am arguing that in any modestly complex world, the laws of the BSA will lose the simplicity-strength contest to laws that do not respect the regularity requirement. We can be sure that the real world is many orders of magnitude more complex than either of my toy worlds, on any reasonable understanding of complexity.

7. *Ceteris Paribus* Redux

A natural move to make in the face of my argument would be to append *ceteris paribus* clauses to (L1)–(L3) to get them to come out true. If such a move worked, it might allow the BSA to help itself to at least (L1)–(L3) while retaining the requirement that they truly report on regularities. Consider (L1); so modified, it becomes

(L1*)   *Ceteris paribus*, if an F-particle strikes a G-particle from the west, the G-particle is moved 1 inch to the east.

I don't see how this move would help with (L4) and (L5): those two laws tell you how Z and Y fields affect the motion of particles, and it makes no sense to

---

[43] My thanks to Benjamin Jantzen for this example.
[44] I owe this example to an anonymous referee.

say that they hold other things being equal. Let's set these last two laws aside and see if the *ceteris paribus* maneuver can help us hold on to (L1)–(L3) as regularities.

Right from the start, of course, we face the Lange/Cartwright trilemma. On the horn we're now considering, the law comes out unverifiable and trivially true. Nothing could count as a counterexample, since it could simply be brushed under the carpet of the *ceteris paribus* clause. But I want to make a different point: so qualified, L1* does not state a regularity. Perhaps it states a rule of thumb, or something like that, but 'other things being equal, F-particles and G-particles ϕ' does *not* say that F-particles and G-particles ϕ. For again, our Z and Y fields make it the case that things are never equal. In toy world, it's not a good idea to go around believing (L1)–(L3) on their own, since they're never true. If you insulate them with *ceteris paribus* clauses, they come out (trivially) true, but then they do not state regularities.

Nor can we appeal to Braddon-Mitchell's 'lossy laws.' Such laws are not true, so the BSA cannot include them in the best system. Even if we waive that problem, it seems to me that the proposal wouldn't go nearly far enough. Braddon-Mitchell's view still respects the 'accuracy' condition: the laws are aiming for the highest degree of truth (that is, the fewest exceptions) compatible with playing the role of the axiom of a best system. From the point of view of accuracy, (L1)–(L3) are not lossy; they're losers. They don't state a for-the-most-part uniformity with a few exceptions. They are never true, not even once.

I argued above that the only way out of the problem of *ceteris paribus* clauses is the hand-coding maneuver, where we simply record each and every exception to the law. But the brute force alternative produces a system of laws that will lose the simplicity/strength contest to (L1)–(L5). The costs to simplicity of entering each exception manually is enormous. Imagine a proposition that stated the free fall equation and went on to specify each and every exception to it as such. That would be the analogue of BSA-1–BSA-6. I don't see how it could beat the analogue of (L1)–(L5), that is, a system of interlocking propositions, each stating a law that is not a regularity but instead works in tandem to make predictions. Nor do I see how such a super-regularity could come anywhere close to living up to the revisions we accepted above, which require that the laws be useful for creatures like us.

I don't think the problem I've identified is one of mere detail. It's not a technical problem that can be overcome with ingenuity. I think it's a central tension, written into the BSA from the beginning: statements of regularities are just not up to the job of being laws.

14

The Alternatives

1. The State of Play

The announced goal of this part of the book is to develop and refine the best
version of the regularity theory. Unfortunately, I think the BSA is doomed. So
this final chapter of Part IV will explore alternative theories that still respect
one animating idea of the BSA: that there are no lawmakers out there, apart
from the individual tiles of the mosaic and their arrangement.

Even though the BSA is flawed, it contains insights that we should develop
and exploit if we can. I've argued that the BSA can mount solid replies to many
of the objections it faces. The real problem is an internal one, and it stems from
adopting the regularity requirement. That requirement conflicts with the role
we need laws to play, as axioms of the best system. But if we free ourselves
from the BSA, we can simply drop this requirement. What happens if we hold
on to the notion of a law as a proposition that functions within a given
scientific framework but stop requiring that the laws state regularities?

For many, such a move would be a non-starter. For it seems to commit us to
an objectionable idealism. The BSA tethers its laws to the world precisely by
making them regularities; if we loose this rope, won't our laws simply float
away? I'll argue that the threat of idealism is not nearly as pressing as it seems.
After working through some versions of projectivism, I'll develop a view I call
the 'rules' theory. Unlike projectivism, the rules theory retains the supervenience of laws on non-nomic facts. Its rules are every bit as anchored to the
real world as the regularities of the BSA. If I'm right, the best competitor to
come out of the present family of views is one that treats laws, not as
regularities, but as rules for prediction and explanation.

2. Conservative Projectivism

I call our first isotope of projectivism 'conservative' because it still treats laws
as a subset of regularities. What singles them out *as laws* is the attitude we
adopt toward them. And it counts as 'projectivism' because it takes this

The Metaphysics of Laws of Nature: The Rules of the Game. Walter Ott, Oxford University Press. © Walter Ott 2022.
DOI: 10.1093/oso/9780192859235.003.0014

attitude to consist in the 'spreading' of the mind's own tendencies on to objects in the world.

This core idea is present in Hume, although it's often submerged in subsequent work in the Humean tradition.[1] For Hume himself, it's a key part of his psychological project of explaining how in fact human beings come to believe in capital-C Causation, with mind-independent necessary connections between distinct existences:

'Tis a common observation, that the mind has a great propensity to spread itself on external objects, and to conjoin with them any internal impressions, which they occasion, and which always make their appearance at the same time that these objects discover themselves to the senses. Thus as certain sounds and smells are always found to attend certain visible objects, we naturally imagine a conjunction, even in place, betwixt the objects and qualities, tho' the qualities be of such a nature as to admit of no such conjunction, and really exist no where... [T]he same propensity is the reason, why we suppose necessity and power to lie in the objects we consider, not in our mind, that considers them; notwithstanding it is not possible for us to form the most distant idea of that quality, when it is not taken for the determination of the mind, to pass from the idea of an object to that of its usual attendant.[2]

Two things are important here. The first is that, in Hume's context, the projection or spreading of the mind on to external objects does not present us with a logically possible state of affairs whose absence we might sensibly mourn. There is no world in which smells and sounds are 'out there,' nor is there any one in which objects really are necessarily connected. The only connections are there in our minds, and the notion that we might have lived in a world where they were really 'out there' is not a coherent one. Second, the spreading of the mind is a natural propensity, not the result of an occasional error of reasoning. In this respect, the projection is like an optical illusion: even after you know that the two lines are really the same length, the Müller–Lyer illusion persists.

Hume is here speaking of the necessity of causes, but since, as we've seen, his laws just are a subset of causes, we can apply his thoughts in the nomic context

[1] A notable exception: Massimi (2018) is very clear that her version of perspectivalism incorporates Hume's story about projection. Note that perspectivalism still relies on regularities, and so would count in the present taxonomy as a conservative version of projectivism.
[2] T 1.3.14.25.

as well. Someone who allows this 'great propensity' free rein will soon find himself insisting that laws somehow force or require new events of one type to follow prior events of another. There's no mystery why we've struggled to find an account of how laws do that: there is no coherent account to be had. All we can do is explain *why* it feels to us as if the events that unfold around us are bound together in a net of necessary connections.

Neither Mill nor Lewis makes much of the projectivist element of their Humean inheritance. To see it come to the surface, we have to look at two intervening figures: F. P. Ramsey and A. J. Ayer. Sometime in the late 1920s, Ramsey endorses and then abandons a Millian Best System Analysis. Rather than worrying about counterfactuals or nomic stability, Ramsey thinks there are epistemic barriers standing in the BSA's path: 'it is impossible to know everything and organize it in a deductive system.'[3] Ramsey takes laws to be 'variable hypotheticals,' propositions of the form for all x, x is F. But 'when we regard it as a proposition capable of the two cases truth and falsity, we are forced to make it a conjunction, and to have a theory of conjunctions which we cannot express for lack of symbolic power.'[4] I take the lack of 'symbolic power' here just to mean that we cannot explicitly list all of the conjunctions.[5] Ramsey concludes that, since it's not a conjunction, the variable hypothetical 'is not a proposition at all,' and is neither true nor false.[6] Instead, variable hypotheticals are 'rules for judging "If I meet a φ, I shall regard it as a ψ."'[7] The talk of rules here can be misleading. For Ramsey, the point is that the hypothetical is an attitude or approach. Attitudes 'cannot be *negated*' but they 'can be *disagreed* with by one who does not adopt' them.[8]

Ayer argues that Ramsey goes too far. Law statements can be true or false; a law statement can be falsified by a single instance just like any other proposition, even if it can never be decisively confirmed (ranging as it does over all of history). To bar law statements from the realm of propositions is to go too far, on Ayer's view. He recommends that we locate lawhood in the attitudes we adopt toward propositions themselves. A plain old universal generalization and a statement of a law do not differ intrinsically. What makes a generalization a law is only the attitude we take up toward it.[9]

---

[3] 'General propositions and causality' (1929), in Ramsey (1990, 150).
[4] Ramsey (1990, 146), (mis)quoted in Ayer (1956, 160).
[5] I might be wrong about this. I've been unable to find a helpful gloss on this sentence in the secondary literature.
[6] Ramsey (1990, 146).   [7] Ramsey (1990, 149).
[8] Ramsey (1990, 149), italics in original. Ayer (1956, 160) points to same move in Ryle's work.
[9] See Ayer (1956, 165).

I think Ayer is right to resist Ramsey's retreat to the realm of the non-propositional: law statements do at least appear to state the facts, and if there's some way to make sense of that, we shouldn't give up so quickly. The problem with Ayer's proposal is that it doesn't go nearly far enough. It inherits all the problems of the original BSA because it still insists on making laws regularities.

3. Contemporary Projectivism

Our next option is contemporary projectivism, which has been developed by Barry Ward. Ward devises the underdetermination argument against nomic supervenience, also used by primitivists such as Maudlin and Carroll. And Ward rejects both planks of the BSA: laws are not regularities, nor are they axioms of the best system. Unlike the primitivists, Ward feels no compulsion to endorse realism. For him, as for Ramsey, nomic claims are not factual claims at all. Ward provides a nice summary:

> When we speak of laws of nature we speak as if we were describing a feature of the world. But my analysis is expressivist in that law claims are understood as normative claims which express attitudes taken to rules for making predictions and explanations. Thus, on this analysis, the surface appearance of our discourse embodies a projection of the expressed attitude onto the world.[10]

For Ward, the underdetermination scenario shows that the laws do not supervene on non-nomic facts. But that's not because there really are laws, over and above such facts. In underdetermination scenarios, we face a kind of 'normative' uncertainty, an uncertainty about which rules *should* be used to make predictions and explanations in the world in question. There is no further fact, primitive or otherwise, that makes one set of rules the one that ought to be used.[11]

Consider the lesson Ward draws from his scenario. The reason why world *w* is indeterminate with respect to its laws is that it provides no means of choosing among competing explanations. Our uncertainty about the laws is uncertainty about which explanation to prefer. If we want to explain the

---

[10] Ward (2002, 192). Ward goes on to develop a semantics for the view, building on the work of Blackburn (1984) and Gibbard (1990).
[11] Ward (2002, 198).

behavior of the E-particle by pointing to what the M-field does, we need to commit ourselves to what would have happened to the other particles had they entered the field. If the M-field explains the trajectory of E-particles simply because it acts that way on *all* particles, then we expect the other particles to take on the same trajectory.

We've already seen that the Best System Analysis can deploy its story about counterfactuals in this context as in any other. In the present situation, the proponent of the BSA ought simply to respond that world w does indeed underdetermine the laws, and so it's no surprise that it also doesn't tell us which explanation we ought to give for the behavior of the E-particle. So I think the BSA has nothing much to fear from the underdetermination argument.

But there are promising features of projectivism we may want to incorporate in the best version of anti-realism. Opponents of the BSA—from anarchists like Ronald Giere to primitivists like Maudlin—have long complained that it gives us the wrong account of science. Physics, for example, doesn't seem to engage in collecting and summarizing regularities, or at least it doesn't *just* do that. The projectivist lets us accommodate this point without endorsing either extreme by divorcing the whole project from regularities. To see how, we need a richer story account of the relationship between laws, models, and the world. Giere and Ward present an attractive, indirect picture of the way laws apply to the world.[12] Roughly, the idea is that nomic equations allow you to generate models, and the question of the accuracy of those equations is to be settled by the degree of agreement between those models and the actual world. In Ward's terms, differential equations, such as those flowing from Newton's second law, are 'model generating rules': they let you input whatever initial conditions you like and generate a model of how the system that has those conditions will evolve.[13]

---

[12] Giere (1999, 92–3). Giere himself is, of course, in the 'no laws' camp. But that might be because of the way he's understanding the term 'laws.' He prefers to speak of Newton's 'equations' rather than 'laws,' because '[i]nterpreting the equations as laws assumes that the various terms have empirical meaning and that there is an implicit universal quantifier out front' (1999, 92). So he is thinking of 'laws' as regularities, and rejects laws because he rejects the Humean picture. I see no reason why his view is not compatible with projectivism on the point at issue here; I note only that Giere is a realist about causal necessity and so would reject projectivism on that score (1999, 95). Projectivism is still faithful to the 'Humean' part of the BSA, that is, the rejection of mind-independent necessary connections.

[13] Ward (2002, 197). I think we should also take on board Andreas Hüttemann's point, that an equation such as $F = ma$ 'becomes a law statement once it is asserted that this equation is meant to represent the behaviour of physical systems' (2021, 174). In our terms, we might say that an equation doesn't count as a law until it is applied, via models, to the world. The equation itself is not what is confirmed or refuted by experience.

In this chapter, we're searching for the most defensible successor to the BSA. I think projectivism can make a good claim to that title. But since I don't find the underdetermination argument persuasive, I think it's too soon to give up on nomic supervenience. Ward takes the failure of supervenience to follow from the role laws have to play in explanations: they have to generate models. But I think we can take Ward's points about models and explanation and retain any version of nomic supervenience worth having.

4. Laws as Rules

I think the strongest anti-realism will combine the best elements of the revised BSA and projectivism. It should also take on board as many of the virtues of Cartesianism as it can without sinking into realism. Let me set out which bits I want to preserve from each before trying to state what I call the 'rules' account more rigorously.

From the BSA, we can take the idea that law statements should function as axioms of the best system. The system should enable prediction and retrodiction. And we should welcome the revisions to the BSA we've seen, which make lawhood relative to a recognizably human Limited Oracular Perfect Physicist. The rules theory should be just as ecumenical with regard to natural properties as the 'package deal account': which properties are natural, and allowed to figure in laws, should be left up to the relevant science. When we move outside of physics, we may need to appeal to a Limited Oracular Perfect Botanist (LOPB) or Economist (LOPE) to fix the laws.

From the Cartesians, we should take the idea that laws are not statements of regularities, or even any kind of statement that reports on a single fact by itself. Instead, a law is a rule that functions with others of its kind to enable prediction and explanation. So we have gotten rid of the BSA's besetting problem: its attachment to regularities. The rules theory doesn't demand that science, whether actual or ideal, aspire to summarize anything at all. That's not what the laws do.

We should also take on board the projectivist's story about how these rules do their jobs. The rules allow us to generate models, and it's the models that fit the world or don't. A set of rules that generates a model that in turn enables prediction and retrodiction will at the same time allow for contrastive, modal explanations. I argued that the BSA can get us both explanation by unification and the counterfactual element. But many will be unconvinced by my defense of the BSA on this last score, and insist that no regularity theory is entitled to the counterfactuals, or at least not in a sufficiently robust way. To such opponents of the BSA, I advertise the rules theory. For the connection to contrastive, modal explanation grows more naturally out of the rules story. It's in the nature of rules to be indifferent to their input conditions, and hence to allow for the construction of counterfactual scenarios of whatever kind we need in scientific explanation.

What does the result look like? On the rules theory, to say that some proposition or equation L is a law is to say that L is an axiom of the human-LOPP relative best system, where that system need not include any statements of regularities. Its laws will typically be equations that allow for the prediction of the evolution of systems through time.[14] Such a law will be an axiom an ideal *human* observer would use in order to predict, explain, and deal with her world. This sounds very like the revisions to the BSA, but it's also not that far from Ward's projectivism. Ward, recall, takes lawhood to consist in an attitude we take up with regard to an equation. Once we buy the relativity of laws to a human LOPP, we're not so far from agreeing. So I think the rules view counts as an isotope of projectivism, even if that projectivism grows out of the revisions to the BSA. The crucial difference is that, unlike projectivism, the rules theory can still get us the kind of supervenience we ought to care about.

There's more to the rules theory than this bare statement, of course. Let's develop it by testing it against the main objections we've posed for its predecessors. This will help bring out the differences among the views, and lay bare just what someone committed to the rules theory has bought into. I think we'll find that on all these metrics, the rules theory does as well as, or better than, any of its anti-realist fellows.

4.1 The Mismatch Problem

We worried that the laws formulated by the BSA would not match those delivered by science, whether actual or ideal. In answer to this, we accepted the 'Humeanism with a human face' revisions. These put extra constraints on the

---

[14] I take seriously Maudlin's (2007) point that the fundamental laws of physics will turn out to be laws of temporal evolution, or FLOTES. The paradigmatic laws of the rules theory should be FLOTES. I'd like to remain neutral, however, on whether other propositions should count as laws. Jaag and Loew (2020) nominate the Past Hypothesis, which states that at the start of time, the universe was in a very low-entropy state. Chen and Goldstein (forthcoming) point to laws that do not privilege one direction of time and so are not FLOTES. In the interest of attracting as many friends to the rules theory as possible, I think it should remain uncommitted on the lawhood of such claims. Nothing about the rules theory as such bars them from lawhood.

regularities that count as laws, and so 'reverse engineer' lawhood from the standards that our modestly idealized LOPP would use. But the rules theory's 'reverse engineering' of the laws is far more thorough-going than anything suggested by the revisions to the BSA. Even the revised BSA is still subject to the nagging worry that actual or ideal science will not care about summaries of regularities at all. The rules theory escapes all this by providing a different account of what laws do. They are rules for generating models, which play a role in prediction and explanation. If some of these models should accurately predict exceptionless patterns, so much the better. But the laws will still not aspire to summarize regularities.

Giving up the regularity requirement allows a reconciliation with scientific practice. The rules of a completed science acting in concert should, if possible, allow for a deduction of all particular matters of fact. Where probabilistic laws are concerned, the standards of success will be different. But there is no reason to suspect such rules will themselves state these matters of fact. It's no surprise that physics is not conducted merely by collecting and summarizing individual facts but also—at least typically—by formulating relationships among variables. That's what the laws of toy-world from the general argument against the BSA try to do, and plausibly what the laws formulated by physics aim to do, too.

4.2 The Problem of *Ceteris Paribus* Clauses

I argued that this problem is a reflection of the central tension in the BSA. Jettisoning the regularity requirement removes the tension. But doesn't the problem of *ceteris paribus* clauses simply reappear? This is where the Cartesian inheritance pays interest. Taken on its own, no nomic equation (or the predictions derived from the model it generates) will state the facts. But that is because so taking the law is perverse. The Cartesians' laws need no *ceteris paribus* qualifiers because they are implicitly 'web'-qualified: they function only as a group.[15]

---

[15] In this respect, the rules theory is similar to the picture of laws Max Kistler (2006) presents. For Kistler, no law on its own entails a universal generalization (2010, 93). Putative exceptions to a given law can be explained by reference to other laws or to the circumstances in which the properties it concerns are instantiated (2010, 129–30). Kistler's view thus avoids the need to posit a blanket *ceteris paribus* clause within the laws themselves. Note, however, that Kistler's view is a brand of realism: '[o]ur own hypothesis that a law of nature is a universal relation between property instances' (2010, 219).

Let's return to the second toy world I set up in presenting the central tension. The simplest set of laws includes propositions that state relations among variables. Laws (L1)–(L3) are simply false if taken on their own. In terms of the rules theory, taken individually, each law generates a model that fails to correspond to the toy world. But that's perfectly fine. The early modern web approach, running from Descartes through Newton and Berkeley, simply abandons any idea that laws will state truths when taken individually. We have to take them as a group.

Nor do we have to tolerate any mistakes, as Braddon-Mitchell's view must. Our laws are not lossy in the sense that they leave out any data. In our toy world, they let you predict everything that happens with complete accuracy. That's what the laws from the LOPP's-eye point of view would do with the mosaic of the actual world.

So the unit of evaluation is the entire system of laws, that is, the whole set of axioms of a given science. I hasten to add that this is no kind of 'holism': the web of laws approach does not say that each law gets its meaning from its relation to other laws. In toy world, for example, each of (L1)–(L5) has a perfectly clear meaning of its own. Nor is this Duhem/Quine-style confirmation holism, although I have no quarrel with that.[16] The view doesn't claim that no law can be tested independently of the others. It's just that there wouldn't be a point in doing so: the whole idea is that you need all of the laws to generate accurate predictions and explanations.

Suppose we propose an equation L1 as a law that generates a given model M. Now, M fits the real world well enough, except under conditions C. That's a sign that we're either quite wrong or missing another law somewhere that is interacting with L1. (Having dropped the regularity requirement, we can help ourselves to the metaphor of nomic interaction.) The solution is not to add 'except in conditions C' to L1. It's to find L2, or L2–Ln, and make our predictions and explanations using the whole group. If we can't at the moment find such laws, that's a sign that our science is incomplete. We can be sure that the LOPP, if there were one, wouldn't need any *ceteris paribus* clauses. Now, nothing stops us from acknowledging the equations of an incomplete science as laws. Suppose we haven't yet discovered (L5) in our toy example. The other four laws do pretty well, half of the time, in generating models that fit toy world. Pragmatic considerations might mean that we have no alternative but

---

[16] By 'confirmation holism,' I mean the claim that no individual sentence has implications of its own for experience, such that it could be confirmed or disconfirmed. On this kind of holism, the unit of empirical evaluation has to be a theory, not a sentence or proposition.

to go on using them and hope we discover (L5). In the real world, of course, there's no a priori guarantee that the laws we have currently formulated will be included at the idealized end of science. But that's the price we pay for trying to be faithful to the facts. Just what such fidelity comes to is our next topic.

4.3 Truth and Supervenience

The BSA's defense against charges of anthropomorphism and idealism is that their laws are chosen from the ranks of regularities. While the standards of the choice are up to us, and need to be made relative to our own purposes and cognitive limitations, the choice is still a choice *among* fully real things. Having given up the regularity requirement, have we not unmoored the laws altogether from the truth? In fact, I think the rules theory can muster a good reply.[17]

On the rules theory, following Giere and Ward, the connection between laws and the world is an indirect one. The laws let us generate models, and it's the models that 'fit' or fail to fit the world. We can plug values in for the mass of the Earth and the Moon and, using Newtonian equations, generate a model. As Giere points out, we still need to specify the aspects of the world we want the model to fit, and how close the fit needs to be.[18] But once we settle that (or more likely, make an array of pragmatic assumptions that vary with the context of inquiry), our questions have objective answers. If I'm trying to predict the position of the Earth relative to the Moon within a sufficiently wide range, I might not need to worry much about the gravitational interference of other massive bodies in the solar system. If I need a more precise prediction, I'll need to take account of the 'token interference,' the disruption to the actual system caused by other bodies whose behavior is predictable by my equations (again, within whatever degree of error is tolerable).

Consider the ideal gas law, which relates the volume, temperature, and pressure of gases. But it treats those gases as if they were made up of mass-less

---

[17] Ward (2002) gives a number of persuasive replies to the charge of subjectivism on behalf of his own projectivist view. The rules view ought to adopt these same replies, to the degree possible. In particular, Ward notes that projectivism does not say there are no laws, or that all law claims are false; 'Bas van Fraassen's (1989) antirealism entails that all nomic claims are false...[B]ut non-factualism doesn't.... [T]o claim that there are no laws is to deny that there are any rules that ought to be used for prediction and explanation,' and that clearly isn't Ward's view (2002, 211). The rules view can make the same point: unless we are in a diabolically unkind world, there will of course be rules our human LOPP would formulate to enable prediction and explanation.

[18] Giere (1999, 92–3).

particles. Why isn't it just false? If it were reporting on regularities, or on any direct feature of the world such as an N-relation, it would be. But that's not its role. Instead, it allows us to generate models that are close enough approximations to be useful in many situations in the real world. There's often this kind of trade-off between fit and utility. A helpful analogy might be (yet again) the relation between a key and a set of locks. The key might fit no one of these locks perfectly; it might be worn at the edges and hard to use. But the very defects that cause this might be responsible for its fitting the range of locks it does.

Now we can turn to our main question: what would it mean, on the rules theory, to say that the laws are 'true'? Suppose our best system is such that its axioms, together with whatever set of conditions one puts in and counts as 'initial,' is sufficient—via its models—to predict every relevant part of the mosaic. Input all the conditions at t, and our best system accurately predicts the ambient temperature of Bremen at $t+n$. And suppose that the system works symmetrically: you can run the axioms on the conditions at t, and figure out conditions at $t-n$. The proposal on offer is to think of the axioms of the system not as regularity statements but as rules or algorithms. Rules are not true or false in isolation; as we've just seen, their connection to the world is indirect. But that doesn't mean there is no way to evaluate them other than on pragmatic grounds. When these rules issue in models that enable predictions, we have a straightforward way of evaluating them: are the predictions correct?

In the real world, a false prediction might be due to an insufficiently complex model. Insufficient complexity can arise from using too few rules in generating the model. Suppose we apply the laws of free fall to a bit of metal in a magnetic field; we'll get the wrong prediction. (This is what I called 'type interference' above.) The right response is not to chuck out the rules but to use more of them. Alternatively, the problem might be not taking account of other players on the scene ('token interference'). I might not need to use any other rules, but I do have to take account of the interlopers.

There might be multiple sets of rules that generate the same predictions. But it would be premature to conclude that we're doomed to some kind of relativism. Faced with the choice between Ptolemy and Copernicus, both of whom—at least from a terrestrial vantage point—get us the right predictions, our LOPP will prefer the Copernican system. Again, just like the BSA, the rules theory has to hope that nature is kind, and gives us enough data to settle disputes among competing laws. These laws are evaluated not just in light of their predictive power but also in light of the epistemic needs of our recognizably human LOPP.

Imagine a set of algorithms that enabled the accurate prediction of hurricanes. How could one answer the question, 'are the algorithms true?' but by looking to see whether the predictions based on the models they generate are correct or not? And having found them to be correct, what extra element could the algorithms be missing, such that its addition would make them true in some other sense, or 'more' true in this one?[19]

The rules of hurricane prediction might well have the form of (L1)–(L5) above. None of them on its own is true: they do not, individually or collectively, state universally quantified generalizations. But they enable the precise prediction and, symmetrically, the explanation, of hurricanes. I have a hard time seeing what else could be required to call the propositions true: they give you the right predictive and explanatory results. What we really *have* given up is the claim that each law individually has its own truthmaker. But a great many propositions fail this one-to-one truth-to-truthmaker standard, and are not mysterious or suspect for that.

Let me come at this point from another direction. Some proponents of truthmaker theory appeal to a slogan: 'truth supervenes on reality.' As John Bigelow puts it, '[i]f something is true then it would not be possible for it to be false unless either certain things were to exist which don't, or else certain things had not existed which do.'[20] The axioms of the best system, construed as rules, have just this kind of responsiveness to the mosaic: if we keep the initial conditions constant and remove any tile of the mosaic, such that the system generates a false prediction, then the system would no longer count as the 'best.' If the mosaic were different, and if hurricanes behaved differently, the laws wouldn't work. It's that kind of responsiveness to the facts that we want out of a scientific system: a difference in the mosaic must be reflected in a difference in the laws or initial conditions included in the best system. And if that's not enough, I can't see what is.

---

[19] Someone might worry that the rules view amounts to a kind of instrumentalism. Narrowly defined, instrumentalism is a thesis about terms that seem to refer to unobservables, and the sentences that contain them. That issue, prominent in early twentieth-century debates in philosophy of science, is orthogonal to the ones here. I think the rules theory need not commit itself to this kind of instrumentalism about unobservables. It's perfectly consistent with the rules theory that whatever properties feature in its equations should be as real as anyone could want. Alternatively, 'instrumentalism' might refer to any view that takes propositions about laws to be evaluable only in terms of their utility, and not as reports of states of affairs. This is the kind of view the Inquisition took of Copernican theory; see Cardinal Bellarmine's letter to Foscarini of 1615, in Galilei (2008, 146–8). But in the text, I argue that the rules theory can preserve the supervenience of laws on the mosaic, relative to the human LOPP. If that still counts as instrumentalism, it's a very anodyne version of it, and it would equally extend to the BSA. An excellent recent treatment of instrumentalism is to be found in Stanford (2015).

[20] Bigelow (1988, 133).

We can put all this more formally in terms of supervenience. We've seen that contemporary projectivism abandons nomic supervenience. By contrast, the rules theory can claim that laws supervene on the mosaic with just as much right as the BSA or its revisions. I argued above that the best version of supervenience is:

> Human LOPP-relative Nomic HS: the laws of a given world supervene on the mosaic (the spatio-temporal arrangement of properties), and are relative to the epistemic goals and cognitive limitations of the LOPP

where 'supervene' means that there is no way to wiggle what supervenes without wiggling what it supervenes *on*.

Projectivism abandons supervenience on the basis of the underdetermination argument. But I tried to show above that that argument is not persuasive. There's nothing troublesome about what Lewis would call an 'unkind' world. The laws supervene on the mosaic in the sense that you can't wiggle the laws without wiggling the mosaic, provided you keep the LOPP fixed. That's the kind of supervenience worth caring about. And it's just what the rules theory gets us.

4.4 Counterfactuals and Explanation

Do laws as rules support counterfactuals? First we have to clarify: given the web picture, the right question is whether the laws as a whole support counterfactuals. No individual law can be expected to do so, for, as we've seen, the obtaining of another individual law can scupper the outcome of the first. A sphere that is also electrically charged may not be predictable on the basis of the law of gravity alone. But even as a whole, we might worry that these rules don't support counterfactuals, or don't do so robustly enough. If the rules in tandem generate a model that predicts outcome S on the basis of conditions R, we would like to be able to say that had R obtained, S would have obtained. Can we?

Let's work with our two going theories of counterfactuals. On the Lewis/Stalnaker view, evaluating the counterfactual circumscribes the worlds we can choose from: it has to be a world that shares the same laws of nature as ours. Within that set, we look for a world that is similar to the actual world in all its non-nomic respects, changing as little as we must to make R come out true. Are such R-worlds also S-worlds? Of course they are. After all, we chose the R-worlds from among those that have the same laws as our own. And in any such world, conditions R lead to outcome S. When we ask, would Neptune's orbit have been different had Pluto not existed, we need to look at the worlds within the sphere picked out by the rules, find one without Pluto, and ask after its orbit. Now, within that sphere, we also want one that is as all-things-considered close to the actual world as possible. That world will differ from ours in lots of ways, not least in its initial conditions. Still, since by stipulation we're using the same rules, we get the right result: of course we would! Given the laws, Neptune's orbit would have been different had Pluto not existed. That's just what the equations we're using to calculate the orbits tell us.

Supposition theory gets us there even more directly. The rules theory says that the laws of the actual world are the web of rules that would be chosen by our human LOPP. If I'm in possession of these rules, and am asked to consider situation R, which is not actual, all I can do is consult the rules, generate a model, and see what happens. And if I hypothetically add a belief in R to my current stock of beliefs, of course I'll end up with outcome S.

We saw above that many philosophers think case explanation-by-unification is not the aim of science, or at least not its sole aim: explanation of cases also involves counterfactual evaluation. Considering counterfactuals lets us contrast various scenarios, and vary our input conditions as we like. The rules theory is just as well positioned as the BSA to deal with this objection. If I purport to explain the orbit of Neptune in terms of its derangement by the gravitational force of Pluto, I am at a minimum saying that, had Pluto not existed, Neptune's orbit would have been different. And the rules theory lets us say exactly that.

The basic story about counterfactuals is the same one I formulated for the BSA. If anything, it just brings out more clearly the way in which any story about counterfactuals has to invoke the evaluator's set of beliefs and commitments. Still, it is bound to seem like thin soup to some. Given the denial of necessary connections and their agents (N-relations and powers), where is the metaphysical *oomph* that forces S to obtain when R does? It's this point that most awakens my realist sympathies. The story I've given has the air of getting something for nothing. All I've done is bring out the consequences of denying mind-independent connections among distinct existences: anyone willing to go that far is already comfortable in a world not held together by any metaphysical glue. But that's not a story any realist would welcome, since it deflates what the realist takes to be a genuine insight into the structure of modal space. There's now nothing left to the claim that, given the laws, some things *have* to be as they are, beyond our own need to project these laws into non-actual states of affairs if we are to evaluate counterfactuals.

4.5 Nomic Stability

I've argued that the real point of the mirror argument is nomic stability: we want laws that can be used to evaluate a wide array of non-actual scenarios without thereby becoming false within those very scenarios. That's why the bite-the-bullet response is so unsatisfying. Carroll is right: just swiveling the mirror shouldn't change the laws. I argued that even the BSA should accommodate, rather than jettison, the desideratum of nomic stability. And I think it can. Now, some readers will not have been persuaded by my case on behalf of the BSA. To them, I advertise the rules theory as superior in this respect.

The key move the rules theory makes is to deny that laws are statements of regularities. And that move pays dividends here. As rules for making model-based predictions instead, the rules are indifferent to changes in boundary conditions. We're free to input whatever values we like for the variables that feature in the laws. That's as true in the special sciences as in physics. Turchin's population ecology equation can take any number as its input for the population of a given species in an environment, and show us how that model will evolve in respect of population. Freed from the regularity requirement, our laws are now even more clearly stable over differences in inputs and boundary conditions.

An analogy might help here. Suppose you've been playing correspondence chess with an incredibly consistent and unimaginative opponent, perhaps Kim Jong-Un. Over the years, you've learned what we'll call 'Kim's Rules,' which include the ordinary rules of chess plus a complete list of Kim's strategies.[21] Kim's rules are revealed over multiple games, as you see him do different things in different scenarios. If you open with a pawn, he opens with a knight; if you open with a knight, he answers with a pawn. Now, there would be no point in learning a summary of every individual move Kim has ever made. What you need in order to beat the Dear Leader is the ability to predict what's coming next. And the way to do that is to master the rules he's playing by. If you change the initial position of the pieces on the board, you don't thereby make your knowledge of Kim's rules useless. You're testing that knowledge by seeing whether it gets you the right prediction or not.

In much the same way, the laws are equations that together allow us—via the models they generate—to predict the next move *nature* will make. Precisely because the laws are no longer yoked to the particular mosaic we find ourselves in, they can be used to model parts of it and indeed scenarios

---

[21] Thanks to Tzuchien Tho for helping me think through this example.

that have never happened. Varying the initial conditions does not, by itself, falsify the laws: there's a massively large large number of games you can play with Kim that all have different moves, and yet Kim's behavior might be predictable by a single set of rules. By the same reasoning, although you can't falsify the laws without changing the mosaic, changing the mosaic on its own is not enough. Take every world whose behavior is predicted by the rules an idealized physics actually settles on in our own. They are nomically identical, but wildly different in their arrangements of tiles.

5. The Path of Anti-realism

We've now surveyed three main anti-realist views: the BSA, projectivism, and the rules theory. I've argued that each is an improvement over its predecessor. Although the BSA withstands the most common objections, it still suffers from what I called the 'central tension': statements of regularities are not suited to playing the axiom role. The 'Humeanism with a human face' revisions, which make lawhood relative to our epistemic goals and cognitive limitations, acknowledge this, albeit in too limited a way. Once we start making lawhood a matter of what we want and need laws to *do* in our epistemic economy, we can move away some of the barriers facing the original BSA. Some, but not all. We're still left with the problems of *ceteris paribus* clauses, and are still vulnerable to the central tension.

Taking its cue from projectivism, the rules theory abandons the idea that laws are statements of universal generalizations. The rules theory cheerfully ignores the regularity requirement. Equally important, it rejects the idea that the laws have to stand on their own: they function instead only as a web. And I think it produces a better fit with the practices, both ideal and real, of science, by making explicit the connections among laws, models, and predictions. These seem to me decisive advances over any version of the Best System Analysis. What is more, the rules theory recaptures the only kind of supervenience worth having. That sets it apart from its near ancestor, projectivism. The rules of a recognizably human LOPP are sensitive to, and responsible to, the mosaic. The rules theory can claim with some justification to get us laws that are not just useful but true. In short, if we want to deny that there are any cosmic policemen enforcing the laws, if we want to craft a plausible view of scientific practice free of any 'metaphysical glue,' the rules theory is the way to go.

PART V
THE ENDGAME

15

Settling up

1. The Rules

This book has had two aims. First, of course, is working out the best meta-physical story we can tell about the laws of nature. It's up to science to tell us what the laws are, but it's up to philosophy to tell us what *kinds* of things they are and how they do what they do. What does it mean to say that nature works according to laws? How are laws related to the world they describe, predict, or govern?

We've had to take up another task as well: working out the rules that govern the debate over laws. Are there some 'intuitions' or putative desiderata that should be shunted aside or ignored? Are some requirements for lawhood the relic of a pre-scientific worldview, or a theological one? Although I have no universal algorithm to decide such questions, I think we've made some progress.

The first rule I propose is this: *ceteris paribus*, no view should be dismissed on the grounds that it cannot accommodate a claim that the view itself rejects. Here's an example. Suppose we set up a single possible world with just one particle and one field. The particle just travels along at a constant velocity and never encounters the field. We can invent an indefinitely large number of laws, each of which is consistent with the meager 'facts' of that world. Presto! The laws don't supervene on the non-nomic facts. Such a crude version of the underde-termination argument shouldn't be allowed to settle the debate. If you think laws are summaries of regularities, you will rightly be unmoved. The sentence 'Call me Ishmael' doesn't uniquely determine the summary of *Moby Dick*. That doesn't mean the writers of Cliff's Notes are doing Melville's job for him.

We observed the same rule when evaluating nomic stability. Laws ought to be stable over a wide variety of variations. But we shouldn't require that a view preserve the laws even in worlds where those laws, by its own account, cannot hold. For example, the universals view is immanentist: it requires that a given universal be instantiated if that universal is to exist. So we can't run a counter-example like the spin argument, which assumes from the start that a law might hold even when one of the universals joined by the N-relation is not instantiated.

The Metaphysics of Laws of Nature: The Rules of the Game. Walter Ott, Oxford University Press. © Walter Ott 2022.
DOI: 10.1093/oso/9780192859235.003.0015

In my statement of this rule, I smuggled in a *ceteris paribus* clause, and I think for good reason. Other things might not be equal, if we can show that the claim in question is not just any old feature a competing theory builds in, but in fact a requirement of scientific practice. To extend our stability example: suppose someone wanted to reject nomic stability full stop, rather than merely limiting its scope to worlds that share the lawmaking features of the actual world. As we've seen, nomic stability is the real upshot of the mirror argument: changing the position of the mirror shouldn't change the laws. Some Humeans dismiss this intuition as an artifact of the top-down approach, which holds that laws are independent of the mosaic they govern. No doubt that explains part of the appeal of the stability claim. But it doesn't explain it away, for that claim has some purchase in scientific practice, as I'll continue to argue.

Are there other rules to guide us? I opened the book by suggesting that the concept of law is very unlike some other concepts philosophers deal with. It's not the inevitable and natural product of reflection on, say, morality or knowledge. It's a concept introduced to play a specific role in doing science. So that's where we should look to find the criteria we can sensibly hold any view to. These desiderata will then function as first-order rules of the game, constraining the moves we can make.

It's not enough to apply these criteria and see who wins and who loses, although of course we need to do that, too. If multiple views do an equally good job of preserving them, we have the harder task of weighing up the merits and demerits of each. That's what I'll try to do in this chapter. I hasten to add that I don't see this chapter as in any sense the culmination of the others. What really count, I think, are the specific arguments and positions that have been developed and refined in the prior chapters. Much of what we'll need to do here calls for philosophical judgment as much as the diligent crafting and refining of arguments, and I don't claim mine is more attuned to the truth than anyone else's. Even readers (if there be any) who find the arguments of the other chapters persuasive will likely part ways with me here. All any of us can do is honestly weigh up the virtues and vices as we see them.

To begin, then, what are the requirements that survive scrutiny? We started with our 'thin' concept of laws, which is a good place to hunt for respectable desiderata:

(1) A law statement should be specific in that it refers to properties, forces, or phenomena, and not just a way of saying that the universe is orderly;

(2) A law statement should be general in that it doesn't necessarily refer to particular places and times.

Both of these, I think, flow from

   (3) A law statement should play the axiom role relative to whatever science
       counts it as a law.

A law should be a proposition that is fundamental in the sense that it
doesn't follow from other propositions, at least within the same domain of
scientific theory. There is bound to be some elasticity here. Charles's law can,
plausibly, be shown to follow from kinetic theory; and yet there's no movement afoot to strip it of its mantle of lawhood. Much more important is the
fact that playing the axiom role involves permitting operations that allow us to
model regions of the world and indeed counterfactual worlds. I want to predict
the speed at which a ball will return to my hand when I toss it in the air. I plug
in my equations and in effect generate a model of the system defined by hand,
ball, and surrounding air. That model will fit the actual event to as great a
degree as we could like, even if the model represents an isolated system. Laws
of economics—if indeed there are any—should be able to do the same kind of
thing. We want to know what will happen to inflation if a given country raises
interest rates. The only way to figure that out is to plug in the relevant values
(current inflation rate, GDP, and the rest), and see what we think would
happen. So criterion (3) is tightly connected to the idea that

   (4) Laws ought to be indifferent to at least some counterfactual variations
       (nomic stability).

Using laws to generate predictions and consider counterfactual scenarios
requires that laws not crumble when we change a given variable. If we really
did have an economic law that could infallibly predict the effects of raising
interest rates, it would have to be such that you could enter different variables
and generate predictions. This is the lesson at the core of the mirror argument.
A typical law will state a relationship among variables, and ought to be
indifferent over at least a wide range of what particular inputs we use. That
seems to be part of what we want the laws to do: they ought to allow us to
consider nearby counterfactual worlds just as they let us predict unobserved
cases symmetrically in time. In evaluating views on this score, we still have to
be careful not to violate our meta-rule. If a view posits X as the lawmaker, we
can't require its laws to be stable over variations that do not include X.
Descartes's laws, to give an extreme example, are divine volitions. So it
would be absurd to require that his laws hold even in worlds in which there

is no God. In parallel fashion, powers views are entitled to say that the powers of the actual world have to be held constant if the laws of the actual world are to hold in another possible world.

All four of these desiderata have been traced to their origins in the modern period. But a pedigree by itself carries no justificatory weight; the reasons for endorsing (1)–(4) have to come from the roles we expect law statements to play in scientific theories. And I think, on those grounds, they are reasonable places to start. Meeting them is not an all-or-nothing affair; each view is likely to do well on some and less well on others.

From the twentieth-century debate, we can extract another desideratum worth taking account of:

(5) An account of laws should enable us to give a satisfying answer to the problem of *ceteris paribus* clauses.

I argued that (5) doesn't crop up in the early modern debate because the dominant view of laws has an answer to it. It doesn't become a problem until regularity theories get going in the work of Hume. Someone like Descartes, for example, takes the laws to be divine volitions that function in tandem. This is the 'web of laws' approach that later figures in the work of Berkeley, Newton, Mill, and others. I argued that, at least for the pre-Humeans, the web approach can provide a satisfying answer to the problem; more carefully, that approach prevents the problem from arising. But now that we've seen that it *is* a problem, it needs an explicit solution.

Finally, we have to add an extra requirement, one that comes from physics itself. As we've seen, lots of views have trouble with a specific kind of law, namely, the conservation laws. We can hope that such laws will vanish as physics progresses, but I wouldn't want to depend on that hope. Our metaphysics of laws should at least leave open the possibility that there are conservation laws. So I propose adding:

(6) Whatever view of laws we endorse, it must in principle be capable of making sense of conservation laws.

2. Three Axes

We've already winnowed the available views as we moved through the three families. Some fail to meet our desiderata; others labor under internal tensions.

Quite a number remain on the table, however, and still others—the hybrid views I've mentioned along the way—await assessment.

We have three axes along which views of laws can vary:

- the relationship between laws and their truthmakers;
- the top-down/bottom-up divide;
- the realist/anti-realist split, where a view counts as 'realist' just in case it posits some extra feature—primitive laws, powers, N-relations—as shaping the mosaic

I plan to take each axis in turn, and see if we can extend the results by ruling out more positions. I think the arguments so far established have made evaluation of views along the first two axes pretty straightforward, though I'll need to add some arguments to make the case. The third is the tough one.

3. Law Statements to Truthmakers

Our first question concerns the relationship between law statements and truthmakers. There are three options: one-to-one, one-to-many, and many-to-one. It might be that for every true law statement, such as '$F=ma$,' there is a corresponding fact, $F=ma$, that makes it so. Alternatively, a law statement might be true in virtue of a number of independent facts. For example, it might be that the natures of force, mass, and acceleration cooperate, and make it the case that force is the product of the other two. Finally, there might be a single, overarching fact or state of affairs that makes *all* true law statements true. This is the kind of picture we get from positing the maxi-N relation, or a single power, the 'Blower.' This axis cuts across the other two: as long as we have some split between propositions that play the law-role and facts or states of affairs that make them true, we have our choice among the three options.

It would be a surprise, I think, if the universe carved itself so neatly as to conform one-to-one with the laws we need, given our six desiderata. So any view that insists on perfect isomorphism here is less attractive than its competitors. In fact, one way or another, all one-to-one accounts will suffer at the hands of the problem of *ceteris paribus* clauses. That's because any plausible solo candidate for lawhood will need to ignore a huge array of other conditions.

Consider primitivism. The prominent versions of primitivism are one-to-one; for each fundamental law-statement in the Theory of Everything, there is exactly one law. How plausible is it to think that the universe is so neatly constructed as to pair the law-statements even of an idealized science with whatever kind of thing the primitivist thinks answers to them?

James Woodward describes his primitivism as a kind of 'tempered and restricted' realism. He acknowledges that the distinction between initial conditions and laws enshrined in nomic stability is there 'because there are things we want to *do* with [them]—e.g., predict and explain.' A law both 'functions to organize our thinking in a certain way *and*, when applicable, tracks features of the world.'[1] Whatever those features might be, they have to stand one-to-one with our (true) law statements. But that seems to amount to a dubious bet on the final outcome of fundamental physics. I don't see any guarantee that the laws we need to work with are going to walk in lockstep with these features they're meant to track. To insist that they must and will do so seems a far more objectionable kind of anthropomorphism than anything the Humeans contemplate.

The same issue infects the original universals view. We've seen it struggle with the problem of *ceteris paribus* clauses and lose. Recall that the original view promises to go directly from N(F,G) to $\forall x(Fx \to Gx)$. But to go from an N-relation to a regularity requires that we somehow take account of all potential interferers. To include them in the law itself is to produce a law statement that will be all but useless to creatures like us.

There's another force pushing against the one-to-one universals view. Our sixth desideratum requires us to account for conservation laws. But such laws sit awkwardly, to say the least, in any top-down view other than primitivism. In the context of the universals theory, we might have to postulate a third-order N-relation to keep the second-order relations among the universals in line. A more appealing strategy is to hope that conservation laws will emerge organically from the first-order universals themselves, as on the revised view. To say that, of course, is already to give up on a one-to-one match.

The same worries make one-to-one powers views unappealing. Like the other one-to-one views, this one seems to me excessively optimistic about the tightness of fit between the laws we need for a Theory of Everything and the ontology that will underwrite it. Even if that theory's laws reported on all and only powers, it's a substantial further commitment to say that each law statement will report on a single power. Nor are conservation laws any easier to account for. If it's a law that momentum is conserved in isolated systems (even if the only one is the universe itself), how can that law be a power? What

---

[1] Woodward (2018, 167).

is it a power *of*? If it conserves momentum, then it has to be in act whenever collisions happen. And it has to be active at every point in the universe at once.

The original BSA labors under different problems. It, too, is a one-to-one view, identifying each law with a suitable regularity. One upshot of the central tension argument is that this identification cannot work. Pairing law statements with regularities results in a set of laws that will lose the simplicity-strength contest to the rules theory, which rejects that requirement.

I think we can we can say that the considerations from this chapter and previous ones make it pretty clear that the one-to-one view is not going to survive to the final round. In the context of realism, it tacks an extra burden on to an already heavy load. We don't need to insist on such a tight connection between a concept that we need to play a specific set of roles and the world that makes that concept applicable. Now, the realist can welcome the *possibility* that the final Theory of Everything—and even the completed special sciences—will get us a nice isomorphism. But that seems unlikely, and there seems to me no good reason to build such a requirement in at the start. So we have our first result in this chapter: the one-to-one view is out.

4. Top-down/Bottom-up

Our next axis is top-down/bottom-up. Top-down views make the laws independent of what they govern. On all such views, it's not the powers of objects that make events take the course they do; the order of explanation is just the opposite. Our first contemporary contender was primitivism. Although the one-to-one version has been rejected, nothing stops the primitivist from saying that any given law statement has multiple truthmakers, or even that there's just one primitive maxi-law. The real problem with primitivism is the governing dilemma: there's just too little to be said about not only what the laws are, but how they do their jobs. So the only top-down view still in the running is some version of the universals picture.

Unlike its top-down antagonists, the bottom-up position comes in realist and anti-realist flavors. The realist version is of course the powers view; but the Best System Analysis and laws-as-rules theories should also count as bottom-up pictures. Even if it isn't held together with any metaphysical glue, the mosaic fixes the laws, and not the other way around. We can deal with the anti-realist picture in the next section. Here, I want to argue that the best versions of the universals and powers theories very nearly collapse into one. What differences remain, however important in themselves, shouldn't obscure the state of play. The revised universals and revised powers views are both, in the end, bottom-up pictures. Let's slow down and take each in turn.

Two features set the revised universals view apart from its predecessor. First, it refuses to identify individual laws with universals linked by the N-relation; it's not a one-to-one account and so is still in the running. To account for conservation laws and answer the problem of *ceteris paribus* clauses, the view helps itself to a single, indefinitely long state of affairs: a network of universals joined by the N-relation. The laws that figure in any actual or completed science will be reflections of the single, maxi-N-relation. Second, it trades in the contingent N-relation for an internal one, such that any world with both Fs and Gs in it will be a world in which N(F,G) holds. This was the key to solving the mechanism version of the inference problem.

I think these revisions are important advantages; I hope those attracted to the view in the future will consider taking them on board. They de-fang a number of troublesome objections, especially the worry about intra-world contingency. And they make possible an answer to the problem of conservation laws. Now, many philosophers have wondered just how different even the original universals view is from a powers ontology: given immanentism, the N-relation is fully present wherever the universals it connects are. The revisions push it even further in the direction of the powers view. The N-relation is internal, and so flows from the natures of the universals connected. And as we sifted through versions of the powers view, we came closer and closer to the revised universals view. Let's turn, then, to a brief recapitulation of our refinements of the powers view.

We have three varieties of the powers view on the table: the Aristotelian view, the Boylean view, and the 'Blower.' The Aristotelian picture, which has dominated the last few decades among powers theorists, takes powers to be at once intrinsic to their bearers and directed at actual and possible states of affairs. I argued that the best version of the Aristotelian view allows for properties that are not themselves powers. These non-dispositional properties might be either categorical properties or qualitative aspects of properties that are also dispositional. Either way, we need some property or feature of a property that is not itself a disposition.

The chief problem with the Aristotelian view is the 'little souls' objection from the moderns, which comes to the surface in the current debate in the form of the problem of fit. One way to solve it is some kind of power holism, but I argued that that view is no more attractive than its semantic cousin. So I suggested taking a cue from the moderns themselves, and proposed the Boylean theory. On that view, to say that object x has power P is elliptical for something of the form: objects x and y have non-directed, monadic properties F and G.

Those attracted to powers views would do well to see if the Boylean version could be made to work, especially since it survives the attacks of the moderns while the Aristotelian version simply cannot. Sadly, as things stand, I think it's out of the running. The Boylean view has to posit powers as relations that are grounded in the monadic properties of the relata, where those properties exhibit no directedness or *esse-ad*. The Boylean needs those properties to explain the course of events simply of their own nature. It's very hard to imagine how properties devoid of *esse-ad* could do that, unless Locke and Boyle's geometrical model were right.

The way out, I suggested, is to treat the universe as a single 'Blobject,' and posit just one Aristotelian power, the 'Blower.' Now, this still leaves us with the problem of the missing compass needle: how can there be a property whose nature is exhausted by its 'pointing'? If the Boylean view is untenable, any powers view will have to live with that mystery. And I argued that the Blower has important advantages over pluralist powers views, particularly in answering the moderns' little souls objection and its contemporary incarnation in the problem of fit. But it also pays dividends when we ask how well the view meets our sixth desideratum, being able to fund conservation laws.

The one-to-one powers-to-laws view makes such laws pretty mysterious. But the one-to-many position will have trouble, too. How could the distribution of energy, momentum, and all the rest, accord with those laws, if only individual powers are operating? We would have to pack all the relevant 'information' into each power. Each power would be such that, in its exercise, it is precisely calibrated with all the others so that it doesn't disturb the overall amount of whatever quantity is at issue. Someone might say, we know from the problem of fit that every power *already* has to be finely attuned and sensitive to the exercise of all other powers. Once we go for that, why not ride the train to the next stop? Why not just stock each power with whatever resources are necessary to bring about conservation?

That seems like a tough position to believe. Such inflated powers would be even worse than the 'little souls' derided by the moderns, for they would have to have an extra layer of information. They would need something close to omniscience, since they need to take account of the total quantities that need to be conserved at every moment. Cartwright's Nature or God would have to be on the scene to make sure things come out right. But that seems to me like a last resort.

Another option is to invoke higher-order powers that constrain the operations of the first-order ones. We might keep all the individual little powers and just add a meta-power that keeps them in line. That sounds to me too much like Nature or God: how would this higher-order power know where to make adjustments and when, so the relevant quantities are conserved? And as Bird points out, the powers theorist cannot take this route: '[p]roperties are already constrained by their own essences and so there is neither need nor opportunity for higher-order properties to direct which relations they can engage in.'[2] Bird ends up suggesting that conservation laws might be mere features of our way of representing the world and so eliminable. I'm in no position to rule out that possibility. Still, I stand by desideratum six: a powers view should at least leave conceptual space for the possibility that the conservation laws are real features of the world.

The best bet, then, is to argue that there's only one power. We might in practice differentiate the power of charge to weaken or strengthen chemical bonds, or the power of gravity to attract objects as it does; but in reality, these are all aspects of a single power, the Blobject's Blower. So there can be no question of distinct properties being calibrated so as to conserve anything. It's just one of the Blower's features that its exercise conserves whatever quantities or properties the final Theory of Everything says are getting conserved.

I think the Blower does pretty well on our other desiderata, too. Its laws are indifferent to a wide range of initial conditions, namely, any that are consistent with the existence of the Blower in the first place. Following our first meta-rule, it would be absurd to require this position to give us a set of laws that is indifferent to the Blower itself. The claim is just that nothing about the Blower dictates that one starts with this or that set of initial conditions. Nor does the Blower have a particular problem with *ceteris paribus* clauses. When they occur in our own law statements, they mark our ignorance of the Blower.

When we take a step back from the details, the revised powers view seems to converge with the revised universals view. We can see this by reviewing some of the problems we had with each of them. Although I defended both the original and revised universals views from some objections, I didn't succeed in resolving all of the issues. In particular, I wasn't able to say anything substantive about *how* a universal could be such that its nature enabled and even required it to stand in an internal N-relation. This parallels a worry about the Blower: it's just a brute fact that the categorical properties that exist are related in the way the Blower makes them. Whether we choose the maxi-N-relation or the Blower seems to turn on where we want to locate the mystery: is it in the

---

[2] Bird (2007a, 214).

natures of the immanent universals themselves, which fund the N-relation, or is it in the Blower that connects these property instantiations?

Some might find this a depressing result: all this work has generated two realist options that, for all their individual merits and demerits, come to much the same thing, in the grand scheme. I think just the opposite is the case: the arguments we've covered make realism converge on what is essentially a single view. I hadn't anticipated this result when I started. I had expected to defend a version of the powers view.[3] But when we dig into the problems with universals and powers, we find the same themes re-emerging. Solving these problems required us to push the views much closer together than I had guessed.

There's a larger lesson here: if we want to go for realism, we need to go big. Piecemeal N-relations, or powers that knit together just a few properties, produce views that founder on the shoals. They have no good way of accounting for conservation laws, or *ceteris paribus* clauses; and each suffers from particular defects besides. Only the Blower, or the maxi-N-relation, will give the realist a chance of meeting our six desiderata.

5. Realism/Anti-realism

If the considerations so far are persuasive, our options have narrowed considerably. I propose to take the Blower as the best candidate from the realist camp; insofar as it differs from the revised universals view at all, it seems to have at least some advantages. In particular, it leaves aside the inference problem in all of its forms. But partisans of the revised universals view are free to substitute it in what follows.[4] Let's see how the best realist view, whichever one we prefer, stacks up against its anti-realist competitors.

Much about the regularity theory is attractive, particularly when we look at Lewis's version and the friendly amendments we adopted along the way. I argued that it emerges unscathed from the battles with the typical objections. In particular, I argued the BSA's laws can support counterfactuals and also get us nomic stability. With the realists, I worried that this support and stability

---

[3] As I tentatively ended up doing at the end of my (2009).
[4] Another advantage worth noting is the ease with which the Blower can deal with probabilistic laws. Jonathan D. Jacobs and Robert Hartman (2017) argue that Armstrong's view cannot, on pain of self-contradiction, accommodate probabilistic laws. I'm unsure whether this is an in-principle problem for all universals views, or is peculiar to Armstrong's own formulation. Still, it provides a further, if prima facie, reason to prefer the Blower.

might in some way be counterfeit, or less robust than that provided by realism. But working through the two main theories of counterfactuals going— supposition theory and Lewis/Stalnaker semantics—assuages that worry.

It's true that, in considering counterfactuals, the BSA has to project its own laws onto modal space: in Lewis/Stalnaker terms, we have to limit the worlds we consider to those that share the BSA laws of the actual world. But everyone seems to be in the same boat. There is just no alternative to projection of some kind when considering counterfactuals. I argued that the powers and universals views both get us nomic stability. But they, too, need to project their own laws on to the realm of the possible. Put in terms of the mirror argument, they need to anchor their attitudes to the starred worlds in the unstarred worlds. Let's take the original universals view as an example. Suppose the N-relation is contingent. If we just consider U_1^* on its own, there's no reason to think that L1 is preserved there. There are worlds where the relevant universals are not N-related, so there will be plenty of worlds described just as U_1^* is in which L1 fails. The only reason to take L1 as a law in U_1^* is that it's a variation on U_1, where the universals *are* so related. Even the revised universals view and the Blower have laws that hold only in worlds in which their preferred entities exist. So I don't think that the BSA is any worse off than its competitors, when it comes to counterfactuals and nomic stability.[5]

Unfortunately, the BSA still suffers from the central tension: any plausible candidate for a regularity statement is unlikely to pull its weight as a law. Our third desideratum requires law statements to play the axiom role; the BSA itself requires a maximal balance of strength and simplicity in its laws. I argued that any system of laws consisting only of regularities will lose on those measures to a system free of that requirement. Nor can the original BSA deal with *ceteris paribus* clauses while giving us laws that can play the axiom role.

If we want to preserve the idea that made Lewis choose Hume as his inspiration, namely, the rejection of mind-independent powers or necessary connections, then we'd be best off giving up the regularity requirement. I've argued that such a move doesn't have the dire consequences one might think, and pays dividends in other ways.

For example, the rules view can exploit the early modern web of laws approach. Each law is implicitly *ceteris paribus* qualified only in the sense that the laws have to be used together to generate the right prediction and explanation. This came out especially clearly in the second toy world. Our

---

[5] For a broadly similar point, see Roberts (2008, 339).

world isn't a toy world, of course, but I see no reason to think that it's relevantly different in kind. And here it's worth noticing that this web of laws innovation was right there in Descartes and his successors. In this respect, the rules view takes over some of the attractive features of Cartesianism while jettisoning the divine intervention Descartes himself requires.

Although clearly inspired by projectivism, the rules theory lets us preserve the only kind of supervenience worth having. If you keep the LOPP steady, you can't wiggle the laws-cum-rules without wiggling the mosaic, but you can change the mosaic without thereby changing the laws. That's the asymmetric dependence of laws on the mosaic that we're after. So the rules theory doesn't land us in an objectionable form of instrumentalism. There's a straightforward sense in which the rules that the final theory will use are true: they predict and explain every tile of the mosaic. And they are sensitive to the mosaic in the way any kind of supervenience worth caring about requires.

6. The Hybrids

Although I think the rules view is more appealing than any other anti-realist proposal, I still feel the pull of realism. Heather Demarest has suggested marrying a powers ontology to the Best System Analysis.[6] Might we make our own refined versions of each, and have all the virtues of realism without the vices?

In fact, once we give up on a one-to-one marriage of law statements and powers, something like the BSA view of lawhood becomes attractive. Consider a view like Alexander Bird's, on which the laws supervene on the nature and distribution of powers. The BSA, one might argue, is worth adding to the powers view because it can do much more than the bare supervenience claim to explain why some propositions are laws and others not. Bird, of course, recognizes that some special features must be invoked to sort nomic from non-nomic truths.[7] We might, with Bird, try hand-coding these features, and I see no in principle objection to that. But, Demarest might counter, why not avail ourselves of all the work that has already gone into the BSA? If the BSA in fact produces a concept of laws that fits with science as it is actually practiced, we might graft it on to a powers theory. Leaving the laws simply to supervene on

---

[6] See Demarest (2017).

[7] Bird settles on the following definition of a law: '[t]he laws of a domain are the fundamental, general explanatory relationships between kinds, quantities, and qualities of that domain, that supervene upon the essential natures of those things' (2007a, 201).

the mosaic leaves them unsystematized: there is no justification for demanding that the laws fit together as axioms of a system.[8]

Much turns on just how the BSA-cum-powers marriage is to be effected. Consider Demarest's formulation:

> The basic laws of nature at w are the axioms of the simplest, most informative, true systematization of all w-potency-distributions, where a w-potency-distribution is a possible distribution of only potencies appearing in w.[9]

This is a massive inflation of the best system's job description. It's not enough that it systematize all the non-modal facts of our world; it now must cover all worlds in which any of the powers of the actual world figure. Samuel Kimpton-Nye suggests a clarification: on his view, the best version of powers-cum-BSA restricts the relevant worlds to those that instantiate all and only the powers found at the actual world.[10]

Even so, we now face a non-trivial skeptical problem. The mosaic of our own world is, at least in principle, open to empirical investigation. That, trivially, is not true of other possible worlds. Nor does our evidence from the actual world tell us what would happen in these other worlds. Powers have modal profiles, that is, they behave differently in different circumstances. The modal profile of fire includes burning paper in certain conditions. That much we can know from experience, despite the recrudescence of the problem of *ceteris paribus* conditions. But it is a crucial part of being a power that no finite list of actual events exhausts that power's modal profile.

Although I'm sure Demarest's 'potency-BSA' can come up with a response to this epistemic worry, it's enough to suggest we experiment with a different way to bring about the marriage of powers and BSA laws. The key question is: *what* is getting systematized? If the laws have to directly systematize powers, we have an epistemic problem on our hands. But suppose we leave the truthmaking relationship between the laws and the mosaic alone: it is the mosaic itself that makes the laws true. The distribution of properties is what is getting summarized. I've given independent reasons above to reject pan-dispositionalism, so there's no cost, in my view, to going with dualism or dual-aspect theory. The mosaic will consist of the distribution of powers and categorical properties (or, if you prefer, the qualitative aspects of metaphysically

---

[8] Samuel Kimpton-Nye (2017, 133) makes this sort of argument in favor of Demarest's proposal, as opposed to Bird's.

[9] Demarest (2017, 49), quoted in Kimpton-Nye (2017, 134).   [10] See Kimpton-Nye (2017).

neutral properties, which are also dispositional). On this view, a proposition such as '$F=ma$' might earn its keep as a law by functioning as an axiom of the best system and reporting a regularity. That's what makes it a law. So far, so Humean. But what make the *mosaic* the way it is—namely, such as to be describable by '$F=ma$'—are the powerful properties that are instantiated in it.

To breed the best powers-BSA hybrid, we should keep the BSA's job description steady: the best system systematizes the facts of the actual world. That is the 'user-facing' side of laws: law statements have to perform a function within a scientific practice. The other side, the metaphysical side, is where powers get invoked.[11] Unfortunately, this marriage of the BSA and the powers view still requires that its laws report regularities, even if those regularities are underwritten by powers. So even the best BSA-powers hybrid will still suffer from the central tension. We're stuck with the main problem that made us abandon the BSA in the first place.

In my view, the best hybrid will substitute the rules theory about laws for the BSA's. We can say that the rules view tells us what the laws are; but *why* the laws are as they are is a separate question. Why is the mosaic of our world faithful to the rules? The powers theory can be brought in to give us an answer. Unlike the anti-realist theories, it aspires to give a metaphysical explanation for the laws alongside the scientific ones. And for all the reasons I've given, we should also trade the original, Aristotelian story about powers for the Blower.

Note that this final hybrid—rules-plus-the-Blower—isn't making the Blower into the truthmaker for the laws. It isn't saying that laws can be paired with the power in any way at all, neither one-to-many, one-to-one, or many-to-one. The job of truthmaker for the laws is already occupied by the mosaic itself. Instead, the Blower is on the scene to do a different job: to glue the tiles of the mosaic together.

7. Decisions

I think we've arrived at the two most powerful and defensible positions on the table: the rules theory, and its realist hybrid, rules-cum-Blower. I've tried my best to follow the rules of the game, and in particular not to allow a consideration a view simply rejects to count against it unless that consideration can claim a solid foundation in scientific practice. Even though no argument in

---

[11] Neil Williams (2019, 220) defends a view similar to the one I suggest here; he calls it the 'Best System Blueprint' account.

this book is immune from objection—this is philosophy, after all—I feel mildly confident that our winnowing process has been correct. What none of the arguments made so far can do for us is choose between the realist and anti-realist camps. Is the mosaic all there is, or is there some deep metaphysical element such as the Blower making it the way it is?

From one perspective, the question isn't important. For our two surviving competitors agree on the best way to cash out the concept of a law and the best way to understand how it can function as it does in our epistemic economy. The rules theory and its Blower-hybrid have a clear advantage, I think, over any competitor on that score.

Let me extend the chess analogy. Again suppose you're playing correspondence chess with someone you think is Kim Jong-un. His moves arrive by telegram, you communicate your moves, and the game goes on. Over time, you work out your theory of Kim's rules: the combination of the rules of chess and the strategies from which Kim never departs.

One day, you read in the papers that Kim Jong-un has never played chess. In fact, you have good reason to suspect that nobody at all is playing chess with you. The telegrams just keep turning up. What happens next depends on your interests. If you want to keep playing, you shrug your shoulders and go on. It really makes no difference either way.

From another point of view, the question could hardly be more pressing. Is there in fact anything ensuring that events unfold over time as they do? Before writing this book, I would have insisted that there had to be. I fully expected to deal quickly with the anti-realist option and then settle on some version of the powers view. But then I worked through the details of the available realisms. Only after seeing the exorbitant prices realist views exact did I begin to have doubts. We seem always in the business of positing something mysterious—the N-relation, a power with *esse-ad*—to explain the facts. At some point, we have to resist the temptation of unexplained explainers, divine or otherwise, and just live with the world as it is.

In particular, I find the Aristotelian powers view implausible, partly because it is so very hard to fully spell out. I think the moderns' 'little souls' objection cannot be answered, because I just don't know what it could mean for each power to have enough 'information' to behave itself properly around all the others. That leaves the Blower and the Blobject. This is a real improvement over the original Aristotelian view. And although in many respects it converges with the revised universals view, it wins that contest—albeit narrowly—on points.

If the Blower exists, so much the better. But I have to confess, I find it simply not believable. The Blower beats out all realist competitors and it can provide metaphysical backing for the tiles of the mosaic. Still, the cost, to my mind, is exorbitant: a single, massively complex power, with all the mysteries of *esse-ad* thrown in. Whether it exists or not, the telegrams will keep coming, and moves will have to be made.

References

Airaksinen, Timo. 2010. 'Berkeley and Newton on Gravity in Siris.' In *George Berkeley: Religion and Science in the Age of Enlightenment*, ed. Silvia Parigi. New York: Springer.

Anon. 1819. *Enquiry Respecting the Relation of Cause and Effect*. Edinburgh: James Ballantyne.

Aquinas, St Thomas. 1997. *Basic Writings of St. Thomas Aquinas*, ed. Anton C. Pegis. 2 vols. Indianapolis: Hackett Publishing Company.

Aristotle. 1984. *The Complete Works of Aristotle: The Revised Oxford Translation*, ed. J. A. Barnes and W. D. Ross. Vol. 1. 2 vols. Princeton: Princeton University Press.

Armstrong, David Malet. 1981. 'The Causal Theory of the Mind.' In *The Nature of Mind and Other Essays*, 16–31. Ithaca, NY: Cornell University Press.

Armstrong, David Malet. 1983. *What is a Law of Nature?* Cambridge: Cambridge University Press.

Armstrong, David Malet. 1993. 'The Identification Problem and the Inference Problem.' *Philosophy and Phenomenological Research* 53 (2): 421–2.

Armstrong, David Malet. 1997. *A World of States of Affairs*. Cambridge: Cambridge University Press.

Armstrong, David Malet. 2005. 'Four Disputes About Properties.' *Synthese* 144 (3): 309–20.

Armstrong, David Malet. 2010. *Sketch for a Systematic Metaphysics*. Oxford: Oxford University Press.

Ayer, Alfred Jules. 1956. 'What is a Law of Nature?' *Revue internationale de philosophie* 10 (36, 2): 144–65.

Backmann, Marius, and Alexander Reutlinger. 2014. 'Better Best Systems—Too Good to be True.' *Dialectica* 68 (3): 375–90.

Bacon, Francis. 2000. *The New Organon*, ed. Lisa Jardine and Michael Silverthorne. Cambridge: Cambridge University Press.

Barfoot, Michael. 1990. 'Hume and the Culture of Science in the Early Eighteenth Century.' In *Studies in the Philosophy of the Scottish Enlightenment*, ed. M. A. Stewart, 151–90. Oxford: Oxford University Press.

Barker, Stephen. 2009. 'Dispositional Monism, Relational Constitution and Quiddities.' *Analysis* 69 (2): 242–50.

Barker, Stephen 2013. 'The Emperor's New Metaphysics of Powers.' *Mind* 122 (487): 605–53.

Bauer, William A. 2019. 'Powers and the Pantheistic Problem of Unity.' *Sophia* 58 (4): 563–80.

Bechtel, William, and Robert C. Richardson. 1993. *Discovering Complexity: Decomposition and Localization as Strategies in Scientific Research*. Princeton: Princeton University Press.

Beebee, Helen. 2000. 'The Non-Governing Conception of Laws of Nature.' *Philosophy and Phenomenological Research* 61 (3): 571–94.

Beebee, Helen. 2011. 'Necessary Connections and the Problem of Induction.' *Noûs* 45 (3): 504–27.

Bell, J.S. 1964. 'On the Einstein Podolsky Rosen Paradox.' *Physics* 1 (3): 195–200.

Berenstain, Nora, and James Ladyman. 2012. 'Ontic Structural Realism and Modality.' In *Structural Realism: Structure, Object, and Causality*, ed. Elaine Landry and Dean Rickles, 149–68. New York: Springer.

Berkeley, George. 1901. *The Works of George Berkeley*, ed. A. C. Fraser. 4 vols. Oxford: Clarendon Press.

Berkeley, George. 1975. *Philosophical Works, Including the Works on Vision*, ed. Michael Ayers. London: J. M. Dent & Sons.

Berkeley, George. 1991. *De Motu and the Analyst: A Modern Edition with Notes and Commentary*, ed. Douglas Jesseph. New York: Springer.

Bhaskar, Roy. 1978. *A Realist Theory of Science*. London: Routledge.

Bhogal, Harjit. Forthcoming. 'Nomothetic Explanation and Humeanism about Laws of Nature.' *Oxford Studies in Metaphysics*.

Bigelow, John. 1988. *The Reality of Numbers: A Physicalist's Philosophy of Mathematics*. Oxford: Oxford University Press.

Bigelow, John, Brian Ellis, and Caroline Lierse. 2004. 'The World as One of a Kind: Natural Necessity and Laws of Nature.' In *Readings on Laws of Nature*, ed. John W. Carroll, 141–60. Pittsburgh: University of Pittsburgh Press.

Bird, Alexander. 2005. 'The Ultimate Argument against Armstrong's Contingent Necessitation View of Laws.' *Analysis* 65 (2): 147–55.

Bird, Alexander. 2007a. *Nature's Metaphysics*. Oxford: Oxford University Press.

Bird, Alexander. 2007b. 'The Regress of Pure Powers?' *The Philosophical Quarterly* 57 (229): 513–34.

Bird, Alexander. 2016. 'Overpowering: How the Powers Ontology has Overreached Itself.' *Mind* 125 (498): 341–83.

Bird, Alexander, Brian Ellis, Stathis Psillos, and Stephen Mumford. 2006. 'Looking for Laws.' *Metascience* 15 (3): 437–69.

Black, Robert. 2000. 'Against Quidditism.' *Australasian Journal of Philosophy* 78 (1): 87–104.

Blackburn, Simon. 1984. *Spreading the Word: Groundings in the Philosophy of Language*. Oxford: Clarendon Press.

Blackburn, Simon. 1990. 'Filling in Space.' *Analysis* 50 (2): 62–5.

Boyle, Robert. 1991. *Selected Philosophical Papers of Robert Boyle*, ed. M. A. Stewart. Indianapolis: Hackett Publishing Company.

Braddon-Mitchell, David. 2001. 'Lossy Laws.' *Nous* 35 (2): 260–77.

Braithwaite, R. B. 1968. *Scientific Explanation*. Cambridge: Cambridge University Press.

Brook, Richard J. 1973. *Berkeley's Philosophy of Science*. The Hague: M. Nijhoff.

Brown, Thomas. 1835. *Inquiry into the Relation of Cause and Effect*. 4th edn. London: Henry G. Bohn.

Callender, Craig. 2004. 'Measures, Explanations and the Past: Should "Special" Initial Conditions Be Explained?' *British Journal for the Philosophy of Science* 55 (2): 195–217.

Carnap, Rudolf. 1936. 'Testability and Meaning.' *Philosophy of Science* 3 (4): 419–71.

Carraud, Vincent. 2002. *Causa Sive Ratio: la raison de la cause de Suarez à Leibniz*. Paris: Presses Universitaires de France.

Carroll, John. 1990. 'The Humean Tradition.' *Philosophical Review* 99 (2): 185–219.

Carroll, John. 1994. *Laws of Nature*. Cambridge: Cambridge University Press.

Carroll, John. 2012. Review of John T. Roberts: The Law-Governed Universe. *British Journal for the Philosophy of Science* 63 (4): 895–901.

Carroll, John. 2018. 'Becoming Humean.' In *Laws of Nature*, ed. Walter Ott and Lydia Patton. Oxford: Oxford University Press.

Cartwright, Nancy. 1983a. 'Do the Laws of Physics State the Facts?' In *How the Laws of Physics Lie*. Oxford: Clarendon Press.

Cartwright, Nancy. 1983b. *How the Laws of Physics Lie*. Oxford: Clarendon Press.

Cartwright, Nancy. 1993. 'In Defence of "This Worldly" Causality: Comments on van Fraassen's Laws and Symmetry.' *Philosophy and Phenomenological Research* 53 (2): 423–9.

Cartwright, Nancy. 2009. 'Causal Laws, Policy Predictions and the Need for Genuine Powers.' In *Dispositions and Causes*, ed. Toby Handfield, 138–68. Oxford: Oxford University Press.

Cartwright, Nancy. 2019. *Nature, the Artful Modeler*. Chicago: Open Court.

Chakravartty, Anjan. 2007. *A Metaphysics for Scientific Realism*. Cambridge: Cambridge University Press.

Chakravartty, Anjan. 2017. 'Saving the Scientific Phenomena: What Powers Can and Cannot Do.' In *Putting Powers to Work*, ed. J. D. Jacobs, 24–37. Oxford: Oxford University Press.

Charleton, Walter. 1654. *Physiologia Epicuro-Gassendo-Charletoniana*. London: Tho. Newcomb for Thomas Heath.

Chen, Eddy Keming, and Sheldon Goldstein. Forthcoming. 'Governing Without a Fundamental Direction of Time: Minimal Primitivism about Laws of Nature.' In *Rethinking the concept of Laws of Nature*, ed. Yemima Ben-Menahem. Berlin: Springer.

Choi, Sungho, and Michael Fara. 2018. 'Dispositions.' In *The Stanford Encyclopedia of Philosophy*, ed. Edward N. Zalta, Fall 2018. Metaphysics Research Lab, Stanford University. <https://plato.stanford.edu/archives/fall2018/entries/dispositions/>.

Cobb, Aaron D. 2016. 'Mill's Philosophy of Science.' In *A Companion to Mill*, ed. Christopher Macleod and Dale E. Miller, 234–49. London: Blackwell.

Cohen, Jonathan, and Craig Callender. 2009. 'A Better Best System Account of Lawhood.' *Philosophical Studies* 145 (1): 1–34.

Cohen, Jonathan, and Craig Callender. 2010. 'Special Sciences, Conspiracy, and the Better Best System Account of Lawhood.' *Erkenntnis* 73 (3): 427–47.

Contessa, Gabriele. 2015. 'Only Powers Can Confer Dispositions.' *The Philosophical Quarterly* 65 (259): 160–76.

Corry, Richard. 2019. *Power and Influence: The Metaphysics of Reductive Explanation*. Oxford, UK: Oxford University Press.

Creary, Lewis. 1981. 'Causal Explanation and the Reality of Natural Component Forces.' *Pacific Philosophical Quarterly* 62 (2): 148–57.

Crombie, A. C. 1994. *Styles of Scientific Thinking in the European Tradition: The History of Argument and Explanation Especially in the Mathematical and Biomedical Sciences and Arts*. 3 vols. London: Duckworth.

Cross, Troy. 2005. 'What Is a Disposition?' *Synthese* 144 (3): 321–41.

Cudworth, Ralph. 1837. *The True Intellectual System of the Universe*. 2 vols. New York: Gould and Newman.

Curley, Edwin M. 1969. *Spinoza's Metaphysics: An Essay in Interpretation*. Cambridge, MA: Harvard University Press.

Curley, Edwin M. 1988. *Behind the Geometrical Method: A Reading of Spinoza's Ethics*. Princeton: Princeton University Press.

Demarest, Heather. 2012. 'Do Counterfactuals Ground the Laws of Nature? A Critique of Lange.' *Philosophy of Science* 79 (3): 333–44.

Demarest, Heather. 2017. 'Powerful Properties, Powerless Laws.' In *Putting Powers to Work: Causal Powers in Contemporary Metaphysics*, ed. J. Jacobs, 39–54. Oxford: Oxford University Press.

Descartes, René. 1984. *The Philosophical Writings of Descartes*. Vols 1 and 2, ed. John Cottingham, Robert Stoothoff, and Dugald Murdoch; vol. 3, ed. Cottingham, Stoothoff, Murdoch, and Anthony Kenny. Cambridge: Cambridge University Press.

Descartes, René. 1996. *Oeuvres de Descartes*. 12 vols, ed. Charles Adam and Paul Tannery. Paris: Librarie Philosophique J. Vrin.

Dipert, Randall R. 1997. 'The Mathematical Structure of the World: The World as Graph.' *Journal of Philosophy* 94 (7): 329–58.

Domski, Mary. 2018. 'Laws of Nature and the Divine Order of Things: Descartes and Newton on Truth in Natural Philosophy.' In *Laws of Nature*, ed. Walter Ott and Lydia Patton, 42–61. Oxford: Oxford University Press.

Dorr, Cian. 2010. 'Review of Everything Must Go By Ladyman et Al.' *Notre Dame Philosophical Reviews*. <https://ndpr.nd.edu/reviews/every-thing-must-go-metaphysics-naturalized/>.

Dorst, Chris. 2019a. 'Towards a Best Predictive System Account of Laws of Nature.' *British Journal for the Philosophy of Science* 70 (3): 877–900.

Dorst, Chris. 2019b. 'Humean Laws, Explanatory Circularity, and the Aim of Scientific Explanation.' *Philosophical Studies* 176 (10): 2657–79.

Douven, Igor. 2017. 'Abduction.' In *The Stanford Encyclopedia of Philosophy*, ed. Edward N. Zalta. Metaphysics Research Lab, Stanford University. <https://plato.stanford.edu/archives/sum2017/entries/abduction/>.

Downing, Lisa. 1995. 'Siris and the Scope of Berkeley's Instrumentalism.' *British Journal for the History of Philosophy* 3 (2): 279–300.

Downing, Lisa. 2005. 'Berkeley's Natural Philosophy and Philosophy of Science.' In *The Cambridge Companion to Berkeley*, ed. Kenneth Winkler, 230–65. Cambridge: Cambridge University Press.

Downing, Lisa. 2021. 'Qualities, Powers, and Bare Powers in Locke.' In *Reconsidering Causal Powers: Historical and Conceptual Perspectives*, ed. Benjamin Hill, Henrik Lagerlund, and Stathis Psillos, 186–205. Oxford: Oxford University Press.

Dretske, Fred. 1977. 'Laws of Nature.' *Philosophy of Science* 44: 248–68.

Dretske, Fred. 2002. 'A Recipe for Thought.' In *Philosophy of Mind: Classical and Contemporary Readings*, ed. David J. Chalmers, 491–9. New York: Oxford University Press.

Du Châtelet, Emilie. 1740. *Institutions de Physique*. Amsterdam: Pierre Mortier.

Ducheyne, Steffen. 2009. 'Whewell, Necessity, and the Inductive Sciences: A Philosophical-Systematic Survey.' *South African Journal of Philosophy* 28 (4): 333–58.

Dupleix, Scipion. 1640. *La Physique, ou science des choses naturelles*. Rouen: Louys du Mesnil.

Earman, John, and John Roberts. 1999. '"Ceteris Paribus," There Is No Problem of Provisos.' *Synthese* 118 (3): 439–78.

Edgington, Dorothy. 2008. 'Counterfactuals.' *Proceedings of the Aristotelian Society* 108 (1): 1–21.

Edgington, Dorothy. 2020. 'Indicative Conditionals.' In *The Stanford Encyclopedia of Philosophy*, ed. Edward N. Zalta, Fall 2020. Metaphysics Research Lab, Stanford University. <https://plato.stanford.edu/archives/fall2020/entries/conditionals/>.

Ellis, Brian. 2001. *Scientific Essentialism*. Cambridge: Cambridge University Press.

Ellis, Brian. 2002. *The Philosophy of Nature: A Guide to the New Essentialism*. Montreal: McGill-Queen's University Press.

Ellis, Brian. 2010. 'Causal Powers and Categorical Properties.' In *The Metaphysics of Powers*, ed. Anna Marmodoro, 133–42. London: Routledge.

Ellis, Brian. 2021. 'Causal Powers and Structures.' In *Reconsidering Causal Powers: Historical and Conceptual Perspectives*, ed. Benjamin Hill, Henrik Lagerlund, and Stathis Psillos, 271–83. Oxford: Oxford University Press.

Emery, Nina. 2019. 'Laws and their Instances.' *Philosophical Studies* 176 (6): 1535–61.

Euler, Leonhard. 1750. 'Reflexions sur l'espace et le tems.' *Memoires de l'Académie des Sciences de Berlin*, no. 4: 324–33.

Euler, Leonhard. 1802. *Letters of Euler on Different Subjects in Physics and Philosophy Addressed to a German Princess*, trans. Henry Hunter. 2 vols. London: Murray and Highley.

Fodor, Jerry. 1974. 'Special Sciences (Or: The Disunity of Science as a Working Hypothesis).' *Synthese* 28 (2): 97–115.

Fodor, Jerry A., and Ernest Lepore. 1992. *Holism: A Shopper's Guide*. London: Blackwell.

Forbes, Graeme, and Rachael Briggs. 2017. 'The Growing-Block: Just One Thing After Another?' *Philosophical Studies* 174 (4): 927–43.

Foster, John. 2004. *The Divine Lawmaker: Lectures on Induction, Laws of Nature, and the Existence of God*. Oxford: Oxford University Press.

Friedman, Michael. 1974. 'Explanation and Scientific Understanding.' *Journal of Philosophy* 71 (1): 5–19.

Funkenstein, Amos. 1986. *Theology and the Scientific Imagination*. Princeton: Princeton University Press.

Galilei, Galileo. 2001. *Dialogue Concerning the Two Chief World Systems: Ptolemaic and Copernican*, ed. Stephen Jay Gould and Stillman Drake, trans. Stillman Drake. New York: Modern Library.

Galilei, Galileo. 2008. *The Essential Galileo*, ed. and trans. Maurice A. Finocchiaro. Indianapolis: Hackett.

Garber, Daniel. 1992. *Descartes's Metaphysical Physics*. Chicago: University of Chicago Press.

Garrett, Don. 1997. *Cognition and Commitment in Hume's Philosophy*. Oxford: Oxford University Press.

Gassendi, Pierre. 1658. *Petri Gassendi Opera Omnia in Sex Tomos Divisa*. 6 vols. Lyon: Laurent Anisson and Jean Baptiste Devenet.

Gaukroger, Stephen. 2001. *Francis Bacon and the Transformation of Early-Modern Philosophy*. Cambridge: Cambridge University Press.

Gibbard, Allan. 1990. *Wise Choices, Apt Feelings: A Theory of Normative Judgment*. Cambridge, MA: Harvard University Press.

Giere, Ronald N. 1999. *Science Without Laws*. Chicago: University of Chicago Press.

Glennan, Stuart. 2009. 'Mechanisms.' In *The Oxford Handbook of Causation*, ed. Helen Beebee, Christopher Hitchcock, and Peter Menzies. Oxford: Oxford University Press.

Glennan, Stuart. 2017. *The New Mechanical Philosophy*. Oxford: Oxford University Press.

Greaves, Hilary. 2011. 'In Search of (Spacetime) Structuralism.' *Philosophical Perspectives* 25 (1): 189–204.

Greco, John, and Ruth Groff, eds. 2012. *Powers and Capacities in Philosophy: The New Aristotelianism*. London: Routledge.

Grene, Marjorie, and David Depew. 2004. *The Philosophy of Biology: An Episodic History*. Cambridge: Cambridge University Press.

Gundersen, Lars. 2002. 'In Defence of the Conditional Account of Dispositions.' *Synthese* 130 (3): 389–411.

Hájek, Alan. 2014. 'Probabilities of Counterfactuals and Counterfactual Probabilities.' *Journal of Applied Logic* 12 (3): 235–51.

Hall, Ned. 2004. 'Two Mistakes About Credence and Chance.' *Australasian Journal of Philosophy* 82 (1): 93–111.

Hall, Ned. 2011. Review of *Laws & Lawmakers: Science, Metaphysics, and the Laws of Nature*, by Marc Lange. *Notre Dame Philosophical Reviews*, September. <https://ndpr.nd.edu/news/laws-lawmakers-science-metaphysics-and-the-laws-of-nature/>.

Hall, Ned. 2015. 'Humean Reductionism about Laws of Nature.' In *A Companion to David Lewis*, ed. Barry Loewer and Jonathan Schaffer, 262–77. Oxford: Wiley & Sons.

Halpin, John F. 2003. 'Scientific Law: A Perspectival Account.' *Erkenntnis* 58 (2): 137–68.

Handfield, Toby. 2005. 'Armstrong and the Modal Inversion of Dispositions.' *Philosophical Quarterly* 55 (220): 452–61.

Harper, William. 2012. 'Newton, Huygens, and Euler: Empirical Support for Laws of Motion.' In *Interpreting Newton: Critical Essays*, ed. Andrew Janiak. Cambridge: Cambridge University Press.

Harrison, Peter. 2002. 'Voluntarism and Early Modern Science.' *History of Science* 40 (1): 63–89.

Harrison, Peter. 2019. 'Laws of God or Laws of Nature? Natural Order in the Early Modern Period.' In *Science Without God? Rethinking the History of Scientific Naturalism*, ed. Peter Harrison and Jon H. Roberts. Oxford: Oxford University Press.

Hattab, Helen. 2007. 'Concurrence or Divergence? Reconciling Descartes's Physics with his Metaphysics.' *Journal of the History of Philosophy* 45 (1): 49–78.

Hattab, Helen. 2018. 'Early Modern Roots of the Philosophical Concept of a Law of Nature.' In *Laws of Nature*, ed. Walter Ott and Lydia Patton, 18–41. Oxford: Oxford University Press.

Healey, Richard. 2008. Review of *The Metaphysics Within Physics*, by Tim Maudlin. *Notre Dame Philosophical Reviews*, February. <https://ndpr.nd.edu/news/the-metaphysics-within-physics/>.

Healey, Richard. 2017. *The Quantum Revolution in Philosophy*. Oxford: Oxford University Press.

Heil, John. 2003. *From an Ontological Point of View*. Oxford: Oxford University Press.

Heil, John. 2012. *The Universe as we Find it*. Oxford: Oxford University Press.

Helmholtz, Hermann von. 1995. *Science and Culture: Popular and Philosophical Essays*, ed. David Cahan. Chicago: University of Chicago Press.

Hempel, Carl G. 1988. 'Provisoes: A Philosophical Problem Concerning the Inferential Function of Scientific Laws.' In *The Limits of Deductivism*, ed. A. Grünbaum and W. Salmon, 19–36. Berkeley: University of California Press.

Henninger, Mark Gerald. 1989. *Relations: Medieval Theories 1250–1325*. Oxford: Oxford University Press.

Henry, John. 1994. ' "Pray Do Not Ascribe That Notion to Me": God and Newton's Gravity.' In *The Books of Nature and Scripture: Recent Essays on Natural Philosophy, Theology and Biblical Criticism in the Netherlands of Spinoza's Time and the British Isles of Newton's Time*, ed. James E. Force and Richard H. Popkin, 123–47. International Archives of the History of Ideas/Archives Internationales D'Histoire Des Idées. Dordrecht: Springer Netherlands.

Henry, John. 2004. 'Metaphysics and the Origins of Modern Science: Descartes and the Importance of Laws of Nature.' *Early Science and Medicine* 9 (2): 73–114.

Hicks, Michael Townsen. 2017. 'Dynamic Humeanism.' *British Journal for the Philosophy of Science* 69 (4): 983–1007.

Hildebrand, Tyler. 2013. 'Can Primitive Laws Explain?' *Philosopher's Imprint* 13 (1): 1–15.

Hildebrand, Tyler. 2016. 'Natural Properties, Necessary Connections, and the Problem of Induction.' *Philosophy and Phenomenological Research* 96 (3): 668–89.

Hildebrand, Tyler, and Thomas Metcalf. Forthcoming. 'The Nomological Argument for the Existence of God.' *Noûs*.

Hill, Benjamin. 2021. 'The Ontological Status of Causal Powers: Substances, Modes, and Humeanism.' In *Reconsidering Causal Powers: Historical and Conceptual Perspectives*, ed. Benjamin Hill, Henrik Lagerlund, and Stathis Psillos, 121–48. Oxford: Oxford University Press.

Hobbes, Thomas. 1839. *The English Works of Thomas Hobbes of Malmesbury*. Vol. 1. 11 vols. London: J. Bohn.

Hume, David. 1999. *An Enquiry Concerning Human Understanding*, ed. Tom L. Beauchamp. Oxford: Oxford University Press.

Hume, David. 2000. *A Treatise of Human Nature*, ed. David Fate Norton and Mary J. Norton. Oxford: Oxford University Press.

Hüttemann, Andreas. 1998. 'Laws and Dispositions.' *Philosophy of Science* 65 (1): 121–35.

Hüttemann, Andreas. 2021. 'The Return of Causal Powers?' In *Reconsidering Causal Powers: Historical and Conceptual Perspectives*, ed. Benjamin Hill, Henrik Lagerlund, and Stathis Psillos, 168–85. Oxford: Oxford University Press.

Huygens, Christiaan. 1669. 'Extrait d'une lettre de M. Hugens à l'auteur du journal.' *Journal des Sçavans* 18 (March): 28–32.

Ingthorsson, R. D. 2015. 'The Regress of Pure Powers Revisited.' *European Journal of Philosophy* 23 (3): 529–41.

Jaag, Siegfried, and Christian Loew. 2020. 'Making Best Systems Best for Us.' *Synthese* 197: 2525–50.

Jackson, Frank. 1998. *From Metaphysics to Ethics: A Defence of Conceptual Analysis*. New York: Oxford University Press.

Jacobs, Jonathan D. and Robert J. Hartman. 2017. 'Armstrong on Probabilistic Laws of Nature.' *Philosophical Papers* 46 (3): 373–387.

Jacobson, Anne Jaap. 1986. 'Causality and the Supposed Counterfactual Conditional in Hume's Enquiry.' *Analysis* 46 (3): 131–3.

Janiak, Andrew. 2008. *Newton as Philosopher*. Cambridge: Cambridge University Press.

Jantzen, Benjamin C. 2014. *An Introduction to Design Arguments*. Cambridge: Cambridge University Press.

Jolley, Nicholas. 2002. 'Occasionalism and Efficacious Laws in Malebranche.' In *Renaissance and Early Modern Philosophy*, ed. Peter French and Howard Wettstein, 245–57. New York: Blackwell.

Jolley, Nicholas. 2019. 'Malebranche, Occasionalism, and the Janus Faces of Laws.' In *Occasionalism: From Metaphysics to Science*, ed. M. Camposampiero, M. Priarolo, and E. Scribano, 127–45. Turnhout, Belgium: Brepols Publishers.

Katz, Victor J. 1998. *A History of Mathematics*. Reading, MA: Addison-Wesley.

Kaufman, Dan. 2006. 'Locks, Schlocks, and Poisoned Peas: Boyle on Actual and Dispositive Qualities.' In *Oxford Studies in Early Modern Philosophy Volume 3*, ed. Daniel Garber and Steven Nadler, 153–98. Oxford: Clarendon Press.

Kim, Jaegwon. 1998. *Mind in a Physical World: An Essay on the Mind–Body Problem and Mental Causation*. Cambridge, MA: MIT Press.

Kimpton-Nye, Samuel. 2017. 'Humean Laws in an UnHumean World.' *Journal of the American Philosophical Association* 3 (2): 129–47.

Kistler, Max. 2006. *Causation and Laws of Nature.* London: Routledge.

Kitcher, Philip. 1989. 'Explanatory Unification and the Causal Structure of the World.' In *Scientific Explanation,* ed. Philip Kitcher and Wesley C. Salmon, 410–505. Minneapolis: University of Minnesota Press.

Kripke, Saul. 1982. *Wittgenstein on Rules and Private Language.* Cambridge, MA: Harvard University Press.

Ladyman, James, Don Ross, John Collier, and David Spurrett. 2007. *Every Thing Must Go: Metaphysics Naturalized.* Oxford: Oxford University Press.

Lange, Marc. 1993. 'Natural Laws and the Problem of Provisos.' *Erkenntnis* 38 (2): 233–48.

Lange, Marc. 2002. *An Introduction to the Philosophy of Physics: Locality Fields, Energy, and Mass.* London: Blackwell.

Lange, Marc. 2009. *Laws and Lawmakers: Science, Metaphysics, and the Laws of Nature.* Oxford: Oxford University Press.

Lange, Marc. 2013. 'Grounding, Scientific Explanation, and Humean Laws.' *Philosophical Studies* 164 (1): 255–61.

Lange, Marc, Jim Woodward, Barry Loewer, and John W. Carroll. 2011. 'Counterfactuals All the Way down? Marc Lange: Laws and Lawmakers: Science, Metaphysics, and the Laws of Nature.' *Metascience* 20 (1): 27–52.

Langton, Rae, and David Lewis. 1998. 'Defining "Intrinsic."' *Philosophy and Phenomenological Research* 58 (2): 333–45.

Lehoux, Daryn. 2006. 'Laws of Nature and Natural Laws.' *Studies in History and Philosophy of Science Part A* 37 (4): 527–49.

Leibniz, Gottfried Wilhelm. 1989. *Leibniz: Philosophical Essays,* trans. Roger Ariew and Daniel Garber. 1st edn. Indianapolis: Hackett Publishing Company.

Leibniz, Gottfried Wilhelm, and Antoine Arnauld. 2016. *The Leibniz–Arnauld Correspondence: With Selections from the Correspondence with Ernst, Landgrave of Hessen-Rheinfels,* trans. Stephen Voss. New Haven: Yale.

Leibniz, G. W., and Samuel Clarke. 1956. *The Leibniz–Clarke Correspondence,* ed. H. G. Alexander. London: The Philosophical Library.

Lewis, David K. 1979. 'Counterfactual Dependence and Time's Arrow.' *Noûs* 13 (4): 455–76.

Lewis, David K. 1983. 'New Work for a Theory of Universals.' *Australasian Journal of Philosophy* 61 (4): 343–77.

Lewis, David K. 1986a. *On the Plurality of Worlds.* Oxford: Wiley-Blackwell.

Lewis, David K. 1986b. *Philosophical Papers.* Vol. 2. Oxford: Oxford University Press.

Lewis, David K. 1994. 'Humean Supervenience Debugged.' *Mind* 103 (412): 473–90.

Lewis, David K. 1997. 'Finkish Dispositions.' *Philosophical Quarterly* 47 (187): 143–58.

Lewis, David K. 1999. *Papers in Metaphysics and Epistemology: Volume 2.* Cambridge: Cambridge University Press.

Lewis, David K. 2009. 'Ramseyan Humility.' In *Conceptual Analysis and Philosophical Naturalism,* ed. David Braddon-Mitchell and Robert Nola, 203–22. Cambridge, MA: MIT Press.

Lewis, Peter J. 2016. *Quantum Ontology: A Guide to the Metaphysics of Quantum Mechanics.* New York: Oxford University Press.

Lipton, Peter. 1999. 'All Else Being Equal.' *Philosophy* 74 (288): 155–68.

Locke, John. 1975. *An Essay Concerning Human Understanding,* ed. Peter H. Nidditch. Oxford: Oxford University Press.

Loewer, Barry. 1996. 'Humean Supervenience.' *Philosophical Topics* 24 (1): 101–27.

Loewer, Barry. 2007. 'Laws and Natural Properties.' *Philosophical Topics* 35 (1/2): 313–28.

Loewer, Barry. 2012. 'Two Accounts of Laws and Time.' *Philosophical Studies* 160 (1): 115–37.

Loewer, Barry. 2019. 'Humean Laws and Explanation.' *Principia: An International Journal of Epistemology* 23 (3): 373–85.

Loewer, Barry. Forthcoming a. 'The Package Deal Account of Laws and Properties.' *Synthese.*

Loewer, Barry. Forthcoming b. 'What Breathes Fire into the Equations.'

LoLordo, Antonia. 2006. *Pierre Gassendi and the Birth of Early Modern Philosophy.* Cambridge: Cambridge University Press.

LoLordo, Antonia. 2019. 'Mary Shepherd on Causation, Induction, and Natural Kinds.' *Philosopher's Imprint* 19 (52): 1–14.

Lowe, E. J. 2005. *The Four-Category Ontology: A Metaphysical Foundation for Natural Science.* Oxford: Clarendon Press.

Lowe, E. J. 2011. 'How Not to Think of Powers: A Deconstruction of the "Dispositions and Conditionals Debate."' *The Monist* 94 (1): 19–33.

Lowe, E. J. 2012. 'Mumford and Anjum on Causal Necessitarianism and Antecedent Strengthening.' *Analysis* 72 (4): 731–5.

McCracken, Charles J. 1983. *Malebranche and British Philosophy.* Oxford: Oxford University Press.

Mach, Ernst. 1893. *The Science of Mechanics*, trans. Thomas J. McCormack. Chicago: Open Court.

Machamer, Peter K., Lindley Darden, and Carl F. Craver. 2000. 'Thinking about Mechanisms.' *Philosophy of Science* 67 (1): 1–25.

McKitrick, Jennifer. 2003a. 'A Case for Extrinsic Dispositions.' *Journal of Philosophy* 81 (2): 155–74.

McKitrick, Jennifer. 2003b. 'The Bare Metaphysical Possibility of Bare Dispositions.' *Philosophy and Phenomenological Research* 66 (2): 349–69.

McKitrick, Jennifer. 2005. 'Are Dispositions Causally Relevant?' *Synthese* 144 (3): 357–71.

McKitrick, Jennifer. 2009. 'Dispositional Pluralism.' In *Debating Dispositions: Issues in Metaphysics, Epistemology and Philosophy of Mind*, ed. Gregor Damschen, Robert Schnepf, and Karsten Stueber, 186–203. Berlin: de Gruyter.

McKitrick, Jennifer. 2018. *Dispositional Pluralism.* Oxford: Oxford University Press.

McKitrick, Jennifer. 2021. 'Resurgent Powers and the Failure of Conceptual Analysis.' In *Reconsidering Causal Powers: Historical and Conceptual Perspectives*, ed. Benjamin Hill, Henrik Lagerlund, and Stathis Psillos, 241–70. Oxford: Oxford University Press.

Malebranche, Nicolas. 1958. *Oeuvres complètes*, ed. A. Robinet. 20 vols. Paris: Vrin.

Malebranche, Nicolas. 1992. *Treatise on Nature and Grace*, trans. Patrick Riley. Oxford: Clarendon Press.

Malebranche, Nicolas. 1997a. *Dialogues on Metaphysics and on Religion.* Cambridge: Cambridge University Press.

Malebranche, Nicolas. 1997b. *The Search after Truth*, ed. Paul Olscamp and Thomas Lennon. Cambridge: Cambridge University Press.

Martin, C. B. 1993a. 'The Need for Ontology: Some Choices.' *Philosophy* 68 (266): 505–22.

Martin, C. B. 1993b. 'Power for Realists.' In *Ontology, Causality and Mind: Essays in Honour of D. M. Armstrong*, ed. John Bacon, Keith Campbell, and Lloyd Reinhardt, 175–85. Cambridge: Cambridge University Press.

Martin, C. B. 1994. 'Dispositions and Conditionals.' *Philosophical Quarterly* 44 (174): 1–8.

Martin, C. B. 1997. 'On the Need for Properties: The Road to Pythagoreanism and Back.' *Synthese* 112 (2): 193–231.

Martin, C. B. 2007. *The Mind in Nature*. Oxford, UK: Oxford University Press.

Massimi, Michela. 2017. 'Laws of Nature, Natural Properties, and the Robustly Best System.' *The Monist* 100 (3): 406–21.

Massimi, Michela. 2018. 'A Perspectivalist Better Best System Account of Lawhood.' In *Laws of Nature*, ed. Walter Ott and Lydia Patton, 139–57. Oxford: Oxford University Press.

Maudlin, Tim. 2007. *The Metaphysics Within Physics*. Oxford: Oxford University Press.

Mayr, Ernst. 1988. *Toward a New Philosophy of Biology: Observations of an Evolutionist*. Cambridge, MA: Harvard University Press.

Mellor, D. H. 2000. 'The Semantics and Ontology of Dispositions.' *Mind* 109 (436): 757–80.

Menzies, Peter. 1993. 'Laws of Nature, Modality, and Humean Supervenience.' In *Ontology, Causality, and Mind*, ed. J. K. Bacon, K. Campbell, and L. Reinhardt, 195–224. Cambridge: Cambridge University Press.

Mill, John Stuart. 1973. *The Collected Works of John Stuart Mill*, ed. J. M. Robson. 32 vols. Toronto: University of Toronto Press.

Miller, Elizabeth. 2015. 'Humean Scientific Explanation.' *Philosophical Studies* 172 (5): 1311–32.

Millican, Peter. 2002. 'Hume's Sceptical Doubts Concerning Induction.' In *Reading Hume on Human Understanding*, ed. Peter Millican, 107–74. Oxford: Oxford University Press.

Milton, John R. 1998. 'Laws of Nature.' In *The Cambridge History of Seventeenth-Century Philosophy*, ed. Michael Ayers and Garber, Daniel, 1: 680–701. Cambridge: Cambridge University Press.

Molnar, George. 1999. 'Are Dispositions Reducible?' *Philosophical Quarterly* 49 (194): 1–17.

Molnar, George. 2003. *Powers: A Study in Metaphysics*, ed. Stephen Mumford. Oxford: Oxford University Press.

Mumford, Stephen. 1998. *Dispositions*. Oxford University Press.

Mumford, Stephen. 1999. 'Intentionality and the Physical: A New Theory of Disposition Ascription.' *Philosophical Quarterly* 49 (195): 215–25.

Mumford, Stephen. 2004. *Laws in Nature*. London: Routledge.

Mumford, Stephen. 2006. 'The Ungrounded Argument.' *Synthese* 149 (3): 471–89.

Mumford, Stephen. 2013. 'The Power of Power.' In *Powers and Capacities in Philosophy: The New Aristotelianism*, ed. Ruth Groff and John Greco, 9–24. London: Routledge.

Mumford, Stephen. 2018. 'Laws and their Exceptions.' In *Laws of Nature*, ed. Walter Ott and Lydia Patton, 205–20. Oxford: Oxford University Press.

Mumford, Stephen, and Rani Lill Anjum. 2011. *Getting Causes From Powers*. Oxford: Oxford University Press.

Mumford, Stephen, and Rani Lill Anjum. 2018. *What Tends to Be: The Philosophy of Dispositional Modality*. London: Routledge.

Nadler, Steven M. 1993. 'Occasionalism and General Will in Malebranche.' *Journal of the History of Philosophy* 31 (1): 31–47.

Newton, Isaac. 1999. *The Principia: The Authoritative Translation and Guide: Mathematical Principles of Natural Philosophy*, trans. I. Bernard Cohen, Anne Whitman, and Julia Budenz. 1st edn. Berkeley: University of California Press.

Newton, Isaac. 2004. *Philosophical Writings*, ed. Andrew Janiak. Cambridge: Cambridge University Press.

Nida-Rümelin, Martine. 1996. 'Pseudonormal Vision: An Actual Case of Qualia Inversion?' *Philosophical Studies* 82 (2): 145–57.

Oakley, Francis. 1961. 'Christian Theology and the Newtonian Science: The Rise of the Concept of the Laws of Nature.' *Church History* 30 (4): 433–57.

Oakley, Francis. 2019. 'The Rise of the Concept of Laws of Nature Revisited.' *Early Science and Medicine* 24 (1): 1–32.

Ott, Walter. 2009. *Causation and Laws of Nature in Early Modern Philosophy.* Oxford: Oxford University Press.

Ott, Walter. 2015. 'Locke and the Real Problem of Causation.' *Locke Studies* 15: 53–77.

Ott, Walter. 2017. '"Archetypes Without Patterns": Locke on Relations and Mixed Modes.' *Archiv für Geschichte der Philosophie* 99 (3): 300–25.

Ott, Walter. 2018. 'Leges Sive Natura: Bacon, Spinoza, and a Forgotten Concept of Law.' In *Laws of Nature*, ed. Walter Ott and Lydia Patton, 62–79. Oxford: Oxford University Press.

Ott, Walter. 2019. 'Berkeley's Best System: An Alternative Approach to Laws of Nature.' *Journal of Modern Philosophy* 1 (1): 1–13.

Ott, Walter. 2021. 'The Case Against Powers.' In *Causal Powers in Science: Blending Historical and Conceptual Perspectives*, ed. Stathis Psillos, Benjamin Hill, and Henrik Lagerlund, 149–67. Oxford: Oxford University Press.

Place, Ullin T. 1999. 'Intentionality and the Physical: A Reply to Mumford.' *Philosophical Quarterly* 49 (195): 225–31.

Platt, Andrew. 2011. 'Divine Activity and Motive Power in Descartes's Physics.' *British Journal for the History of Philosophy* 19 (4): 623–646.

Prior, E. W. 1985. *Dispositions.* New York: Humanities Press.

Prior, Elizabeth, Robert Pargetter, and Frank Jackson. 1982. 'Three Theses about Dispositions.' *American Philosophical Quarterly* 19 (3): 251–7.

Psillos, Stathis. 2002. *Causation and Explanation.* Montreal: McGill-Queen's University Press.

Psillos, Stathis. 2006. 'What Do Powers Do When They Are Not Manifested?' *Philosophy and Phenomenological Research* 72 (1): 137–56.

Psillos, Stathis. 2009. 'Regularity Theories.' In *The Oxford Handbook of Causation*, ed. Helen Beebee, Christopher Hitchcock, and Peter Menzies, 132–57. Oxford: Oxford University Press.

Psillos, Stathis. 2014. 'Regularities, Natural Patterns and Laws of Nature.' *Theoria* 29 (1): 9–27.

Psillos, Stathis. 2018. 'Laws and Powers in the Frame of Nature.' In *Laws of Nature*, ed. Walter Ott and Lydia Patton, 80–107. Oxford: Oxford University Press.

Psillos, Stathis. 2021. 'The Inherence and Directedness of Powers.' In *Reconsidering Causal Powers: Historical and Conceptual Perspectives*, ed. Benjamin Hill, Henrik Lagerlund, and Stathis Psillos, 45–67. Oxford: Oxford University Press.

Psillos, Stathis, and Eirini Goudarouli. 2019. 'Principles of Motion and the Absence of Laws of Nature in Hobbes's Natural Philosophy.' *HOPOS: The Journal of the International Society for the History of Philosophy of Science* 9 (1): 93–119.

Putnam, Hilary. 1981. *Reason, Truth and History.* Cambridge: Cambridge University Press.

Ramsey, Frank Plumpton. 1990. 'General Propositions and Causality.' In *Philosophical Papers*, ed. D. H. Mellor, 145–63. Cambridge: Cambridge University Press.

Régis, Pierre-Sylvain. 1996. *L'Usage de la raison et de la foi.* Paris: Fayard.

Reid, Thomas. 1788. *Essays on the Active Powers of Man.* London: John Bell, G. G. J. & J. Robinson.

Rickles, Dean. 2016. *The Philosophy of Physics.* Cambridge: Polity Press.

Roberts, John T. 1998. 'Lewis, Carroll, and Seeing Through the Looking Glass.' *Australasian Journal of Philosophy* 76 (3): 426–38.

Roberts, John T. 2008. *The Law Governed Universe*. Oxford: Oxford University Press.

Robinson, Howard. 1982. *Matter and Sense: A Critique of Contemporary Materialism*. Cambridge: Cambridge University Press.

Rosenberg, Alex. 1993. 'Hume's Philosophy of Science.' In *The Cambridge Companion to Hume*, ed. David Fate Norton, 64–89. Cambridge: Cambridge University Press.

Rouhault, Jacques. 1671. *Traité de physique*. Paris: Charles Savreux.

Roux, Sophie. 2001. 'Les Lois de la nature à l'âge classique: la question terminologique.' *Revue de synthèse* 4 (2): 531–76.

Ruby, Jane E. 1986. 'The Origins of Scientific "Law."' *Journal of the History of Ideas* 47 (3): 341–59.

Russell, Bertrand. 1953. 'On the Notion of Cause.' *Proceedings of the Aristotelian Society* 13: 171–96.

Ryle, Gilbert. 1949. *The Concept of Mind*. London: Hutchinson & Co.

Salmon, Wesley C. 1984. *Scientific Explanation and the Causal Structure of the World*. Princeton, NJ: Princeton University Press.

Salmon, Wesley C. 1989. *Four Decades of Scientific Explanation*. Minneapolis: University of Minnesota Press.

Salusbury, Thomas. 1641. *Mathematical Collections and Translations*. Vol. 1. 2 vols. London: William Leybourne.

Sankey, Howard, ed. 1999. *Causation and Laws of Nature*. Dordrecht: Kluwer Academic Publishers.

Scarre, Geoffrey. 1998. 'Induction and Scientific Method.' In *The Cambridge Companion to Mill*, ed. John Skorupski, 112–38. Cambridge: Cambridge University Press.

Schaffer, Jonathan. 2016. 'It is the Business of Laws to Govern.' *Dialectica* 70 (4): 577–88.

Schliesser, Eric. 2020. 'Hume's Newtonianism and Anti-Newtonianism.' In *The Stanford Encyclopedia of Philosophy*, ed. Edward N. Zalta, Winter 2008. Metaphysics Research Lab, Stanford University. <https://plato.stanford.edu/archives/win2008/entries/hume-newton/>.

Schmaltz, Tad M. 1997. 'Spinoza's Mediate Infinite Mode.' *Journal of the History of Philosophy* 35 (2): 199–235.

Schneider, Susan. 2007. 'What Is the Significance of the Intuition That Laws of Nature Govern?' *Australasian Journal of Philosophy* 85 (2): 307–24.

Schrenk, Markus. 2007a. 'Can Capacities Rescue Us from Ceteris Paribus Laws?' In *Dispositions in Philosophy and Science*, ed. B. Gnassounou and M. Kistler, 221–48. London: Ashgate.

Schrenk, Markus. 2007b. *The Metaphysics of Ceteris Paribus Laws*. Berlin: Ontos Verlag.

Schrenk, Markus. 2014. 'Better Best Systems and the Issue of CP-Laws.' *Erkenntnis* 79 (10): 1787–99.

Schrödinger, Erwin. 1935. 'Discussion of Probability Relations Between Separated Systems.' *Mathematical Proceedings of the Cambridge Philosophical Society* 31 (4): 555–563.

Scriven, Michael. 1959. 'Truisms as the Grounds for Historical Explanation.' In *Theories of History*, ed. Patrick Gardiner, 443–75. New York: The Free Press.

Segal, Aaron. 2014. 'Causal Essentialism and Mereological Monism.' *Philosophical Studies* 169 (2): 227–55.

Sergeant, John. 1696. *The Method to Science*. London: W. Redmayne.

s'Gravesande, Willem. 1747. *Mathematical Elements of Natural Philosophy*. London: W. Innys, T. Longman, and T. Shewell.

Shank, Michael H. 2019. 'Naturalist Tendencies in Medieval Science.' In *Science Without God? Rethinking the History of Scientific Naturalism*, ed. Peter Harrison and Jon H. Roberts. Oxford: Oxford University Press.

Shepherd, Mary. 2020. *Mary Shepherd's Essays on the Perception of an External Universe*, ed. Antonia LoLordo. Oxford: Oxford University Press.

Shoemaker, Sydney. 1994. 'Phenomenal Character.' *Noûs* 28 (1): 21–38.

Shoemaker, Sydney. 2003. *Identity, Cause, and Mind*. Oxford: Oxford University Press.

Sider, Theodore. 1992. 'Tooley's Solution to the Inference Problem.' *Philosophical Studies* 67 (3): 261–75.

Sider, Theodore. 2020. *The Tools of Metaphysics and the Metaphysics of Science*. Oxford: Oxford University Press.

Skow, Bradford. 2016. *Reasons Why*. Oxford: Oxford University Press.

Slavov, Matias. 2013. 'Newton's Law of Universal Gravitation and Hume's Concept of Causality.' *Philosophia Naturalis* 50 (2): 277–305.

Snyder, Laura J. 2019. 'William Whewell.' In *The Stanford Encyclopedia of Philosophy*, ed. Edward N. Zalta. Metaphysics Research Lab, Stanford University. <https://plato.stanford.edu/archives/spr2019/entries/whewell/>.

Spinoza, Baruch. 1926. *Opera*. 4 vols, ed. Carl Gebhardt. Heidelberg: C. Winter.

Spinoza, Baruch. 1985. *The Collected Works of Spinoza, Volume I*, ed. and trans. Edwin Curley. Princeton: Princeton University Press.

Spinoza, Baruch. 1994. *A Spinoza Reader: The Ethics and Other Works*, trans. Edwin Curley. 1st edn. Princeton: Princeton University Press.

Spinoza, Baruch. 2002. *Spinoza: The Complete Works*, ed. Michael L. Morgan, trans. Samuel Shirley. Indianapolis: Hackett.

Stalnaker, Robert C. 1968. 'A Theory of Conditionals.' In *Studies in Logical Theory* (American Philosophical Quarterly Monographs 2), ed. Nicholas Rescher, 98–112. Oxford: Blackwell.

Stalnaker, Robert C. 1984. *Inquiry*. Cambridge, MA: MIT Press.

Stalnaker, Robert C. 2019. *Knowledge and Conditionals: Essays on the Structure of Inquiry*. Oxford: Oxford University Press.

Stanford, P. Kyle. 2015. 'Instrumentalism: Global, Local, and Scientific.' In *The Oxford Handbook of Philosophy of Science*, ed. Paul Humphreys, 318–36. Oxford: Oxford University Press.

Starr, William. 2019. 'Counterfactuals.' In *The Stanford Encyclopedia of Philosophy*, ed. Edward N. Zalta. Metaphysics Research Lab, Stanford University. <https://plato.stanford.edu/archives/fall2019/entries/counterfactuals/>.

Steinle, Friedrich. 2002. 'Negotiating Experiment, Reason and Theology: The Concept of Laws of Nature in the Early Royal Society.' In *Wissensideale und Wissenskulturen in der Frühen Neuzeit*, ed. Wolfgang Detel and Claus Zittel, 197–212. Berlin: Akademie Verlag.

Stoneham, Tom, and Angelo Cei. 2009. '"Let the Occult Quality Go": Interpreting Berkley's Metaphysics of Science.' *European Journal of Analytic Philosophy* 5 (1): 73–91.

Swinburne, Richard. 1980. 'Properties, Causation, and Projectibility: Reply to Shoemaker.' In *Applications of Inductive Logic*, ed. L. J. Cohen and M. Hesse, 313–20. Oxford: Oxford University Press.

Swoyer, Chris. 1982. 'The Nature of Natural Laws.' *Australasian Journal of Philosophy* 60 (3): 203–23.

Tooley, Michael. 1977. 'The Nature of Laws.' *Canadian Journal of Philosophy* 7 (4): 667–98.

Tugby, Matthew. 2013. 'Platonic Dispositionalism.' *Mind* 122 (486): 452–80.

Tugby, Matthew. 2016. 'Universals, Laws, and Governance.' *Philosophical Studies* 173 (5): 1147–63.

Turchin, Peter. 2001. 'Does Population Ecology Have General Laws?' *Oikos* 94 (1): 17–26.

van Fraassen, Bas C. 1989. *Laws and Symmetry*. Oxford: Oxford University Press.

van Fraassen, Bas C. 1993. 'Armstrong, Cartwright, and Earman on Laws and Symmetry.' *Philosophy and Phenomenological Research* 53 (2): 431–44.

Venn, John. 1889. *The Principles of Empirical or Inductive Logic*. London: Macmillan.

Vetter, Barbara. 2014. 'Dispositions without Conditionals.' *Mind* 123 (489): 129–56.

Ward, Barry. 2002. 'Humeanism without Humean Supervenience: A Projectivist Account of Laws and Possibilities.' *Philosophical Studies* 107 (3): 191–218.

Westfall, Richard S. 1971. *Force in Newton's Physics*. New York: Elsevier.

Westfall, Richard S. 1978. *The Construction of Modern Science: Mechanisms and Mechanics*. Cambridge: Cambridge University Press.

Whewell, William. 1860. *On the Philosophy of Discovery, Chapters Historical and Critical*. London: J. W. Parker and Son.

Wigner, Eugene. 1967. *Symmetries and Reflections*. London: Indiana University Press.

Williams, Neil. 2009. 'The Ungrounded Argument Is Unfounded: A Response to Mumford.' *Synthese* 170 (1): 7–19.

Williams, Neil. 2010. 'Puzzling Powers: The Problem of Fit.' In *The Metaphysics of Powers: Their Grounding and their Manifestations*, ed. Anna Marmodoro, 84–105. London: Routledge.

Williams, Neil. 2011. 'Dispositions and the Argument from Science.' *Australasian Journal of Philosophy* 89 (1): 71–90.

Williams, Neil. 2019. *The Powers Metaphysic*. Oxford: Oxford University Press.

Wilson, Jessica M. 2014. 'No Work for a Theory of Grounding.' *Inquiry* 57 (5–6): 535–79.

Wilson, Mark. 2013. 'What is "classical mechanics", anyway?' In *The Oxford Handbook of Philosophy of Physics*, ed. Robert Batterman, 43–106. Oxford: Oxford University Press.

Wilson, Mark. 2017. *Physics Avoidance: And Other Essays in Conceptual Strategy*. Oxford: Oxford University Press.

Winkler, Kenneth P. 1989. *Berkeley: An Interpretation*. Oxford: Oxford University Press.

Woodward, James. 2003. *Making Things Happen: A Theory of Causal Explanation*. Oxford: Oxford University Press.

Woodward, James. 2013. 'Laws, Causes, and Invariance.' In *Metaphysics and Science*, ed. Stephen Mumford and Matthew Tugby, 48–72. Oxford: Oxford University Press.

Woodward, James. 2018. 'Laws: An Invariance-Based Account.' In *Laws of Nature*, ed. Walter Ott and Lydia Patton, 158–80. Oxford: Oxford University Press.

Wootton, David. 2015. *The Invention of Science: A New History of the Scientific Revolution*. New York: Harper Collins.

Yablo, Stephen. 1992. 'Mental Causation.' *Philosophical Review* 101 (2): 245–80.

Yates, David. 2016. 'Is Powerful Causation an Internal Relation?' In *The Metaphysics of Relations*, ed. Anna Marmodoro and David Yates, 138–56. Oxford: Oxford University Press UK.

Zilsel, Edgar. 1942. 'The Genesis of the Concept of Physical Law.' *The Philosophical Review* 51 (3): 245–79.

Index

For the benefit of digital users, table entries that span two pages (e.g., 52–53) may, on occasion, appear on only one of those pages.

actualism
  Megaric 145n.14, 192
  non-Megaric 146, 151, 192
Anjum, Rani Lill 35n.85, 141n.2
anthropomorpism
  and Best System Analysis 219–22
  see also ratbag idealism
Aquinas, St. Thomas 12, 155
Aristotelianism 23–8, 38, 67–8, 142;
    see also Bacon, Francis; Spinoza,
    Baruch; powers
Aristotle 21–2, 58n.51
Armstrong, D.M. 85–90, 160, 173–4,
    177–8, 230, 238–42; see also circularity
    objection; inference problem; modal
    inversion; quiddities
Arnauld, Antoine 45–6
axiom role
  and single law 117
  as desideratum 277
  defined 15–17

Backmann, Marius 248–50
Bacon, Francis 24, 124–9, 131–2, 137,
    144–5, 215
Bacon, Roger 22
Barker, Stephen 115n.21, 154n.35, 157n.45,
    173n.35
Beebee, Helen 74, 78, 99n.25, 201–2, 236,
    245–6
Berkeley, George
  and gravity 58–60
  and instrumentalism 54, 58n.50
  and laws as divine volitions 30–1,
    57–8
  and materialism 106
  and Newton 51, 58–60
  and web-of-laws approach 59–61
  divergence from Descartes 58
  on explanation 38, 239

Best Systems Analysis see
    anthropomorphism; Better Best
    System, central tension;
    counterfactuals; explanation; Lewis,
    David; mismatch problem; phony
    constant; ratbag idealism; stability of
    laws, supervenience
Better Best System 222n.16, 250n.41
Bigelow, John 114, 184, 268
biology, laws of 20n.37, 247–8, 271
Bird, Alexander 36n.87, 113, 137, 139,
    152–3, 158, 162–3, 283–4, 287–8
Blackburn, Simon 161
Blobject 186–7, 283
Blower 187–90, 194, 196–7, 279, 282–5,
    289–91
Boyle, Robert 28, 62n.60, 190–4
Braddon-Mitchell, David 248–9,
    256, 265
Braithwaite, R.B. 245
Brown, Thomas 239n.8

Callender, Craig see Better Best System
Cambridge changes 191
Carnap, Rudolf 162
Carroll, John 71–6, 79n.27; see also mirror
    argument
Cartwright, Nancy 4n.3, 32, 141n.5, 143,
    167n.21, 182–3, 215
  see also ceteris paribus clauses, trilemma of
causation
  and explanation 38
  and god(s) see occasionalism
  composition of causes (Mill) 215
  four kinds of 124
  geometrical model of 194–6
  Hume and 258–9
  laws as causes 43–5; see also Descartes,
    René; Malebranche, Nicolas
  mind-body 42

central tension
  defined 202–3, 251–5
  in Best Systems Analyses 218–19, 250–1,
    264–6, 272, 281, 286
  in Hume 209–13
  in Mill 216
ceteris paribus clauses
  and Blower 188–9, 284
  and dispositions 34–5, 137–9
  as desideratum 278
  in Best System Analysis 247–51,
    255–6, 286
  in Mill 215
  in moderns 31–3, 60, 138–9, 278
  in Rules theory 264–6, 286–7
  in universals theory 116–17, 280, 282
  trilemma of 31–3, 114, 139, 248–50, 256
  see also web of laws approach
Chakravartty, Anjan 167–9, 170n.25,
    175n.41, 184n.11
Charleton, Walter 22n.41, 135n.44
Chen, Eddy Keming 15n.17, 81–2, 263n.14
Cicero 12
circularity objection 238–42
Clarke, Samuel 53
classical mechanics 15–16, 18–19
Cohen, Jonathan see Better Best System
conservation
  of energy 19–20
  of motion 63
  laws of 20, 88, 114–15, 118–19, 132, 181,
    278, 282–4
Contessa, Gabriele 88–90
Corry, Richard 141–2, 144–5, 185n.17
Coulomb's law 20, 32
counterfactuals
  and Best Systems Analyses 230–4
  and context 83
  and nomic explanation 17–18, 241–2
  and Rules theory 269–71
  and stability of laws 237–8
  laws' support of 39, 88
  Lewis/Stalnaker theory of 231–3, 237–8,
    242, 269–70, 272
  supposition theory of 230–1, 233–4, 238,
    242, 270, 285–6
Creary, Lewis 32n.72
Cross, Troy 163n.7
Cudworth, Ralph 28–9

Depew, David 20n.37
Descartes, René
  and intentionality of powers 151–2
  and the problem of ceteris paribus
    clauses 33–7
  and web of laws approach 35
  epistemic difficulties faced by 37–8
  merits of position on laws 38–41
  nomic stability in 40
  on ontology of laws 27–31
  on primary and secondary causes 29,
    42, 45n.8
design arguments 244
dispositionalism, varieties of
  dualism 177–8
  dual-sided 176–7
  neutral monism 176–7
  pan-dispositionalism 172–6
dispositions
  as higher-order properties 156–7
  thick vs. thin 36, 88–9
  intrinsic vs. extrinsic 147–8, 192–4
  irreducibility of 162–3, 189–94
  see also finks; masks; dispositionalism,
    varieties of
Dorst, Chris 220–1, 241n.16, 250
Dretske, Fred 85, 103–4
Dupleix, Scipion 23

Earman, John 32n.73, 247–8
Edgington, Dorothy 230–1, 233–4
Ellis, Brian 91–2, 114, 150–1,
    164–71, 184
entanglement 172–3, 181, 223–7
epiphenomenalism 95–6, 111
esse-ad see intentionality
eternalism 246–7
Euclid 21–2
Euler, Leonhard 132–3
exclusion argument 156–7
explanation
  as unification or assimilation 17,
    39–40
  case 238–42, 269–71
  case vs. whole 17
  deductive-nomological model 81
  from laws 16–18
  scientific vs. metaphysical 38,
    239–41

whole 97–8, 145, 242–4
  see also counterfactuals; Newton, on
  mathematical vs. physical distinction;
  transitivity principle

Fibonacci sequence 57–8
finks 138, 162–3, 244
FLOTEs (fundamental laws of temporal
  evolution) 15n.17, 263n.14
force(s)
  composition of 59–60, 143, 215
  in Helmholtz 135–6
  in Mill 208–9
  see also dispositions; gravity; powers
forms *see* Bacon, Francis
functions 155–7, 159

Galilei, Galileo 11n.1, 24, 31, 47
Gassendi, Pierre 22n.41
Gaukroger, Stephen 124
ghost 91, 94–5, 111, 171
Giere, Ronald 4n.3, 261, 266
Glennan, Stuart 4n.3, 141–2, 233
Giffen goods 21
god(s) *see* Berkeley, George; Cudworth,
  Ralph; Descartes, René; occasionalism;
  Reid, Thomas
Goldstein, Sheldon 15n.17, 81–2, 263n.14
Goudarouli, Eirini 63
governing dilemma 79–82, 85, 93
gravity
  in Berkeley 58–60
  in Helmholtz 133
  in Hume 203–9
  in Locke 194
  in Mumford 137–8
  in Newton 47–8, 52–6
  see also inverse square law
Grene, Marjorie 20n.37
grizzly bears 249
growing block 246n.31
Gundersen, Lars 163n.7

Hall, Ned 220, 225–6
Handfield, Toby 112–13
Harré, Rom and E.F. Madden 142
Healey, Richard 72n.9, 77n.25
Heil, John 150–1
Helmholtz, Hermann von 135–7

Hempel, Carl 31, 32n.73, 247–8; see also
  ceteris paribus clauses
Henry, John 11n.2, 12n.5, 25n.50, 27n.56
Hicks, Michael 221
Hildebrand, Tyler 97–100, 243
Hobbes, Thomas 62–3
holism
  about confirmation 35, 265
  about powers 183–6, 282–3; *see also*
    structure without stuffing
  semantic 115, 184–5
Hume, David
  and Newton 203–4, 207
  instrumentalist readings of 205–6
  on definition of 'law of nature' 204–5
  on gravity 203–9
  on necessary connections 194–5, 223, 286
  on probability 206, 210–11
Hume's revenge 34–5
Huygens, Christiaan 15–16
hybrid views of laws 287–9

ideal gas law 21
immanentism 87–8
immutability, divine 33, 42
inference problem
  defined 102–9
  legitimacy version of 104–7
  mechanism version of 107–9, 113,
    117–19, 282
inference to the best explanation 95,
  99–100
Ingthorsson, R.D. 173nn.35,36, 175n.41
inheritance principle 102–3, 107–10
intensionality 152–3
intentionality 151–5, 159–60, 180–9, 283
interference, token 31–2, 59
interference, type see *ceteris paribus* clauses
intra-world variance of laws 99–101
intrinsicality 147
inverse square law 15 *see also* gravity
iron vs. oaken laws 116–17

Jackson, Frank 165
Jesseph, Douglas 59
Jolley, Nicholas 44–5

Kaufman, Dan 191n.25
Kepler, Johannes 16, 24, 128–9

Kim, Jaegwon *see* exclusion argument
Kim Jong-un, chess with 271–2, 290–1
Kripkenstein 229n.33

Lange, Marc 75n.21, 240–1
Lange/Cartwright trilemma see *ceteris paribus* clauses, trilemma of
laws to truthmakers
   defined 5, 279–81
   in primitivism 69, 279–80
   one-to-one 39, 69, 85–6, 116–18, 132, 139, 157, 188
   one-to-many 80–1, 132–5, 139–40, 283
   many-to-one 116–18, 188–9
Leibniz, Gottfried Wilhelm von
   on Malebranche 45–7
   on Newton 49–50, 52–3, 55
   on relations 154
   on why laws cannot execute themselves 28–9
Lewis, David
   and anti-realism 201
   and central tension 218–19
   on intrinsicality 147
   on natural properties 222–3
   on universals view 102
   see also anthropomorphism; counterfactuals, Lewis/Stalnaker theory of; mismatch problem; ratbag idealism; supervenience
Lewis, Peter J. 72n.9
Lierse, Caroline 114, 184
Lipton, Peter 34–5
little souls argument 151–3, 180–9, 282–3, 290
locations, causal and nomic relevance of 91–2, 150–1, 176
Locke, John 133, 195–6
Loewer, Barry 78, 222–3, 239; *see also* package deal account; supervenience
LoLordo, Antonia 22n.41
LOPP (limited oracular perfect physicist) 225–8, 230, 263–5, 267, 270, 287; *see also* supervenience
lossy laws 248–9, 256, 265
Lowe, E.J. 138n.57, 162–3

Mach, Ernst 213n.38
masks 162–3, 244

McKitrick, Jennifer 87n.3, 147–8, 156n.42, 163n.10, 167n.20, 174, 191n.26
Malebranche, Nicolas
   and divine volitions 30–1
   autonomy reading of 42–7, 53, 79–80
   intervention reading of 42–7
   on divine concursus argument 44
   see also little souls argument; occasionalism
many-to-one concept of laws *see* laws to truthmakers
Martin, C.B. 148n.21, 162, 173–4, 176–7
Massimi, Michela 225n.24, 258n.1
Maudlin, Tim 76–8, 80, 227–8, 242–3, 263n.14; *see also* FLOTEs
Mayr, Ernst 20n.37
mechanism, new 4n.3, 141–2
Meinongianism 112, 160
Mill, John Stuart
   and instrumentalism 207–8
   on composition of forces and causes 215–16
   on laws as regularities 213
   on laws as dispositions 137
   on laws as summaries 213–14
   on gravity 207
   see also web of laws approach; force
Miller, Elizabeth 223n.20
Millican, Peter 205–6
Milton, John R. 12, 23n.44
mimicking argument 89–90, 95
mirror argument 71–8, 235–8, 276–8, 286
mismatch problem 219–23, 263–4
modal inversion 112–14, 145–6
models see Giere, Ronald; Rules theory; Ward, Barry
monads 185
Molnar, George 152–3, 160, 177, 191
monism see Blobject; Blower
Mumford, Stephen
   and little souls argument 152–3
   and quiddities 92–5
   and ungrounded argument 165–7
   on laws as powers 136–40

natural properties 97–8
necessitation relation (universals theory) 85–7, 99–101, 117; *see also* inference problem; modal inversion; relations

necessity of fit *see* little souls
necessity of laws 112–13
Newton, Isaac
  against the Cartesians 48–9
  and occasionalism 47–8, 52–4
  and web-of-laws approach 55
  first law of motion 18
  inverse square law and 15
  laws as axioms 47, 55
  on ancients vs. moderns 23
  on mathematical vs. physical
    distinction 48–56, 206–7
  second law of motion 59, 261
  see also forces, composition of; gravity
Nida-Rümelin, Martine 186n.19
Noether, Emmy 19–20
nomic profiles, wide and narrow 93–4, 111

occasionalism 26–7, 42, 45; *see also*
  Descartes, René; Malebranche,
  Nicolas; Reid, Thomas; Newton, Isaac
one-to-many concept of laws *see* laws to
  truthmakers
one-to-one concept of laws *see* laws to
  truthmakers

package deal account of laws 221n.9,
  222n.16, 224n.22, 246n.28,
  250n.41, 262
pan-dispositionalism *see* dispositionalism,
  varieties of
Past Hypothesis 221n.12, 263n.14
phony constant 220, 250, 253–4
Place, Ullin T. 152–3
powers
  and argument from science 164–8
  essentiality of manifestations to 144
  invariance of manifestations 144–5
  ungrounded argument 165–7
  see also dispositions; regress arguments
prescriptivity of laws 13
price and demand, law of 21
primitivism
  and counterfactuals 69
  and conservation laws 69–71
  and laws-to-truthmakers 279–80
  and web-of-laws approach 68–9
  not entertained by moderns 28–9, 42
  see also underdetermination argument

probabilistic laws 80n.32, 264, 285n.4
problem of fit *see* little souls
projectivism 227, 257–63, 269, 272
Psillos, Stathis 63, 110n.12, 231n.37,
  233n.43, 245–6
Ptolemy 12, 22, 26

quantum mechanics 18–20, 71–2, 223–7;
  see also entanglement; spin;
  Schrödinger equation
quiddities 90–6, 111, 171; *see also* ghost;
  share; swap
Quine, Willard van Orman 35

Ramsey, Frank Plumpton 213n.38, 230–1,
  259–60; *see also* counterfactuals,
  supposition theory of
ratbag idealism 219–20, 225; *see also*
  anthropomorphism
realism
  defined by stipulation 4–5, 201
  evaluated 285–7
reducibility, kinds of 149–50
reflective equilibrium 2–3
Régis, Pierre-Sylvain 14, 195
regress arguments
  against pan-dispositionalism 159,
    173–4, 183
  and extrinsic powers 193
  for powers 168–70
regularity
  and universals theory 86
  defined 11, 244–7
  in early modern figures 12–13, 25, 61
  partial regularity 213
  see also Best System Analysis; Braithwaite,
    R.B.; Hume, David; inference problem;
    Lewis, David; Mill, J.S.
Reid, Thomas 26–7
relations
  and intentionality 155
  higher-order 114–15, 158
  internal vs. external 107–10, 115–16,
    118–19, 153–5
Reutlinger, Alexander 248–50
Roberts, John 16nn.20,21, 32n.73, 71–6,
  79n.27, 247–8
Rouhault, Jacques 22n.41
Roux, Sophie 13–14

Ruby, Jane 12
Rules theory
  and counterfactuals 269–71
  and LOPP 262
  and models 262–3, 265–7
  and truth 267–9
  and web-of-laws approach 262, 286–7
  defined 262–72
  explanation in 269–71
  see also *ceteris paribus* clauses;
    supervenience
Russell's paradox 88

Schneider, Susan 75, 79–80, 236n.1
Schrenk, Marcus 34–5
Schrödinger equation 19nn.32,33
Schrödinger, Erwin 224n.21, 225n.23
Sergeant, John 23–4
s'Gravesande, Willem 14, 44–5
Shanghai maglev train 142–3
Shaffer, Jonathan 105–7
share 91, 94–5, 111
Shepherd, Mary 22n.41, 134–5
Shoemaker, Sydney 90–2, 161, 170, 175–6,
  192–3
Sider, Theodore 106n.8, 119n.26
simplicity 219–22; *see also*
    anthropomorphism; ratbag idealism
Slavov, Matias 205–6
spin 20, 71–2
Spinoza, Baruch 128–32, 213
stability of laws
  and Best System Analysis 235–8
  and primitivism 71
  and Rules theory 271–2
  and universals theory 87–8
  as desideratum 275, 277–8
  defined 40, 83–4
  in Descartes 40
  see also mirror argument
Stalnaker, Robert *see* counterfactuals,
    Lewis/Stalnaker theory of
structure without stuffing 158–9, 183
supervenience of laws on mosaic
  Human LOPP-relative 226–7, 229,
    268–9
  moderns' denial of 67
  in Best Systems Analysis 223–30
  in Rules theory 266–9

swap 91, 94–5, 111
Swoyer, Chris 109n.11

thin concept of laws 14–21, 26, 61–3
Theory of Everything ('TOE') 139–40,
  222–3, 279–81, 284
time 246–7
tofu, magical powers of 89, 92
Tooley, Michael 71–2, 85
top-down/bottom-up contrast 3–5, 24–5,
  61–3, 132, 281–5
transitivity principle 240–2
Tugby, Matthew 87n.4, 159n.50
Turchin, Peter 20n.37, 247–8, 271

underdetermination argument 76–8,
  227–30, 260–1
ungrounded argument *see* powers
universals *see* Armstrong, D.M.: immanentism;
    necessitation relation; quiddities

van Fraassen, Bas 102–3, 221–2, 266n.17
volitions, divine
  structure of 30
  in Descartes 29–31, 36
  in Malebranche 43–5

Ward, Barry 227–30, 260–2
web-of-laws approach
  and universals theory 117–18
  in Berkeley 60
  in Descartes 35–6, 39
  in Mill 5–7
  in moderns generally 63
  in Newton 55
  see also *ceteris paribus* clauses
Wigner, Eugene 26n.51
Williams, Neil 164n.12, 165–7, 167n.20,
  180–9, 289n.11
Wilson, Jessica 166n.17
Wilson, Mark 53n.31
Woodward, James 26n.51, 70–1, 236n.3,
  242n.17, 280
Wootton, David 11n.1

Yablo, Stephen 156–7
Yorick, skull of 156

Zilsel, Edgar 12